PARTIAL DIFFERENTIAL EQUATIONS

AN INTRODUCTION

PARTIAL DIFFERENTIAL EQUATIONS

AN INTRODUCTION

BERNARD EPSTEIN

PROFESSOR OF MATHEMATICS
GRADUATE SCHOOL OF SCIENCE
YESHIVA UNIVERSITY

ROBERT E. KRIEGER PUBLISHING COMPANY

HUNTINGTON, NEW YORK   1975

Original edition 1962
Reprint 1975

Printed and Published by
Robert E. Krieger Publishing Company, Inc.
645 New York Avenue
Huntington, New York 11743

© Copyright 1962 by McGraw-Hill Book Company
Reprinted by arrangement

Printed in the United States of America

Library of Congress Cataloging in Publication Data

Epstein, Bernard.
    Partial differential equations.

    Reprint of the ed. published by McGraw-Hill, New York,
in series: International series in pure and applied
mathematics.
    Bibliography: p.
    1. Differential equations, Partial. I. Title.
[QA374.E65 1975]   515'.353   75-11905
ISBN 0-88275-330-4

PREFACE

It has been the purpose and hope of the author in writing this book to help fill a serious need for introductory texts on the graduate level in the field of partial differential equations. The vastness of the field and— even more significantly—the absence of a comprehensive basic theory have been responsible, we believe, for the comparative scarcity of introductory books dealing with this subject. However, the importance of this field is so tremendous that the difficulties and pitfalls awaiting anyone who seeks to write such a book should be looked upon as a provocative and stimulating challenge. We hope that we have achieved some measure of success in meeting this challenge.

Any book dealing with a subject possessing substance and vitality is bound to reflect the particular interests and prejudices of the author. Even if the field is well organized and has been worked out with a considerable degree of completeness, the author's inclinations will be reflected in the manner in which the subject matter is presented. When, in addition, the subject is as extensive and incompletely developed as that here under consideration, they will also be reflected in the choice of material. We are well aware that many important topics are presented briefly or not at all. However, we are consoled by the thought that in the writing of a book of moderate size the omission of much significant material was inevitable, and by the hope that our presentation will be such as not only to interest the student in the topics presented here, but also to stimulate him to pursue some of them, as well as topics not touched upon here, in other books and in the research journals.

Throughout this book the stress has been on existence theory rather than on the effective determination of solutions of specific classes of problems. It is hoped that the presentation will complement usefully

v

any text which emphasizes the more "practical" or "applied" aspects of the subject.

A word is in order concerning the intimate relationship between physics and the theory of differential equations, both ordinary and partial. Physics has certainly been the richest source of problems in this field, and physical reasoning has often been an invaluable guide to the correct formulation of purely mathematical problems and to the successful development of techniques for solving such problems. In this connection we would strongly urge every prospective student of differential equations (indeed, every prospective student of mathematics) to read, and deliberate on, the splendid preface to the Courant-Hilbert masterpiece, "Methods of Mathematical Physics." Although little is said in the following pages concerning the physical origins of many of the mathematical problems which are discussed, the student will find that his understanding of these problems will be heightened by an awareness of their physical counterparts.

The author has good reason to follow the tradition of acknowledging gratefully his wife's unselfish aid in the seemingly endless task of typing successive drafts of the manuscript. More important than the typing, however, was her constant encouragement to carry the writing task through to its completion. It is hoped that her encouragement was directed to a worthwhile objective.

Bernard Epstein

CONTENTS

TERMINOLOGY AND BASIC THEOREMS

For convenience we list here a few terms, notations, and theorems that will be used frequently.

A *domain* is an open connected set (in the plane or in a higher-dimensional euclidean space); an equivalent definition is that a domain is an open set which cannot be expressed as the union of two disjoint nonvacuous open sets.

The "Kronecker delta" δ_{ij} assumes the values 1 and 0, according as the indices i, j are equal or unequal.

$\bar{S}$ denotes the closure of the set S.

A *disc* is the set of all points (in the plane) satisfying an inequality of the form $(x - x_0)^2 + (y - y_0)^2 < r^2$; a *circle* is the boundary of a disc. The *unit disc* is defined by the inequality $x^2 + y^2 < 1$, and the *unit circle* is its boundary.

When dealing with a curve we denote arc length, measured from some fixed point of the curve, by the letter s.

The symbols $\in$, $\subset$, $\cup$, $\cap$, are used with their customary set-theoretic significance: $a \in A$ means that a belongs to (is an element of) the set A, and $A \subset B$ means that every element which belongs to A also belongs to B (A is a subset of B). Note that $A \subset B$ does not imply that A is a *proper* subset of B. $A \cup B$ and $A \cap B$ denote the union and intersection, respectively, of the sets A, B.

The *distance* between two sets is the minimum distance between a pair of points, one from each set. If both sets are closed and at least one is bounded, then the minimum is actually attained.

The Heine-Borel theorem: Given a compact (i.e., bounded and closed) set S in a euclidean space and an open covering of S (i.e., a collection of open sets whose union contains S), then it is possible to extract from this covering a finite number of open sets which suffice to cover S.

A real-valued function which is defined and continuous on a compact set is uniformly continuous, is bounded above and below, and attains its maximum and minimum values.

A function defined in a domain D is said to be of class C^n if all partial derivatives of order up to and including the nth exist and are continuous throughout D.

If the functions $f(x,y)$, $g(x,y)$ are of class C^n in a neighborhood of (x_0,y_0), and if the Jacobian $f_x g_y - f_y g_x$ does not vanish at that point, then the equations $\xi = f(x,y)$, $\eta = g(x,y)$ can be solved in a sufficiently small neighborhood of (x_0,y_0) for x and y in terms of ξ and η, say $x = \phi(\xi,\eta)$, $y = \psi(\xi,\eta)$, and the functions $\phi(\xi,\eta)$, $\psi(\xi,\eta)$ are also of class C^n.

1. SOME PRELIMINARY TOPICS

Before entering on the subject of partial differential equations, it seems appropriate to devote a chapter to some concepts and theorems with which the reader is perhaps not yet acquainted. The reader may choose to study the topics covered in this chapter as the need for them arises later, rather than beginning by going through this chapter (or those parts of it which he has not previously encountered) systematically.

1. Equicontinuous Families of Functions

A basic difficulty that besets many mathematical investigations is the fact that there exists no simple extension to families of functions of the Bolzano-Weierstrass theorem, which asserts (in one of its several possible formulations) that from every bounded sequence of real numbers it is possible to extract a convergent subsequence. If, instead of a sequence of numbers, we consider a sequence of functions $\{f_n(x)\}$ defined on a fixed interval, say $0 \leq x \leq 1$, then, from the hypotheses that each function $f_n(x)$ is continuous on this interval and that these functions are uniformly bounded [i.e., there exists a positive number M independent of x and n such that $|f_n(x)| < M$], it does *not* follow that it is possible to extract a subsequence convergent throughout the interval. (Cf. Exercise 1.) However, there is a certain more stringent condition of continuity, which is frequently found to be satisfied by sequences (or families) of functions encountered in problems of analysis, and which suffices, when taken together with the condition of uniform boundedness, to assure the existence of a *uniformly* convergent subsequence; this more stringent condition will now be explained.

A basic theorem of analysis asserts that a function $f(x)$ continuous on a compact (bounded and closed) set S is uniformly continuous on S;

1

that is, given any positive number ϵ, there exists a positive number δ such that whenever the conditions $x_1 \in S$, $x_2 \in S$, $|x_1 - x_2| < \delta$ are satisfied, the inequality $|f(x_1) - f(x_2)| < \epsilon$ holds. Given any finite number of continuous functions defined on S and any positive number ϵ, it is possible to find a single number δ which will suffice for each of these functions, for one can determine a suitable δ for each of these functions and then choose the smallest of these numbers. However, this method fails for an infinite family of functions, for there may not exist a positive lower bound to the corresponding set of numbers δ. (This occurs in the family of functions considered in Exercise 1.) We may describe this situation as a lack of "over-all" uniformity, despite the uniform continuity of each individual function of the family. We are thus led to formulate the following more restrictive concept of uniform continuity, which refers to a *family* of functions, not to a single function.

Definition 1. A family of functions defined on a set S of real numbers[1] is said to be "equicontinuous" if for every positive number ϵ there exists a positive number δ such that, for every function $f(x)$ of the family and every pair of numbers x_1, x_2 contained in S and satisfying the inequality $|x_1 - x_2| < \delta$, the inequality $|f(x_1) - f(x_2)| < \epsilon$ holds. (It should be emphasized, for clarity, that each function of an equicontinuous family is uniformly continuous. Also it may be noted that no restriction of boundedness or closure is imposed on S.)

A simple example of an equicontinuous family is furnished by any set of functions defined and continuous on a fixed interval (open or closed) and having, at all points of this interval, a first derivative whose absolute value never exceeds some fixed number C; for then, by the theorem of mean value, we have, for any function $f(x)$ of the family and any two numbers x_1, x_2 of the interval, the inequality

$$|f(x_1) - f(x_2)| = |f'(\xi)(x_1 - x_2)| \leq C|x_1 - x_2| \qquad (x_1 < \xi < x_2) \quad (1)$$

so that, for any given ϵ, the choice $\delta = \epsilon/C$ will suffice. (Note that in Exercise 1 the derivatives of the functions under consideration are not uniformly bounded.)

We now state the following striking theorem, which accounts for the important role played in analysis by the concept of equicontinuity.

Theorem 1. Ascoli Selection Theorem. Let F be an infinite, uniformly bounded, equicontinuous family of functions defined on a

---

[1] For simplicity, we formulate the definition only for functions of one real variable; the extension to more general classes of functions is quite straightforward. Similarly, the theorem of this section is stated only in the one-dimensional case, the extension to functions of any (finite) number of variables being clear.

finite closed interval S: $a \leq x \leq b$. Then from every sequence $\{f_n(x)\}$ chosen from F it is possible to select a uniformly convergent subsequence.

Proof. Select any countable dense subset S_1 of S, such as the set of all rational numbers in S, and enumerate them: $r_1, r_2, r_3, \ldots$. Let a sequence $\{f_n(x)\}$ be selected from F. Then the sequence $\{f_n(r_1)\}$ satisfies the hypothesis of the Bolzano-Weierstrass theorem, and so we may select a subsequence $\{f_{n1}(x)\}$ of the original sequence which converges at the point r_1. By applying the preceding argument to $\{f_{n1}(x)\}$ we obtain a sequence $\{f_{n2}(x)\}$ which converges at r_2 and also at r_1 (for any subsequence of a convergent sequence is also convergent, and has the same limit). Repeating this argument, we obtain further sequences $\{f_{n3}(x)\}$, $\{f_{n4}(x)\}$, $\ldots$, each of which is a subsequence of the preceding one, and such that the kth sequence converges at $r_1, r_2, \ldots, r_k$. In order to obtain a single sequence which will converge at all the points of S_1, we employ the "diagonalization procedure," originally employed by Cantor to demonstrate the uncountability of the set of real numbers. Consider the sequence $\{f_{nn}(x)\}$ formed by taking the first function of the first subsequence, the second function of the second subsequence, etc. This last sequence is convergent at each point of S_1, for it is evidently a subsequence of the first sequence $\{f_{n1}(x)\}$ and, aside perhaps from the first $k - 1$ terms, a subsequence of the sequence $\{f_{nk}(x)\}$, $k = 2, 3, \ldots$.

It now remains to show that the sequence $\{f_{nn}(x)\}$ converges throughout S, and that the convergence is uniform. Given any $\epsilon > 0$, choose $\delta > 0$ such that, for every function f in F, and a fortiori for every function of the sequence $\{f_{nn}(x)\}$, the inequality $|x_1 - x_2| < \delta$ implies that $|f(x_1) - f(x_2)| < \epsilon$. Now we select a *finite* subset S_2 of S_1 such that each point of S differs by less than δ from at least one point of S_2. (This can be accomplished, for example, by dividing S into adjoining segments, each of length not exceeding δ, and selecting one point of S_1 in each of these segments.) We next determine a positive integer N so large that, for $n, m > N$, the inequality $|f_{nn}(y) - f_{mm}(y)| < \epsilon$ holds at each point y of S_2. Then for any point x of S we choose $y \in S_2$ such that $|x - y| < \delta$, and we obtain, for $n, m > N$, the chain of inequalities

$$|f_{nn}(x) - f_{mm}(x)| \leq |f_{nn}(x) - f_{nn}(y)| + |f_{nn}(y) - f_{mm}(y)| \\ + |f_{mm}(y) - f_{mm}(x)| < 3\epsilon \quad (2)$$

Since the index N has been chosen independently of x, (2) implies the uniform convergence of the sequence $\{f_{nn}(x)\}$, and the proof is complete.

Two brief remarks may be helpful in clarifying the significance of the two hypotheses (uniform boundedness and equicontinuity). First, the proof of the existence of a subsequence of the original sequence which converges at all points of a preassigned countable dense set requires only the existence of a pointwise bound (*not* a uniform bound) on the family F, and does not involve equicontinuity. Secondly, the proof given above may be easily modified to establish the following corollary, which may be left as a simple exercise.

COROLLARY.   Let F be an infinite family of functions defined on an open interval S: $a < x < b$, equicontinuous on every closed subinterval, and bounded at some point ξ, $a < \xi < b$ (cf. Exercise 4); then from every sequence $\{f_n(x)\}$ chosen from F it is possible to select a subsequence which converges uniformly on every compact subset of S.

The important Montel selection theorem of the theory of analytic functions is closely related to (the two-dimensional version of) this corollary; the essential point of the proof of this theorem consists in showing that a family of analytic functions uniformly bounded in a domain is equicontinuous in every compact subset of the domain.

EXERCISES

1. Consider the sequence of functions $\{\sin n\pi x\}$ on the interval $0 \le x \le 1$, $n = 1$, $2, \ldots$; these functions are uniformly bounded on this interval, for $|\sin n\pi x| \le 1$. Prove that there does not exist a subsequence which converges uniformly at each point of the interval.

2. Consider the sequence of functions $\{x^n\}$ on the interval $0 \le x \le 1$, $n = 1$, $2, \ldots$; in contrast to the preceding exercise, this sequence (and hence every subsequence) converges throughout the interval, but the limit function is discontinuous at the end point $x = 1$.   Prove directly from the definition that this sequence of functions is not equicontinuous.   (This fact also follows, of course, from the selection theorem.)   Show that in any *smaller* interval, $0 \le x \le a < 1$, the above functions are equicontinuous, in agreement with the fact that the limit function is continuous in this smaller interval.

3. Prove that a sequence of continuous functions which converges uniformly on a compact set forms an equicontinuous family.

4. Show that a family of functions equicontinuous on any bounded set S and bounded at one point of S is uniformly bounded on S.

2. The Weierstrass Approximation Theorem

In many branches of analysis there are theorems whose proofs have to be given in two parts: First the theorem is proved subject to certain additional hypotheses, and then it is shown, by the use of suitable approximation techniques, that the additional hypotheses may be dropped.   To cite only one example, we may mention the Riemann-Lebesgue lemma, of fundamental importance in the theory of Fourier series and integrals, which asserts that, for any function $f(x)$ which is (absolutely) integrable over an interval I, finite or infinite, the quantity $\int_I f(x)e^{i\lambda x}\,dx$ approaches zero as the real parameter λ becomes infinite. It is a simple matter to prove this theorem under the additional assumptions that the interval I is finite and the function $f(x)$ is continuously differentiable, for in this case an integration by parts yields the desired result immediately. One can then extend the proof to continuous functions by using the fact that a continuous function can be suitably approximated by continuously differentiable functions; then, similarly, one uses the fact that any integrable function can be suitably approximated by continuous functions; finally, the restriction to a finite interval is easily dropped.

One of the most important and striking approximation theorems is the following, which will be used subsequently a number of times.

Theorem 2. Weierstrass Approximation Theorem. Let $f(x_1,x_2,$ $\ldots,x_n)$ be defined and continuous on any compact set R. Given any positive ϵ, there exists a polynomial $P(x_1,x_2,\ldots,x_n)$ such that the inequality

$$|f(x_1,x_2,\ldots,x_n) - P(x_1,x_2,\ldots,x_n)| < \epsilon \qquad (3)$$

holds at all points of R.

Proof. For simplicity, we consider the case $n = 2$; the modifications required for any other value of n will be clear from the proof to be presented. (Cf. Exercise 6.) First we make the additional assumption that the set R is a rectangle and that f vanishes at all boundary points of R. We extend the function f over the entire plane by assigning it the value zero at all points outside R. Clearly the extended function is uniformly continuous, not only in R, but also over the entire plane. Consider the one-parameter family of functions:

$$f_t(x_1,x_2) = \iint\limits_{-\infty}^{\infty} f(\xi_1,\xi_2)P(\xi_1 - x_1,\ \xi_2 - x_2,\ t)\,d\xi_1\,d\xi_2 \qquad (t > 0) \quad (4)$$

where

$P(\xi_1 - x_1,\ \xi_2 - x_2,\ t)$ (or, for brevity, P_t)
$$= (\pi t)^{-1}\exp\{-t^{-1}[(\xi_1 - x_1)^2 + (\xi_2 - x_2)^2]\}$$

On account of the uniform continuity of f we can choose $\delta > 0$, independent of x_1 and x_2, such that $|f(\xi_1,\xi_2) - f(x_1,x_2)| < \epsilon/3$ whenever $(\xi_1 - x_1)^2 + (\xi_2 - x_2)^2 < \delta^2$. Taking account of the fact that the right side of (4) has the value one when the function f is replaced by the constant function $f(x_1,x_2) \equiv 1$, we obtain the chain of inequalities

$$|f_t(x_1,x_2) - f(x_1,x_2)| = \left| \iint_{-\infty}^{\infty} (f(\xi_1,\xi_2) - f(x_1,x_2)) P_t \, d\xi_1 \, d\xi_2 \right|$$

$$\leq \iint_{-\infty}^{\infty} |f(\xi_1,\xi_2) - f(x_1,x_2)| \, P_t \, d\xi_1 \, d\xi_2 \leq \frac{\epsilon}{3} \iint_{\Delta} P_t \, d\xi_1 \, d\xi_2$$

$$+ 2M \iint_{\Delta'} P_t \, d\xi_1 \, d\xi_2 < \frac{\epsilon}{3} \iint_{-\infty}^{\infty} P_t \, d\xi_1 \, d\xi_2$$

$$+ \frac{2M}{\pi t} \int_0^{2\pi} \int_{\delta}^{\infty} e^{-r^2/t} r \, dr \, d\theta = \frac{\epsilon}{3} + 2M e^{-\delta^2/t} \quad (5)$$

[Here $M = \max |f|$, while Δ and Δ' denote the disc

$$(\xi_1 - x_1)^2 + (\xi_2 - x_2)^2 < \delta^2$$

and its complement, respectively.] We now choose t (independent of x_1 and x_2) such that $2M \exp(-\delta^2/t) < \epsilon/3$. With this choice of t, (5) yields the inequality

$$|f_t(x_1,x_2) - f(x_1,x_2)| < \frac{2\epsilon}{3} \quad (6)$$

(This inequality holds everywhere, not only in R.) We now write the Taylor expansion of the function e^{-u} in the form

$$e^{-u} = 1 - \frac{u}{1!} + \frac{u^2}{2!} - \cdots + \frac{(-u)^N}{N!} + R_N(u) \quad (7)$$

choosing N so large that $|R_N| < \epsilon \pi t/3MA$ whenever $|u| \leq \rho^2/t$, A and ρ denoting the area and diameter, respectively, of R. (That N can be so chosen follows either from the elements of the theory of analytic functions or from Taylor's theorem with remainder.) Letting

$$u = t^{-1}[(\xi_1 - x_1)^2 + (\xi_2 - x_2)^2]$$

and $$P(x_1,x_2) = (\pi t)^{-1} \iint_R f(\xi_1,\xi_2) \sum_{k=0}^{N} \frac{(-u)^k}{k!} \, d\xi_1 \, d\xi_2$$

we obtain

$$|f_t(x_1,x_2) - P(x_1,x_2)| \leq (\pi t)^{-1} \iint_R |f(\xi_1,\xi_2)| \cdot |R_N(u)| \, d\xi_1 \, d\xi_2$$

$$\leq \frac{M}{\pi t} \cdot \frac{\epsilon \pi t}{3MA} \cdot A = \frac{\epsilon}{3} \quad (8)$$

This is valid at all points of R. Combining (6) and (8), we obtain

$$|f(x_1,x_2) - P(x_1,x_2)| < \frac{2\epsilon}{3} + \frac{\epsilon}{3} = \epsilon \quad (9)$$

By inspection, it is clear that $P(x_1,x_2)$ is a polynomial, so that the proof of the theorem is complete, subject to the restrictions concerning R and f that were made at the beginning.

To eliminate these restrictions we require the following lemma.

LEMMA. LEBESGUE EXTENSION THEOREM. If a function $f(x_1,x_2,$ $\ldots,x_n)$ is defined and continuous on a compact set R, it is possible to extend f continuously to any larger set.

Momentarily accepting this lemma, we now choose a (closed) rectangle R' containing the compact set R in its interior, and extend f by defining it to be zero on the boundary of R'. Clearly f is still continuous under this extension, and its domain of definition is still compact. We invoke the lemma to extend f continuously to all of R'. Now the preceding argument is applicable, so that f can be approximated within ϵ throughout R', and a fortiori throughout R, by a polynomial.

It remains to prove the lemma. Again we may restrict attention to the case of two independent variables. First, we consider the very simple particular case of extending a continuous function defined on the boundary of a square to the interior. We merely assign to the center of the square the mean of the values at the four vertices, and then define the function along each line segment connecting the center to a boundary point by linear interpolation. It should be noted that this method of extending the given function assigns to each interior point a value between the minimum and maximum values which are assigned on the boundary. Now let a continuous function f be defined on any compact set R, and let R be contained in the interior of a square S whose boundary we denote by Γ. We then extend f continuously to the compact set $R \cup \Gamma$ by defining f to vanish everywhere on Γ. To prove the lemma it will suffice to show that f can be extended continuously to all of S, for if this can be done, we can accomplish the continuous extension of f to the entire plane, and hence to any specified set in the plane, by defining f to vanish everywhere outside Γ. Let G denote the set of points inside Γ which do not belong to R. Since G is open (and nonvacuous), it is possible to construct a network of equally spaced horizontal and vertical lines sufficiently fine that at least one square of this network lies, together with its boundary, entirely in G. Then the original network is refined by adding horizontal and vertical lines midway between those originally constructed. Those squares (if any) of the finer network which lie (together with their boundaries) in G but whose interiors are disjoint from the square or squares previously selected from the original network are now determined. By repeatedly refining the network and selecting squares, we evidently break G down into a countable union of closed squares whose interiors are disjoint. Let V be defined as the set of all points of G which appear as vertices of this collection of squares. We define f at each point Q of V as the minimum value of f at those points of $R \cup \Gamma$ which are closest to Q. (Since these points constitute a compact set, this definition is meaningful.) It is readily seen from the manner in which the squares were chosen that f has been defined at the vertex of each square, and perhaps also at a *finite* number of additional points on the boundary of each square. Let f be defined along the boundary of each square by linear interpolation between successive points of the set V, and then let f be defined inside each square by the method described earlier in this paragraph. Then f has evidently been defined throughout G and is continuous there, but it must still be shown that this function possesses the proper behavior near the boundary of G. Let any boundary point of G, say T, be selected, and let a positive number ϵ be given. Then $\delta(>0)$ can be so chosen that, for all points T' of $R \cup \Gamma$ whose distance from T does not exceed δ, the inequality $|f(T) - f(T')| < \epsilon$ holds. For any point P of V whose distance from T is less than $\frac{1}{2}\delta$, it is evident that those points of $R \cup \Gamma$ closest to P all lie within a distance less than δ from T, so that $|f(T) - f(P)| < \epsilon$. If, finally, P lies within a distance less than $\frac{1}{4}\delta$ from T, it is readily seen that P lies inside or on the boundary of a square such that all the points of V on the boundary of this square lie within a distance less than $\frac{1}{2}\delta$ from T. From the manner in which f was defined on the boundary and inside each square, it follows that $|f(T) - f(P)| < \epsilon$, and the proof is complete.

EXERCISES

5. Carry out in detail the proof of the Riemann-Lebesgue lemma sketched at the beginning of this section, for any function absolutely integrable over the real axis (in the Riemann sense).

6. Modify the proof given in the text to apply equally well to any value of n.

7. Prove the Lebesgue extension theorem in the one-dimensional case. (This case is decidedly simpler than in higher dimensions.)

8. Carry out in detail the proof of the Weierstrass theorem in one dimension which is outlined here: By the preceding exercise, we may assume that the function f is defined on a closed interval, rather than on an arbitrary compact set. By uniform continuity, f can be approximated uniformly by a polygonal function (i.e., a function which is continuous and sectionally linear, with only a finite number of "corners"). This polygonal function can be expressed as a finite sum of polygonal functions, each having only one "corner." Each such function, in turn, can be expressed as the sum of a linear function and a function of the form constant $\cdot |x - a|$. It therefore suffices to prove that the function $|x|$ can be uniformly approximated by polynomials on the interval $-1 \le x \le 1$. To do this, consider the identity $|x| = [1 - (1 - x^2)]^{\frac{1}{2}}$ and the Taylor series of the function $(1 - u)^{\frac{1}{2}}$ about the point $u = 0$. (This proof is due to Lebesgue.)

9. Prove the following extension of the Weierstrass theorem: If $f(x)$ is of class C^n on the interval $a \le x \le b$ [i.e., $f^{(n)}(x)$ exists and is continuous on the interval], then, given any $\epsilon > 0$, there exists a polynomial $P(x)$ such that $|f^{(k)}(x) - P^{(k)}(x)| < \epsilon$, $k = 0, 1, 2, \ldots, n$.

10. Let $f(x)$ be continuous and strictly increasing in the interval $a \le x \le b$. Show that, in this case, one can impose the additional requirement that the approximating polynomial should also be increasing. *Hint:* Approximate f by a suitable function of class C^1, and use Exercise 9.

11. A family of functions $\{f_t\}$ defined on a fixed domain D (in any number of dimensions) is said to be "concentrated" at Q, where Q is some point of D, if the following conditions are satisfied: (a) $f_t \ge 0$ throughout D, (b) $\int_D f_t = 1$, (c) $\int_\Delta f_t \to 1$ as $t \to 0$ (or some other limit) for each neighborhood Δ of Q. Prove that if the function g is bounded in D and continuous at Q, then $\int_D g f_t \to g(Q)$. (It is assumed that g is such that the integrals $\int_D g f_t$ exist. It is worth noting, however, that $\int_D g$ need not exist. For example, if the integrals are taken in the Lebesgue sense (cf. Sec. 6) it suffices that g be bounded and measurable, even though $\int_D g$ may then fail to exist if the measure of D is infinite.) The functions P_t used in the present section constitute a sequence concentrated at (x_1, x_2).

12. Let $\{p_n(x)\}$, $n = 0, 1, 2, \ldots$, be a sequence of polynomials orthonormal with respect to an interval $a \le x \le b$ [i.e., $\int_a^b p_n(x)p_m(x)\,dx = \delta_{nm}$], and let p_n be of degree n. (For $a = -1$, $b = +1$, the p_n are, aside from constant factors, identical with the Legendre polynomials.) Let $f(x)$ be continuous in the afore-mentioned interval, and let the coefficients f_n be defined as follows: $f_n = \int_a^b f(x)p_n(x)\,dx$. Prove that the series $\sum_{n=0}^{\infty} f_n^2$ converges and that its sum is equal to $\int_a^b f^2(x)\,dx$.

As in the case of Fourier series, there is no assurance that the series $\sum_{n=0}^{\infty} f_n p_n(x)$ actually converges unless suitable additional conditions are imposed on $f(x)$.

3. The Fourier Integral

In the theory of Fourier series one investigates the relationship between a periodic function $f(x)$ and its formal expansion in exponential functions,

$$f(x) \sim \sum_{-\infty}^{\infty} a_n \exp \frac{2\pi i n x}{L} \tag{10}$$

where L is the period of $f(x)$ and the coefficients a_n are given by

$$a_n = \frac{1}{L} \int_0^L f(\xi) \exp\left(-\frac{2\pi i n \xi}{L}\right) d\xi \tag{11}$$

We assume that the reader is acquainted with the elements of this theory, and shall devote the present section to a brief discussion of the

Fourier integral, which can be looked upon as the limiting case of (10) obtained by letting L become infinite. Given any finite interval of the real axis, the number of integers n for which n/L lies in this interval increases with L; in the limit as $L \to \infty$, one would expect (10) to assume a form in which the quantity n/L assumes all real values, so that the summation should be replaced by an integration. The following theorem renders this heuristic remark more precise.

Theorem 3. Fourier Integral Theorem. Let $f(x)$ be defined for all real x and let the integral[1] $\int_{-\infty}^{\infty} |f(x)|\, dx$ exist. Then the integral

$$g(\lambda) = \int_{-\infty}^{\infty} f(x) e^{-i\lambda x}\, dx \tag{12}$$

exists for all real λ, and the function $g(\lambda)$ so defined is continuous. For any value of x, say ξ, at which the derivative $f'(x)$ exists, the value of the function may be recovered from $g(\lambda)$ by the formula[2]

$$f(\xi) = \frac{1}{2\pi} \lim_{A \to \infty} \int_{-A}^{A} g(\lambda) e^{i\lambda\xi}\, d\lambda \tag{13}$$

Proof. The existence of the integral (12) for all real values of λ follows from the absolute integrability of $f(x)$ and the boundedness of $|e^{-i\lambda x}|$. (Since λ and x are real, $|e^{-i\lambda x}| \equiv 1$.) The continuity of $g(\lambda)$ is easily proved in the manner suggested in Exercise 14.

To establish (13) we need the following lemma, known as the Fourier single-integral theorem.

LEMMA. Under the same hypotheses as above,

$$\lim_{A \to \infty} \int_{-\infty}^{\infty} \frac{f(x) - f(\xi)}{x - \xi} \sin A(x - \xi)\, dx = 0 \tag{14}$$

Temporarily assuming this lemma, we establish (13) as follows:[3]

$$\frac{1}{2\pi} \int_{-A}^{A} g(\lambda) e^{i\lambda\xi}\, d\lambda = \frac{1}{2\pi} \int_{-A}^{A} \left[\int_{-\infty}^{\infty} f(x) e^{-i\lambda x}\, dx \right] e^{i\lambda\xi}\, d\lambda$$

$$= \frac{1}{2\pi} \int_{-\infty}^{\infty} f(x) \left(\int_{-A}^{A} e^{-i\lambda(x-\xi)}\, d\lambda \right) dx = \frac{1}{\pi} \int_{-\infty}^{\infty} f(x) \frac{\sin A(x - \xi)}{x - \xi}\, dx \tag{15}$$

---

[1] In the general theory it is essential to employ the Lebesgue integral, but for the applications to be made here it suffices to interpret all integrals in the Riemann sense.

[2] As may be shown by a simple example (cf. Exercise 17), the integral $\int_{-\infty}^{\infty} g(\lambda) e^{i\lambda\xi}\, d\lambda$ may fail to exist; the *symmetric* approach to infinity of the limits of integration indicated in (13) is therefore essential. If $\int_{-\infty}^{\infty}$ does not exist but $\lim_{A \to \infty} \int_{-A}^{A}$ does, this limit is referred to as the "Cauchy principal value" of the divergent integral $\int_{-\infty}^{\infty}$.

[3] Note that for finite A the two iterated integrals appearing in (15) are absolutely convergent and, hence, equal to each other; the passage to the limit $A \to \infty$ is performed *after* the interchange of order of integration.

Subtracting from (15) the equality

$$f(\xi) = \frac{1}{\pi} \int_{-\infty}^{\infty} f(\xi) \frac{\sin A(x - \xi)}{x - \xi} dx$$

(cf. Exercise 15), we obtain

$$\frac{1}{2\pi} \int_{-A}^{A} g(\lambda) e^{i\lambda\xi} d\lambda - f(\xi) = \frac{1}{\pi} \int_{-\infty}^{\infty} \frac{f(x) - f(\xi)}{x - \xi} \sin A(x - \xi) dx \quad (16)$$

We now obtain (13) by applying the above lemma to (16).

It remains to establish the lemma. The existence of the limit appearing in (14) is readily seen as follows. Let two numbers x_1, x_2 be chosen such that $x_1 < \xi < x_2$. The integral

$$\int_{x_1}^{x_2} \frac{f(x) - f(\xi)}{x - \xi} \sin A(x - \xi) dx$$

is well defined on account of the assumption that $f'(\xi)$ exists, for this implies that the function $[f(x) - f(\xi)]/(x - \xi)$, if defined as $f'(\xi)$ for $x = \xi$, is continuous at ξ. The integral

$$\int_{x_2}^{\infty} \frac{f(x)}{x - \xi} \sin A(x - \xi) dx$$

exists, since the integrand is dominated by the function $|f(x)|/(x_2 - \xi)$. As for

$$\int_{x_2}^{\infty} \frac{f(\xi)}{x - \xi} \sin A(x - \xi) dx$$

it is reduced by the substitution $t = A(x - \xi)$ to the integral

$$f(\xi) \int_{A(x_2 - \xi)}^{\infty} \frac{\sin t}{t} dt$$

which is known to exist. (Cf. Exercise 15.) Thus, the integral

$$\int_{x_2}^{\infty} \frac{f(x) - f(\xi)}{x - \xi} \sin A(x - \xi) dx$$

exists, and the same statement is, of course, true for the integral

$$\int_{-\infty}^{x_1} \frac{f(x) - f(\xi)}{x - \xi} \sin A(x - \xi) dx$$

Having established the convergence of the integral appearing in (14), we now consider its behavior for large A. Given $\epsilon > 0$, we impose on the quantities x_1 and x_2, which were introduced previously, the additional conditions

$$x_1 < \xi - 1 \qquad\qquad x_2 > \xi + 1$$

$$\int_{-\infty}^{x_1} |f(x)| \, dx < \epsilon \qquad\qquad \int_{x_2}^{\infty} |f(x)| \, dx < \epsilon$$

The possibility of satisfying the two latter conditions is assured by the existence of $\int_{-\infty}^{\infty} |f(x)|\, dx$. Splitting the integral appearing in (14) into the sum of the following five integrals,

$$I_1 = \int_{-\infty}^{x_1} \frac{f(x)}{x - \xi} \sin A(x - \xi)\, dx$$

$$I_2 = \int_{-\infty}^{x_1} \frac{-f(\xi)}{x - \xi} \sin A(x - \xi)\, dx$$

$$I_3 = \int_{x_1}^{x_2} \frac{f(x) - f(\xi)}{x - \xi} \sin A(x - \xi)\, dx$$

$$I_4 = \int_{x_2}^{\infty} \frac{f(x)}{x - \xi} \sin A(x - \xi)\, dx$$

$$I_5 = \int_{x_2}^{\infty} \frac{-f(\xi)}{x - \xi} \sin A(x - \xi)\, dx$$

we immediately see that $|I_1| < \epsilon$, $|I_4| < \epsilon$, regardless of the value of A. Employing the substitution $t = A(x - \xi)$ used previously, we see that for all sufficiently large values of A the inequalities $|I_2| < \epsilon$, $|I_5| < \epsilon$ hold. Finally, by the Riemann-Lebesgue lemma (cf. Sec. 2), $I_3 \to 0$ as $A \to \infty$, so that $|I_3| < \epsilon$ for A sufficiently large. Hence, the inequalities

$$|I_1 + I_2 + I_3 + I_4 + I_5| \leq |I_1| + |I_2| + |I_3| + |I_4| + |I_5| < 5\epsilon \quad (17)$$

hold for all sufficiently large values of A, and this assertion is equivalent to (14).

The function $g(\lambda)$ associated with $f(x)$ according to (12) is termed the "Fourier transform" of $f(x)$, and will be denoted by the symbol $\mathfrak{F}(f)$.

EXERCISES

13. Derive Theorem 3 *formally* by performing suitable limiting operations in (10) and (11).

14. Prove, under the assumption that $\int_{-\infty}^{\infty} |f(x)|\, dx$ exists, that the function $g(\lambda)$ defined by (12) is continuous, and that $g(\lambda) \to 0$ as λ becomes infinite. *Hint:* Divide the real axis into three parts such that the integrals $\int f(x) e^{-i\lambda x}\, dx$ over the two semi-infinite intervals are small, and apply the Riemann-Lebesgue lemma to the integral over the finite interval.

15. (*a*) Prove the trigonometric identity

$$\tfrac{1}{2} + \cos u + \cos 2u + \cdots + \cos nu = \frac{\sin (n + \tfrac{1}{2})u}{2 \sin \tfrac{1}{2}u} \qquad (n = 1, 2, \ldots) \quad (18)$$

(*b*) Prove that the function $1/u - 1/2 \sin \tfrac{1}{2}u$ is continuous at $u = 0$ if defined to be equal to zero there.

(c) Use (a) and (b) together with the Riemann-Lebesgue lemma to show that[1]

$$\frac{1}{\pi} \int_{-\infty}^{\infty} \frac{\sin kt}{t} \, dt = \begin{cases} 1 & \text{if } k > 0 \\ 0 & \text{if } k = 0 \\ -1 & \text{if } k < 0 \end{cases} \tag{19}$$

(In particular the existence of the integral must be shown.)

16. Modify the proof given in the text to show that (13) holds under the weaker hypotheses that the left- and right-hand limits

$$f(\xi^-) = \lim_{\substack{h \to 0 \\ h > 0}} f(\xi - h) \qquad f(\xi^+) = \lim_{\substack{h \to 0 \\ h > 0}} f(\xi + h)$$

and the left- and right-hand derivatives

$$\lim_{\substack{h \to 0 \\ h > 0}} \frac{f(\xi - h) - f(\xi^-)}{-h} \qquad \lim_{\substack{h \to 0 \\ h > 0}} \frac{f(\xi + h) - f(\xi^+)}{h}$$

all exist and that $f(\xi) = \frac{1}{2}[f(\xi^+) + f(\xi^-)]$.

17. Determine a function $f(x)$ such that $\int_{-\infty}^{\infty} |f(x)| \, dx$ exists and (13) holds for all values of ξ, yet such that for some value of ξ the integral $\int_{-\infty}^{\infty} g(\lambda)e^{i\lambda\xi} \, d\lambda$ [in contrast to $\lim_{A \to \infty} \int_{-A}^{A} g(\lambda)e^{i\lambda\xi} \, d\lambda$] fails to exist.

18. By explicitly evaluating the Fourier transform of the function $e^{-|x|}$ and then employing Theorem 3, evaluate the integral

$$\int_{-\infty}^{\infty} \frac{\cos ax}{1 + x^2} \, dx$$

19. Evaluate the integrals appearing in Exercises 15 and 18 by employing complex integration (residues).

20. Prove, under suitable hypotheses, that $\mathcal{F}(f'(x)) = i\lambda\mathcal{F}(f)$, and, more generally,

$$\mathcal{F}(f^{(n)}(x)) = (i\lambda)^n \mathcal{F}(f) \tag{20}$$

21. Prove, under suitable hypotheses on the functions $f(x)$ and $g(x)$, that

$$\mathcal{F}(f)\mathcal{F}(g) = \mathcal{F}(f * g) \tag{21}$$

where $f * g$, the "convolution" or "faltung" of f and g, is given by

$$f * g = \int_{-\infty}^{\infty} f(\xi)g(x - \xi) \, d\xi = \int_{-\infty}^{\infty} g(\xi)f(x - \xi) \, d\xi = g * f \tag{22}$$

4. The Laplace Transform

Closely related to the Fourier transform is that named after Laplace, which will now be discussed briefly. Applications of both these transforms will be made in Chap. 8 in connection with the equation of heat conduction.

---

[1] The function defined by (19) is termed the "signum function," denoted sgn k.

Let $f(x)$, assumed defined for $x \geq 0$, be such that for some value of the complex parameter s the integral

$$\tilde{f}(s) = \int_0^\infty e^{-sx} f(x) \, dx \tag{23}$$

converges (not necessarily abolutely).[1] [To give several simple examples, the integral (23) exists for all values of s if $f(x) = e^{-x^2}$, for any value of s such that Re $(s - c) > 0$ if $f(x) = e^{cx}$, but for no value of s if $f(x) = e^{x^2}$.] Then the function $\tilde{f}(s)$ thus defined for those values of s for which the integral (23) exists is called the "Laplace transform" of $f(x)$. A most striking fact about integral (23) is that the set of values of s for which it exists [that is, the domain of the function $\tilde{f}(s)$] is always of a very simple form, as described by the following theorem.

Theorem 4. For any given function $f(x)$, the integral (23) converges for (1) no values of s, or (2) all values of s, or (3) all values of s whose real part exceeds a certain real number α [the "abscissa of convergence" of $f(x)$], and, perhaps, for some or all values of s whose real part equals α. In cases 2 and 3 the function $\tilde{f}(s)$ is analytic for all s and in the half plane Re $s > \alpha$, respectively.

Proof. It is readily seen that the first part of the theorem can be reformulated in the following more concise form: If Re $s_1 >$ Re s_0 and if the integral (23) converges for $s = s_0$, it also converges for $s = s_1$. While this statement is trivial if "converges" is replaced by "converges absolutely" [for, since $x \geq 0$, the inequality $|e^{-s_1 x} f(x)| \leq |e^{-s_0 x} f(x)|$ must hold], it is conceivable that when the integral converges only conditionally for $s = s_0$ it might fail to converge for $s = s_1$. (Cf. Exercise 22.) To rule out this possibility, recourse is made to integration by parts, as follows. Let $g(A) = \int_0^A e^{-s_0 x} f(x) \, dx$; by hypothesis, $\lim\limits_{A \to \infty} g(A)$ exists. Then, through integration by parts,[2] we obtain

$$\int_0^A e^{-s_1 x} f(x) \, dx = e^{-(s_1 - s_0)A} g(A) + (s_1 - s_0) \int_0^A e^{-(s_1 - s_0)x} g(x) \, dx \tag{24}$$

Since Re $(s_1 - s_0) > 0$ and $\lim\limits_{A \to \infty} g(A)$ exists, the quantity $e^{-(s_1 - s_0)A} g(A)$ vanishes as A becomes infinite. Since $g(x)$ is continuous (being an integral) and approaches a finite limit as $x \to \infty$, it is bounded in absolute value, say $|g(x)| < M$. Therefore, $\int_0^\infty e^{-(s_1 - s_0)x} g(x) \, dx$ is dominated by

---

[1] However, it is necessary to assume that $\int_0^A |f(x)| \, dx$ exists for every finite A. [Cf. the expression following (25).]

[2] The integration by parts is valid even without the assumption that $f(x)$ is continous; it suffices that $f(x)$ be absolutely integrable on every finite interval $0 \leq x \leq A$. (Cf. preceding footnote.)

the convergent integral $\int_0^\infty |e^{-(s_1-s_0)x}|M\,dx$, and is consequently itself convergent. The right-hand side, and hence the left-hand side, of (24) has been shown to possess a finite limit as A becomes infinite. The first part of the theorem is thus proved.

Turning to the assertion concerning analyticity, we begin by showing that, for any fixed (finite) value of A, the integral $\int_0^A e^{-sx}f(x)\,dx$, which for convenience we denote by $F_A(s)$, is analytic for all values of s. For any complex numbers s and h ($h \neq 0$), we readily obtain

$$\frac{F_A(s+h) - F_A(s)}{h} + \int_0^A e^{-sx}xf(x)\,dx = \int_0^A e^{-sx}\frac{e^{-hx}-1+hx}{h}f(x)\,dx$$

$$= h\int_0^A e^{-sx}\left(-\frac{x^2}{2!} + \frac{hx^3}{3!} - \cdots\right)f(x)\,dx \quad (25)$$

For $|h| < 1$ we immediately find that the right side of (25) is dominated by the expression

$$|h|e^{|s|A}\left(\frac{A^2}{2!} + \frac{A^3}{3!} + \cdots\right)\int_0^A |f(x)|\,dx$$

It follows that the left side of (25) must approach zero with h, and hence that $\dfrac{dF_A(s)}{ds}$ exists and equals $-\int_0^A e^{-sx}xf(x)\,dx$, a result that could have been obtained *formally* by differentiation under the integral.

Next we replace s_1 in (24) by s and subtract the resulting equation from (23). We thus obtain

$$F_A(s) - \tilde{f}(s) = e^{-(s-s_0)A}g(A) - (s - s_0)\int_A^\infty e^{-(s-s_0)x}g(x)\,dx \quad (26)$$

(It is understood, of course, that s is again subjected to the restriction Re $(s - s_0) > 0$.) This immediately yields the inequality

$$|F_A(s) - \tilde{f}(s)| \le Me^{-A\,\text{Re}\,(s-s_0)} + Me^{-A\,\text{Re}\,(s-s_0)}\frac{|s - s_0|}{\text{Re}\,(s - s_0)} \quad (27)$$

Given any $\epsilon > 0$, choose positive numbers δ, R, and A_0 such that the inequalities

$$\text{Re}\,(s - s_0) > \delta \qquad \frac{|s - s_0|}{\text{Re}\,(s - s_0)} < R \qquad Me^{-A_0\delta}(1 + R) < \epsilon \quad (28)$$

are all satisfied. From (27) and (28) we then conclude that the analytic functions $F_A(s)$ converge *uniformly* to $\tilde{f}(s)$, as $A \to \infty$, in the domain defined by the first two inequalities of (28); by a basic theorem on analytic functions, $\tilde{f}(s)$ is itself analytic at s. This completes the proof.

It may happen [and usually does for the functions $f(x)$ most commonly encountered in specific problems] that the function $\tilde{f}(s)$ can be defined analytically in a region larger than the half plane of convergence. Thus, if $f(x) \equiv 1$ the abscissa of convergence is given by $\alpha = 0$, but the transform $\tilde{f}(s) = 1/s$ can be continued analytically over the entire plane except for the origin. It may happen, furthermore, that $\tilde{f}(s)$ becomes multiple-valued when continued analytically. (Cf. Sec. 8-2.)

If we formally set $s = i\lambda$ and define $f(x)$ to be zero for negative values of x, (23) goes over into the Fourier transform (12). The connection between the two types of transform is rendered more precise by the following theorem.

Theorem 5. Laplace Inversion Formula. Let $f(x)$ be such that the integral (23) is absolutely convergent[1] for some real value of s, say $s = \sigma$, and let $f(x) \equiv 0$ for $x < 0$. Then at any value of x, say $x = \xi$, at which $f(x)$ is differentiable,

$$f(\xi) = \frac{1}{2\pi i} \int_{\sigma_1 - i\infty}^{\sigma_1 + i\infty} \tilde{f}(s) e^{s\xi} \, ds \tag{29}$$

where the integral is defined as a Cauchy principal integral (cf. Sec. 3), taken along the vertical line Re $s = \sigma_1$ in the complex plane, σ_1 being any real number exceeding σ.†

Proof. For any s on the path of integration we may write $s = \sigma_1 + i\lambda$, λ real. Then (23) assumes the form

$$\tilde{f}(s) = \tilde{f}(\sigma_1 + i\lambda) = \int_{-\infty}^{\infty} [e^{-\sigma_1 x} f(x)] e^{-i\lambda x} \, dx \tag{30}$$

Replacing $f(x)$ in (12) by $e^{-\sigma_1 x} f(x)$, we obtain, in place of (13), the inversion formula

$$e^{-\sigma_1 \xi} f(\xi) = \frac{1}{2\pi} \int_{-\infty}^{\infty} \tilde{f}(\sigma_1 + i\lambda) e^{i\lambda \xi} \, d\lambda \tag{31a}$$

or $$f(\xi) = \frac{1}{2\pi i} \int_{\lambda = -\infty}^{\lambda = \infty} \tilde{f}(\sigma_1 + i\lambda) e^{(\sigma_1 + i\lambda)} d(\sigma_1 + i\lambda) \tag{31b}$$

Setting $\sigma_1 + i\lambda = s$, we find that (31b) reduces to (29), thus completing the proof.

EXERCISES

22. Construct a pair of continuous functions $g(x)$, $h(x)$ such that (a) $|h(x)| \leq |g(x)|$, (b) $h(x)/g(x) > 0$ (for all x such that $g(x) \neq 0$), (c) $\int_0^\infty g(x) \, dx$ converges, but (d) $\int_0^\infty h(x) \, dx$ diverges.

---

[1] This assumption can be weakened, but it is introduced in order to permit an immediate application of Theorem 3.

† In particular, then, the integral (29) equals zero for any negative value of ξ.

23. Let $\tilde{f}(s)$ exist for a value s_0 whose real part equals α, the abscissa of convergence. Prove that $\lim \tilde{f}(s)$ exists and equals $\tilde{f}(s_0)$ if s approaches s_0 within a "wedge" (Stolz region) defined by the inequalities $-\theta \leq \arg (s - s_0) \leq \theta$, θ being any positive constant less than $\pi/2$. [This theorem is closely related to Abel's theorem, which states that if the series $a_1 + a_2 + \cdots$ converges to a, then the sum of the series $a_1 z + a_2 z^2 + \cdots$ approaches a if z approaches the value one within a wedge defined by the inequalities $-\theta \leq \arg (1 - z) \leq \theta$, $0 < \theta < \pi/2$.]

24. Determine a function whose Laplace transform converges *conditionally* for all values of s.

25. Determine the proper analogues of Theorems 4 and 5 for the "bilateral Laplace transform" $\int_{-\infty}^{\infty} e^{-sx} f(x)\, dx$, where $f(x)$ is now defined for all real values of x.

26. Prove, under suitable assumptions on $f(x)$, that $\widetilde{f'} = s\tilde{f}(s) - f(0^+)$. This formula, which is the analogue of (20), plays a fundamental role in the application of the Laplace transform to the solution of linear differential equations, both ordinary and partial.

27. In analogy with Exercise 21, prove, under suitable hypotheses on the functions $f(x)$ and $g(x)$, the convolution formula

$$\tilde{f}(s)\tilde{g}(s) = \widetilde{f * g},\ f * g = \int_0^x f(\xi)g(x - \xi)\, d\xi \tag{32}$$

28. Let $f(x)$ be periodic (for $x \geq 0$). Prove that $\tilde{f}(s)$, as defined by (23), exists at least for Re $s > 0$, but not necessarily in a larger half plane. On the other hand, show that $\tilde{f}(s)$ can be analytically continued to the entire plane except perhaps for simple poles on the imaginary axis.

5. Ordinary Differential Equations

In the customary introductory course in ordinary differential equations, methods are taught for solving various classes of such equations (linear with constant coefficients, Riccati, Clairaut, etc.), but the question of existence and uniqueness is usually disregarded. This question may be formulated as follows: Under what conditions does a given differential equation possess solutions, and when solutions do exist, what conditions may be imposed in order to assure uniqueness? A large part of the theory of differential equations deals with this question, and the basic result of this part of the theory is given by the following theorem.[1]

Theorem 6.   Cauchy-Picard.   Let $f(x,y)$ be defined and continuous in a closed rectangle $|x - x_0| \leq a$, $|y - y_0| \leq b$, and let $f(x,y)$ satisfy, in this rectangle, a Lipschitz condition with respect to y; that is, a positive constant k is assumed to exist such that the inequality

$$|f(x,y_1) - f(x,y_2)| \leq k|y_1 - y_2|$$

holds throughout this rectangle. Let $M = \max |f(x,y)|$ in the afore-

---

[1] A thorough mastery of this theorem and its proof is urged. The proof provides a comparatively simple, yet typical, illustration of the method of successive approximation, which is a powerful and frequently employed technique in analysis.

mentioned rectangle and let $c = \min (a,b/M)$. Then there exists for $|x - x_0| \leq c$ a function $g(x)$ such that $|g(x) - y_0| \leq b$, $g(x_0) = y_0$, and

$$\frac{dg(x)}{dx} \equiv f(x,g(x)) \tag{33a}$$

i.e., $g(x)$ is a solution of the differential equation

$$\frac{dy}{dx} = f(x,y) \tag{33b}$$

Finally, the solution is unique; i.e., if $h(x)$ also satisfies the differential equation (33b) in the afore-mentioned interval $|x - x_0| \leq c$, and if $h(x_0) = y_0$, then $h(x) \equiv g(x)$.

Proof. The existence portion of the theorem will be proved by actually constructing a solution. Let a sequence of functions $\{g_n(x)\}$, $n = 0, 1, 2, \ldots$, be defined for $|x - x_0| \leq c$ as follows:

$$g_0(x) \equiv y_0 \tag{34a}$$
$$g_n(x) = y_0 + \int_{x_0}^{x} f(\xi,g_{n-1}(\xi)) \, d\xi \qquad (n \geq 1) \tag{34b}$$

First, it must be shown that the definition of the functions $g_n(x)$ provided by (34b) is meaningful. Assume that $g_j(x)$ is defined and continuous for $|x - x_0| \leq c$ and satisfies in this interval the inequality $|g_j(x) - y_0| \leq b$. Then the right-hand side of (34b) is meaningful (and continuous) when $n - 1$ is replaced by j, and we obtain for $g_{j+1}(x)$ the estimate

$$|g_{j+1}(x) - y_0| \leq M|x - x_0| \leq Mc \leq b \tag{35}$$

so that $g_{j+1}(x)$ satisfies the conditions imposed above on $g_j(x)$. Since, in particular, $g_0(x)$ satisfies these conditions, the induction is complete.

Next it will be shown that the sequence of functions $\{g_n(x)\}$ is uniformly convergent. Replacing n in (34b) by $n + 1$ and subtracting, we obtain

$$g_{n+1}(x) - g_n(x) = \int_{x_0}^{x} [f(\xi,g_n(\xi)) - f(\xi,g_{n-1}(\xi))] \, d\xi \tag{36}$$

From (36) and the hypothesis that $f(x,y)$ satisfies a Lipschitz condition, we readily obtain the inequality[1]

$$|g_{n+1}(x) - g_n(x)| \leq k \int_{x_0}^{x} |g_n(\xi) - g_{n-1}(\xi)| \, d\xi \qquad (n \geq 1) \tag{37a}$$

while for $n = 0$ the appropriate inequality is the following:

$$|g_1(x) - g_0(x)| \leq kM(x - x_0) \tag{37b}$$

---

[1] For convenience, we restrict x to the interval $x_0 \leq x \leq x_0 + c$; the reasoning employed obviously applies, with trivial modifications, to the interval $x_0 - c \leq x \leq x_0$ as well.

By an easy induction (cf. Exercise 29), we readily obtain from (37a) and (37b) the inequality

$$|g_{n+1}(x) - g_n(x)| \leq \frac{k^{n+1}M(x - x_0)^{n+1}}{(n + 1)!} \tag{38}$$

and, a fortiori,

$$|g_{n+1}(x) - g_n(x)| \leq \frac{M(kc)^{n+1}}{(n + 1)!} \tag{39}$$

Therefore, the series

$$\sum_{n=0}^{\infty} [g_{n+1}(x) - g_n(x)] \tag{40}$$

is dominated by the series of *constant* terms

$$\sum_{n=0}^{\infty} \frac{M(kc)^{n+1}}{(n + 1)!}$$

which converges for all values of M, k, and c, and so series (40) converges uniformly (as well as absolutely). This is equivalent to the assertion that the sequence of partial sums

$$S_m(x) = \sum_{n=0}^{m} [g_{n+1}(x) - g_n(x)] \qquad (m \geq 0) \tag{41}$$

is uniformly convergent. The sum on the right[1] is readily seen to equal $g_{m+1}(x) - g_0(x)$, or $g_{m+1}(x) - y_0$. Therefore, the sequence of functions $\{g_n(x)\}$ is uniformly convergent to a limit, say $g(x)$, and from (34b) we obtain

$$g(x) = y_0 + \lim_{n \to \infty} \int_{x_0}^{x} f(\xi, g_n(\xi)) \, d\xi \tag{42}$$

It is now necessary to show that the limiting operation appearing in (42) can be performed under the integral sign; i.e., that the operations $\lim\limits_{n \to \infty}$ and $\int_{x_0}^{x}$ are permutable. This may be done as follows: Since, as shown earlier, $|g_n(x) - y_0| \leq b$ for all n, the same inequality must hold in the limit, namely,

$$|g(x) - y_0| \leq b \tag{43}$$

Since, furthermore, $g(x)$ is the uniform limit of continuous functions $g_n(x)$, it is itself continuous. Therefore, $f(x, g(x))$ is well defined and continuous, and the expression $\int_{x_0}^{x} f(\xi, g(\xi)) \, d\xi$ is meaningful. We may

---

[1] A series, finite or infinite, of the form $\Sigma(u_{n+1} - u_n)$, such as (41), is said to be "telescoping."

therefore rewrite (42) as follows:

$$g(x) = y_0 + \int_{x_0}^{x} f(\xi, g(\xi)) \, d\xi + \lim_{n \to \infty} \int_{x_0}^{x} [f(\xi, g_n(\xi)) - f(\xi, g(\xi))] \, d\xi \quad (44)$$

Taking account once again of the Lipschitz condition, we find that the second integral in (44) is dominated by

$$k|x - x_0| \cdot \max_{|x - x_0| \leq c} |g(x) - g_n(x)|$$

and hence must approach zero as n becomes infinite. Therefore, the last term in (44) vanishes, and we conclude that $g(x)$ satisfies the integral identity

$$g(x) = y_0 + \int_{x_0}^{x} f(\xi, g(\xi)) \, d\xi \quad (45)$$

This equation is equivalent to the statement made above concerning the permutability of the operations $\lim\limits_{n \to \infty}$ and $\int_{x_0}^{x}$ appearing in (42).

Since $g(x)$ and $f(x, y)$ are continuous functions of the indicated variables, it follows that $f(x, g(x))$ is a continuous function of x; the elementary rule for differentiation of an integral with respect to its upper limit is therefore applicable, and we obtain

$$\frac{dg(x)}{dx} \equiv f(x, g(x)) \quad (46)$$

Furthermore, it is clear from (45) [or simply from the fact that $g_n(x_0) = y_0$ for all n] that $g(x_0) = y_0$. Thus the existence portion of the theorem has been proved. It is worth stressing that the fact that the function $g(x)$ is differentiable does not follow by any means from the mere fact that it is the uniform limit of the differentiable functions $g_n(x)$; the differentiability of $g(x)$ and the further fact that $g(x)$ satisfies the identity (46) are established only by having recourse to (45). This example illustrates the usefulness of *integral* representations in establishing properties of *differentiability*.[1]

Uniqueness is now easily established. Suppose that $h(x)$ is also a solution of (33b) satisfying $h(x_0) = y_0$. Then by integration we get

$$h(x) = y_0 + \int_{x_0}^{x} f(\xi, h(\xi)) \, d\xi \quad (45')$$

Subtracting (45') from (45) and exploiting the Lipschitz condition, we obtain

$$|g(x) - h(x)| \leq k \int_{x_0}^{x} |g(\xi) - h(\xi)| \, d\xi \quad (47)$$

---

[1] In a quite different connection, the reader may recall that the fact that a function of a complex variable having *first* derivatives throughout a domain has derivatives of *all* orders throughout the domain is proved with the aid of the Cauchy integral representation for analytic functions.

(As before, we restrict attention, for convenience, to values of $x \geq x_0$.) Let $\mu = \max\limits_{x_0 \leq x \leq x_0 + c} |g(x) - h(x)|$. Then, from (47), we obtain, successively,

$$|g(x) - h(x)| \leq k \int_{x_0}^{x} \mu \, d\xi = k\mu(x - x_0) \tag{48a}$$

$$|g(x) - h(x)| \leq k \int_{x_0}^{x} k\mu(\xi - x_0) \, d\xi = \frac{k^2(x - x_0)^2}{2!} \tag{48b}$$

and, by induction,

$$|g(x) - h(x)| \leq \frac{k^n(x - x_0)^n}{n!} \tag{48c}$$

Letting n become infinite, we conclude from (48c) that $|g(x) - h(x)| \equiv 0$. This completes the proof of uniqueness. (Cf. Exercise 31.)

The hypothesis that $f(x,y)$ satisfies a Lipschitz condition was exploited in establishing both the existence and the uniqueness parts of the above theorem. One might conjecture that, by employing more powerful methods, one might prove this theorem without the Lipschitz condition— that is, merely under the hypothesis that $f(x,y)$ is continuous. It is of interest that the existence can be proved under this weaker hypothesis, but not the uniqueness.

To demonstrate that uniqueness may now fail, consider the differential equation

$$\frac{dy}{dx} = |y|^{1/2} \tag{49}$$

The function $|y|^{1/2}$ is easily seen to satisfy a Lipschitz condition in any rectangle which (including its boundary) lies entirely in the upper (or lower) half plane, but not in any rectangle containing a segment of the x axis in its interior or on its boundary. (Cf. Exercise 32.) Through the point $(0,0)$ there exist, among others (cf. Exercise 33), the following four solutions:

$$y \equiv 0 \qquad y = \tfrac{1}{2}x|x| \qquad y = \tfrac{1}{2}[\max (x,0)]^2 \qquad y = \tfrac{1}{2}[\min (x,0)]^2$$

Turning now to the question of existence under the weaker hypothesis that $f(x,y)$ is continuous, we approximate $f(x,y)$ uniformly in R by a sequence of polynomials $\{P_n(x,y)\}$. (Cf. Sec. 2.) Corresponding to each n we define quantities M_n and c_n analogous to M and c; furthermore, since each $P_n(x,y)$ satisfies a Lipschitz condition in R (cf. Exercise 32), we can, by Theorem 6, associate with each n a (unique) function $G_n(x)$ such that

$$G_n(x_0) = y_0 \qquad \frac{dG_n(x)}{dx} \equiv P_n(x, G_n(x)) \qquad (|x - x_0| \leq c_n) \tag{50a}$$

or, in integral form,

$$G_n(x) = y_0 + \int_{x_0}^{x} P_n(\xi, G_n(\xi)) \, d\xi \tag{50b}$$

Now, the functions $G_n(x)$ are uniformly bounded in the intervals $|x - x_0| \leq c_n$, for, as is evident from (50b),

$$|G_n(x)| \leq |y_0| + c_n M_n \leq |y_0| + (\max c_n)(\max M_n) \tag{51}$$

where M_n denotes the maximum of $|P_n(x,y)|$ in the rectangle R. (Since M_n approaches the quantity M defined in the statement of Theorem 6, the quantity $\max M_n$ exists.) Furthermore, from (50a),

$$\left| \frac{dG_n}{dx} \right| \leq M_n \leq \max M_n \tag{52}$$

so that the functions $G_n(x)$ are equicontinuous.[1] Hence Theorem 1 applies, and we can, therefore, select a subsequence of the sequence $\{G_n(x)\}$ which converges uniformly to a function $G(x)$. Confining attention to this subsequence, we let n become infinite and obtain, from (50b),

$$G(x) = y_0 + \lim_{n \to \infty} \int_{x_0}^{x} P_n(\xi, G_n(\xi)) \, d\xi \tag{53a}$$

Taking account of the uniform convergence of $\{P_n(x,y)\}$ to $f(x,y)$ and of $\{G_n(x)\}$ to $G(x)$, we easily show that (53a) may be rewritten in the form

$$G(x) = y_0 + \int_{x_0}^{x} f(\xi, G(\xi)) \, d\xi \tag{53b}$$

and from this it follows that $G(x_0) = y_0$ and that $G(x)$ satisfies in $|x - x_0| \leq c$ the differential equation (33b).

The above theorem is readily extended to systems of differential equations, as indicated by the following corollary, whose proof, which is a routine modification of the proof of the theorem, is left to the reader as Exercise 34.

COROLLARY. Let the functions $f_i(x, y^{(1)}, y^{(2)}, \ldots, y^{(n)})$ be defined and continuous in a region $|x - x_0| \leq a$, $|y^{(j)} - y_0^{(j)}| \leq b^{(j)}$ $(i,j = 1, 2, \ldots, n)$, and let each of the functions f_i satisfy a Lipschitz condition with respect to each of the variables $y^{(j)}$; i.e., there exists a constant k such that, for each i and each j,

$$|f_i(x, y^{(1)}, y^{(2)}, \ldots, y^{(j-1)}, y_1^{(j)}, y^{(j+1)}, \ldots)$$
$$- f_i(x, y^{(1)}, y^{(2)}, \ldots, y^{(j-1)}, y_2^{(j)}, y^{(j+1)}, \ldots)| \leq k|y_1^{(j)} - y_2^{(j)}| \tag{54}$$

---

[1] Since the quantity c_n varies, in general, with n, the interval of definition of the $G_n(x)$ is not necessarily fixed. However, the reasoning of Sec. 1 holds with only minor modifications, which the reader should supply.

Then for $|x - x_0| \leq c$, where c is some suitably chosen constant, there exist uniquely determined functions $g^{(i)}(x)$ such that $g^{(i)}(x_0) = y_0^{(i)}$ and

$$\frac{dg^{(i)}(x)}{dx} \equiv f_i(x, g^{(1)}(x), g^{(2)}(x), \ldots, g^{(n)}(x)) \tag{55}$$

So far only differential equations and systems of first order have been considered, but equations (or systems of m equations in m unknown functions) of higher order are readily handled. Consider, for example, the differential equation

$$\frac{d^r y}{dx^r} = f\left(x, y, \frac{dy}{dx}, \frac{d^2 y}{dx^2}, \ldots, \frac{d^{r-1} y}{dx^{r-1}}\right) \tag{56}$$

where $r > 1$. This single equation is readily seen to be equivalent to the following system of the form (55) [the so-called "canonical system" associated with (56)]:

$$\begin{aligned}
\frac{dy}{dx} &= y^{(2)} \\
\frac{dy^{(2)}}{dx} &= y^{(3)} \\
&\cdots\cdots\cdots \\
\frac{dy^{(r-1)}}{dx} &= y^{(r)} \\
\frac{dy^{(r)}}{dx} &= f(x, y, y^{(2)}, \ldots, y^{(r)})
\end{aligned} \tag{57}$$

An application of the above corollary shows that a solution of (56) exists and is uniquely determined by prescribing the values of y and its first $(r - 1)$ derivatives for a specified value of x, assuming, of course, that f satisfies the necessary continuity and Lipschitz conditions.

The restriction in the statement of Theorem 6 (and similarly in the corollary) to a sufficiently small interval about the point x_0 is essential. This is readily shown by the following simple example. Consider the differential equation

$$\frac{dy}{dx} = 1 + y^2 \tag{58}$$

Although the function $f(x, y) = 1 + y^2$ is defined over the entire x, y plane, the unique solution passing through the origin, namely, the function $\tan x$, becomes infinite as x approaches either of the values $\pm\pi/2$.

In contrast to the essentially "local" character of Theorem 6 (which must be kept in mind whenever differential equations or systems are solved) is the much more satisfactory situation which exists in the linear case.   If (33b) assumes the form

$$\frac{dy}{dx} = q(x)y + r(x) \tag{59}$$

where $q(x)$ and $r(x)$ are defined and continuous on a finite or infinite interval I, there exists a unique solution of (59) defined on the entire interval I and passing through any prescribed point (x_0,y_0), where x_0 is any (inner) point of I. This follows immediately from the familiar rule for solving (59) with the indicated condition, namely,

$$y = \exp\left[\int_{x_0}^{x} q(\xi)\, d\xi\right] \cdot \left\{y_0 + \int_{x_0}^{x} r(\xi) \exp\left[-\int_{x_0}^{\xi} q(t)\, dt\right] d\xi\right\} \tag{60}$$

However, this fact can be easily established, without making use of (60), by suitably extending the proof of Theorem 6.   The obvious analogue of the above statement also holds when the system (55) is linear, i.e., of the form

$$\frac{dy^{(i)}}{dx} = \sum_{j=1}^{n} q_{ij}(x)y^{(j)} + r^{(i)}(x) \tag{61}$$

In particular, it follows [cf. (56)] that, given $n + 1$ continuous functions $a_0(x)$, $a_1(x)$, . . . , $a_{n-1}(x)$, $b(x)$ defined on an interval I, n constants y_0, y_1, . . . , y_{n-1}, and any (inner) point x_0 of I, there exists a unique function $f(x)$ defined on the entire interval I satisfying there the differential equation

$$\frac{d^n y}{dx^n} = \sum_{j=0}^{n-1} a_j(x)\frac{d^j y}{dx^j} + b(x) \tag{62}$$

and the conditions

$$f^{(i)}(x_0) = y_i \qquad (i = 0, 1, \ldots, n - 1) \tag{63}$$

EXERCISES

29. Prove (38) by induction.

30. Show that an attempt to prove (39) directly by induction, rather than as a consequence of (38), fails.

31. Let $g(x)$ and $G(x)$ be solutions of (33b) defined in a common interval.   Under the hypotheses of Theorem 6, prove that if the inequality $g(x) < G(x)$ holds at one point of the interval, it holds at all points.

32. Prove that if $f(x,y)$ is continuous in the (closed) rectangle R and possesses a bounded (not necessarily continuous) derivative with respect to y throughout the interior of R, then $f(x,y)$ satisfies a Lipschitz condition in R.

33. Determine the totality of solutions of (49) by elementary quadrature, and confirm both the existence and uniqueness parts of Theorem 6.

34. Carry out in detail the proof of the corollary on page 22.

35. Prove that a function $f(x,y)$, defined in a rectangle and satisfying a Lipschitz condition in each variable, is necessarily continuous (in contrast to the fact that continuity in each variable separately does not assure continuity as a function of both variables).

6. Lebesgue Integration

In Chap. 5 it will be necessary to employ Lebesgue's, rather than Riemann's, definition of the integral. For convenience we state here, with utmost brevity and without any proofs, the most significant facts about the Lebesgue integral that are needed for our purposes. For simplicity we discuss the case of one independent variable, but the extension to a greater number of variables offers no essential difficulty.

A set E of real numbers is said to be a "null set," or set of measure zero, if, for every $\epsilon > 0$, it is possible to find a finite or denumerable collection of intervals $\{a_i < x < b_i\}$ whose total length $\Sigma(b_i - a_i)$ does not exceed ϵ, and such that each point of E is contained in at least one of the intervals. Now let an interval $a < x < b$, not necessarily finite, be specified. By a "step function" $s(x)$ we mean a real-valued function assuming constant values $c_1, c_2, \ldots, c_n$ on intervals $(a =) x_0 < x < x_1$, $x_1 < x < x_2, \ldots, x_{n-1} < x < x_n (=b)$, respectively, and by the "integral" $\int_a^b s(x) \, dx$ we mean the sum $\sum_{i=1}^{n} c_i(x_i - x_{i-1})$. [If the interval extends to $-\infty$ (or $+\infty$) the constant c_1 (or c_n) is required to be zero.] If a nondecreasing sequence $\{s_n(x)\}$ of step functions is given [i.e., $s_n(x) \leq s_{n+1}(x)$ for all x and for $n = 1, 2, \ldots$], then their integrals evidently form a nondecreasing sequence of numbers which approach either a finite limit or $+\infty$. A nonnegative function $f(x)$ is said to be "measurable" if there exists a nondecreasing sequence $\{s_n(x)\}$ of step functions which converges almost everywhere on the specified interval to $f(x)$. (By "almost everywhere" we mean everywhere except perhaps for a null set.) The limit of the sequence $\left\{\int_a^b s_n(x) \, dx\right\}$ is independent of the particular sequence of step functions which is employed, and therefore represents a property of the function $f(x)$. If this limit is finite, $f(x)$ is said to be "summable," or integrable, and $\int_a^b f(x) \, dx$ is defined as the limit of the integrals of the step functions. In particular, if the interval $a < x < b$ is finite, any bounded measurable function is integrable, for none of the quantities $\int_a^b s_n(x) \, dx$ can exceed $(b - a) \times \max f(x)$. Given a real-valued function $f(x)$ which assumes both positive and negative values, we express $f(x)$ as the difference $f_+(x) - f_-(x)$, where $f_+(x) = \frac{1}{2}[|f(x)| + f(x)]$ and $f_-(x) = \frac{1}{2}[|f(x)| - f(x)]$. [Note that $f_+(x)$ and $f_-(x)$ are both nonnegative.] Then $f(x)$ is said to be integrable if $f_+(x)$ and $f_-(x)$ are both integrable, and $\int_a^b f(x)\,dx$ is defined as the difference $\int_a^b f_+(x)\,dx - \int_a^b f_-(x)\,dx$. A complex-valued function $f(x)$ is said to be integrable if its real and imaginary parts are both integrable, and the integral $\int_a^b f(x)\,dx$ is defined, of course, as

$$\int_a^b [\operatorname{Re} f(x)]\,dx + i \int_a^b [\operatorname{Im} f(x)]\,dx$$

All the familiar properties of the Riemann integral carry over to the Lebesgue integral, and need not be stated here. Any function which is *absolutely* integrable (cf. Exercise 36) in the Riemann sense is also integrable in the Lebesgue sense, and the integral $\int_a^b f(x)\,dx$ has the same value in both senses. The essential difference in power between the two methods of integration lies in the fact that much stronger convergence theorems hold for the Lebesgue method. In particular, the following theorem, which is not true for Riemann integration, is of major significance in many problems of analysis.

Theorem 7. Riesz-Fischer. Given a sequence $\{f_n(x)\}$ of "quadratically integrable" functions—i.e., $f_n(x)$ is measurable and $|f_n(x)|^2$ is integrable—such that $\lim_{n,m\to\infty} \int_a^b |f_n - f_m|^2\,dx = 0$, there exists a quadratically integrable function $f(x)$ such that $\lim_{n\to\infty} \int_a^b |f - f_n|^2\,dx = 0$.

We also state, without proof, two further theorems concerning Lebesgue integration which will be employed in Chap. 5.

Theorem 8. Fubini. Let $f(x,y)$ be integrable over the rectangle R: $a \leq x \leq b$, $c \leq y \leq d$. (Any or all of the quantities a, b, c, d may be infinite.) Then for almost all x the integral

$$\int_c^d f(x,y)\,dy$$

exists, and the function $g(x)$ so defined is integrable; i.e., the iterated integral

$$\int_a^b \left[\int_c^d f(x,y)\,dy \right] dx$$

exists. Similarly, the iterated integral

$$\int_c^d \left[\int_a^b f(x,y)\,dx \right] dy$$

also exists, and the values of both iterated integrals are the same as that of the integral

$$\int\int_R f(x,y)\, dR$$

Theorem 9. Let $f(x)$ be integrable over the interval $I: a \le x \le b$, and let c be any fixed number in I. Then $F(x) = \int_c^x f(\xi)\, d\xi$ is a continuous function of x, and $F'(x)$ exists and equals $f(x)$ almost everywhere in I.

EXERCISE

36. Show that the integral $\int_{-\infty}^{\infty} \frac{\sin x}{x}\, dx$ does not exist in the Lebesgue sense.

7. Dini's Theorem

This is a simple but striking partial converse to the theorem that the sum of a uniformly convergent series of continuous functions is itself continuous.

Theorem 10. Let the nonnegative functions $f_n(x)$ each be continuous on a bounded closed interval I, and let the series $\sum_{n=1}^{\infty} f_n(x)$ converge everywhere on I to a continuous function $f(x)$. Then the convergence of the series is uniform.

Proof. Let the remainders $\sum_{n=N+1}^{\infty} f_n(x)$ be denoted $R_N(x)$. These functions are continuous, nonnegative, and monotone in N; i.e., for each x of I, the inequalities

$$0 \le R_{N+1}(x) \le R_N(x) \tag{64}$$

hold. Given $\epsilon > 0$, we can associate with each number x_0 of I an open interval $I(x_0)$ containing x_0 and an index N, depending both on ϵ and x_0, such that for all x contained in both I and $I(x_0)$ the inequality

$$R_N(x) < \epsilon \tag{65}$$

holds. By the Heine-Borel theorem, a finite number of the intervals $I(x_0)$ suffice to cover I. If we let N' denote the largest of the indices N associated with this finite collection of intervals, it follows from (64) that the inequality

$$R_M(x) < \epsilon \tag{66}$$

holds for all x and all indices M exceeding N'. This is equivalent to the assertion of the theorem.

2. PARTIAL DIFFERENTIAL EQUATIONS OF FIRST ORDER

1. Linear Equations in Two Independent Variables

Let three functions $a(x,y)$, $b(x,y)$, $c(x,y)$ be defined in some domain D of the x,y plane. By a "solution" of the equation

$$L(z) = az_x + bz_y = c \tag{1}$$

is meant a function $f(x,y)$ defined in all or part of D such that (1) reduces to an identity in the variables x and y when z is replaced by $f(x,y)$. Equation (1) is termed a first-order linear partial differential equation; "first-order" because no derivatives of order higher than the first appear, and "linear" because of the following identity, valid for arbitrary constants α_1, α_2 and arbitrary (differentiable) functions z_1, z_2:

$$L(\alpha_1 z_1 + \alpha_2 z_2) = \alpha_1 L(z_1) + \alpha_2 L(z_2) \tag{2}$$

Our principal objective in this section is to demonstrate the nature of the conditions which suffice to determine uniquely a solution of (1). It will be shown that a very simple and complete theory of this equation exists, the essential feature in its development being the reduction of (1), by means of a suitable change of independent variables, to an especially simple form which can be treated as an ordinary differential equation.

Actually, it is possible to develop a more general theory which applies equally well to the analogue of (1) in any number of independent variables, namely,

$$a_1 z_{x_1} + a_2 z_{x_2} + \cdots + a_n z_{x_n} = c \tag{3}$$

Furthermore, the general theory covers the case that the coefficients a_1, a_2, $\ldots$, a_n, c depend on z as well as on the independent variables

$x_1, x_2, \ldots, x_n$; (3) is then termed "quasi-linear." However, in order to introduce the basic ideas as easily as possible, we confine attention in the present section to (1), deferring the discussion of the general quasi-linear equation (3) to the following section.

By analogy with the theory of ordinary differential equations, as presented in Sec. 1-5, it appears plausible that, under suitable hypotheses on the coefficients a, b, c, a unique solution of (1) is determined by the requirement that the surface $z = f(x,y)$ representing the solution shall contain a specified curve C of the x,y,z space. More precisely, let a curve C be defined by a set of three equations,

$$x = \xi(t) \qquad y = \eta(t) \qquad z = \zeta(t) \qquad (t_1 \le t \le t_2) \tag{4}$$

where the functions $\xi(t)$, $\eta(t)$, $\zeta(t)$ are assumed to be sufficiently differentiable for the purposes of the following discussion. Also, we assume that $\xi'(t)$ and $\eta'(t)$ do not vanish simultaneously for any value of t; this assumption assures that C possesses, for each value of t, a well-defined tangent which, furthermore, is not parallel to the z axis. (The latter condition must be satisfied in order that a surface containing C may possess a nonvertical tangent plane; a vertical tangent plane would, of course, be inconsistent with finite values of z_x and z_y.) We also require that C have a one-to-one projection on the x,y plane; this is necessary in order to assure the single-valuedness of the solution. Now, *assuming* that a solution of (1) exists which contains C, then it should be possible to determine from (1) and (4) the values of z_x and z_y at each point of C, for at any such point z_x and z_y must satisfy, in addition to (1), the equality

$$\zeta'(t) = z_x \xi'(t) + z_y \eta'(t) \tag{5}$$

By elementary algebra (Cramer's rule), (1) and (5) determine uniquely the values of z_x and z_y unless the quantities $\xi'(t)$ and $\eta'(t)$ satisfy the "characteristic condition"[1]

$$\begin{vmatrix} a & b \\ \xi'(t) & \eta'(t) \end{vmatrix} = a\eta'(t) - b\xi'(t) = 0 \tag{6}$$

When (6) holds, (1) and (5) are inconsistent unless the "compatibility condition"

$$a:b:c = \xi'(t):\eta'(t):\zeta'(t) \tag{7}$$

is also satisfied. When (6) and (7) are both satisfied, (1) and (5) admit infinitely many solutions.[2] Furthermore, when (6) is not satisfied at

---

[1] If a and b both vanish at a given point of C, no restriction is imposed by (6); we set aside this degenerate case.

[2] That is, when (1) and (5) are considered as algebraic (not differential) equations in the unknowns z_x and z_y (at a given point of C).

a given point of C, all higher derivatives of z at this point are also (formally) uniquely determinable from (1) and (4) (assuming, of course, that the functions a, b, c, ξ, η, ζ possess derivatives of all orders), whereas, if *both* (6) and (7) are satisfied at all points of C it is possible to derive from (1) and (4) consistent, but not uniquely solvable, equations for the higher derivatives of z. (Cf. Exercise 1.) It may therefore be expected, and we shall prove, that when (6) is never satisfied on C there exists a unique solution of (1) containing C, and that when both (6) and (7) are satisfied everywhere on C there exist infinitely many solutions of (1) containing C.

Rewriting (6) in the form

$$\frac{dy}{dx} = \frac{b}{a} \quad \left(\text{or } \frac{dx}{dy} = \frac{a}{b}\right) \tag{8}$$

we see that the characteristic condition determines at each point of D a unique "characteristic ground direction." A curve lying in D which possesses at each of its points this direction is termed a "characteristic ground curve." These curves are obtained by solving the ordinary differential equation (8); referring to Theorem 1-6, we conclude that, under suitable restrictions on a and b, a unique characteristic ground curve passes through each point of D. [Note that neither the function $c(x,y)$ nor the third rectangular coordinate z is involved in determining these curves.] Clearly, (6) asserts that the (orthogonal) projection of C on the x,y plane possesses (at the projection of the point of C under consideration) the characteristic ground direction, and (7) asserts that when this occurs, the curve C must possess at this point the "characteristic direction" determined by the direction numbers a, b, c. A curve (in x,y,z space) possessing at each point this direction will be termed a "characteristic" of equation (1).

Now let a change of independent variables be introduced; we assume that the functions expressing the new variables in terms of the old,

$$x' = g(x,y) \qquad y' = h(x,y) \tag{9a}$$

and the inverse functions

$$x = \phi(x',y') \qquad y = \psi(x',y') \tag{9b}$$

possess continuous first partial derivatives. Then (1) can be converted into the equation

$$\alpha z_{x'} + \beta z_{y'} = \gamma \tag{10}$$

where　　　　　$\alpha = ag_x + bg_y \qquad \beta = ah_x + bh_y \qquad \gamma = c \tag{11}$

and the characteristic ground direction at any point (x',y') is the image of the characteristic ground direction at the corresponding point (x,y). Although the latter assertion can be proved (cf. Exercise 2) by an elementary computation, it is instructive to prove it without computation by the following argument: Given a noncharacteristic direction at (x,y), equations (1) and (5) *uniquely* determine z_x and z_y, and these, in turn, *uniquely* determine $z_{x'}$ and $z_{y'}$ at the corresponding point (x',y'); hence, the corresponding direction at (x',y') is noncharacteristic. Conversely, by symmetry, each noncharacteristic direction at (x',y') corresponds to a noncharacteristic direction at (x,y), and hence the characteristic ground directions must also correspond.

Now, if it is possible to introduce new variables in such a manner that the quantity β appearing in (10) vanishes identically, then the theory of (10) [and hence that of (1)] becomes, as will be shown soon, an almost obvious modification of the theory presented in Sec. 1-5. According to (11), it suffices to find a (nonconstant) function $h(x,y)$ such that

$$ah_x + bh_y = 0 \tag{12}$$

[If such a function $h(x,y)$ is determined, any function $g(x,y)$ independent of $h(x,y)$ may then be employed in (9a).] We leave the analytical discussion of (12) as Exercise 3, and instead employ here the following geometrical argument. In some sufficiently small neighborhood D' of any given point (x_0,y_0) of D we construct the set of all characteristic ground curves. These will cover D'. Let $h(x,y)$ be defined so as to assume a constant value on each one of these curves, the value of the constant being different for each pair of curves.[1] From the remarks made in the preceding paragraph concerning the correspondence between characteristic ground directions under change of variables, it follows that the characteristic ground curves of (10) will be given by $y' =$ constant. Taking account of (8), or, rather, its analogue

$$\frac{dy'}{dx'} = \frac{\beta}{\alpha} \tag{13}$$

we see that β must vanish. We have therefore proved that (1) can be reduced, by a suitable change of variables, to the form

$$z_x = c(x,y) \tag{14}$$

Now consider the problem of obtaining a solution of (14) which has specified values along a ground curve Γ; i.e., the surface representing the solution is to contain a curve C whose projection is Γ. First, we assume that Γ never possesses the characteristic ground direction; this means that C can be represented parametrically in the form (4) with $\eta(t) \equiv t$. We consider any fixed value of y and interpret (14) as an ordinary

---

[1] For example, draw through (x_0,y_0) the orthogonal trajectory of the family of characteristic ground curves, and associate with each ground curve Γ the arc length along the trajectory between (x_0,y_0) and the intersection with Γ.

differential equation; that is, we solve (14) subject to the condition that $z = \zeta(t) [= \zeta(y)]$ when $x = \xi(t) [= \xi(y)]$. We immediately obtain as the solution

$$z = \zeta(y) + \int_{\xi(y)}^{x} c(X,y) \, dX \tag{15}$$

As is readily apparent from (14), the curve obtained from (15) for each fixed value of y is a characteristic, and so the solution may be generated by the simple procedure of constructing the characteristic through each point of C.

On the other hand, if Γ meets a given horizontal line $y = y_0$ at two distinct points (x_1,y_0) and (x_2,y_0), the values of z which are prescribed at these two points must satisfy the relation[1]

$$z(x_2,y_0) - z(x_1,y_0) = \int_{x_1}^{x_2} c(x,y_0) \, dx \tag{16}$$

In particular, if Γ consists entirely of the line $y = y_0$, or a segment of this line, the value of z at any point determines the value everywhere on Γ, so that if the given curve C is not a characteristic there is no solution, but if C is a characteristic there exist infinitely many solutions containing C. Such solutions may be obtained by constructing the characteristic through each point of any curve C' which intersects C and whose ground curve never intersects any line $y =$ constant more than once.

The foregoing remarks are readily illustrated by the simplest of all possible examples, namely, the equation $z_x = 0$. If C is given by the equations $x = \xi(t)$, $y = t$, $z = \zeta(t)$ $(t_1 \leq t \leq t_2)$ the unique solution is given in the strip $t_1 \leq y \leq t_2$, $-\infty < x < \infty$ by $z = \zeta(y)$. Clearly this surface is generated by passing through each point of C the characteristic, namely, the line parallel to the x axis. However, if C is itself a characteristic there are infinitely many cylindrical surfaces and, hence, infinitely many solutions containing C. Finally, if y, but not z, is constant along C, then the characteristics through C form part of a vertical plane, so that no solution exists.

Returning to the general case of (1), and taking account of the fact that characteristic ground curves and characteristics are preserved under change of (independent) variables, we conclude that (1) admits infinitely many solutions containing the curve C if C is a characteristic, no solution if C is not a characteristic but possesses a characteristic ground curve, and one solution in the remaining case, namely, that the ground curve is noncharacteristic.

Finally, we comment briefly on the previously mentioned fact that from (1) and (4) it is possible to determine formally (i.e., even before

---

[1] Assuming that $c(x,y)$ is defined on the line segment connecting the two points; cf. Exercise 4.

establishing the existence of a solution), at any point P of C at which (6) is not satisfied, the values of all derivatives of the solution, not merely the values of the first derivatives z_x and z_y. In order to carry out this formal procedure it is necessary, of course, that the coefficients a, b, c possess partial derivatives of all orders, and that the functions $\xi(t)$, $\eta(t)$, $\zeta(t)$ which define the curve C possess derivatives of all orders with respect to the parametric variable t. Once the derivatives of the presumed solution z have been determined at P, a formal Taylor series solution

$$z \sim \sum_{m,n=0}^{\infty} c_{mn}(x - x_0)^m (y - y_0)^n$$

can be written; here (x_0, y_0) represents the projection of the point P on the x,y plane, and the coefficients c_{mn} are, aside from numerical factors, equal to the values of the derivatives that have been determined. The question then arises whether this series actually converges [in some neighborhood of (x_0, y_0)] and defines a solution of (1) containing the curve C. As might be expected, the answer can be shown to be affirmative if, in addition to the afore-mentioned assumptions concerning a, b, c, $\xi(t)$, $\eta(t)$, $\zeta(t)$, it is assumed that these functions are analytic in their arguments; by this is meant, of course, that each of the functions a, b, c can be expanded in a Taylor series in some neighborhood of (x_0, y_0), and that each of the functions $\xi(t)$, $\eta(t)$, $\zeta(t)$ can be expanded in a Taylor series about t_0, the value of t corresponding to the point P in the parametric representation (4) of C. This assertion constitutes (the simplest form of) the celebrated Cauchy-Kowalewski theorem.

EXERCISES

1. Obtain formulas for z_x, z_y, z_{xx}, z_{xy}, and z_{yy} from (1) and (5), assuming that (6) is not satisfied. Also show that when (6) and (7) hold, there exist infinitely many solutions of (1) and (5) for the above five quantities.

2. Prove analytically the assertion following (11).

3. Work out the analytical counterpart of the geometrical discussion following (12).

4. The illustrative example presented in the text shows that it is not possible to construct a solution in the entire half plane $y \geq 0$ of the equation $z_x = 0$ with the prescribed conditions: $z = 1 - x^2$ on the right half of the parabola $y = x^2$, $z = |1 - x^2|$ on the left half of the same parabola. However, show that if the portion $y \geq 1$ of the y axis is deleted, it is possible to obtain a unique solution defined in the remaining portion of the half plane. Explain why this does not conflict with the illustrative example.

2. Quasi-linear Equations

For ease in geometrical visualization, we shall first consider the quasi-linear equation in two independent variables; that is to say, we shall consider the coefficients a, b, c in (1) to depend on z as well as on x and y. We assume these coefficients to be defined in some domain G of x,y,z space, and to satisfy conditions adequate to permit the following arguments to be applied.[1] As in the preceding section, we define a characteristic to be a curve whose tangent has at each point the direction numbers a, b, c. Through each point (x_0,y_0,z_0) of G there exists a unique characteristic, which can be obtained by solving the system

$$\frac{dx}{ds} = a(x,y,z) \qquad \frac{dy}{ds} = b(x,y,z) \qquad \frac{dz}{ds} = c(x,y,z) \qquad (17)$$

with initial conditions $x(0) = x_0$, $y(0) = y_0$, $z(0) = z_0$; the solution will give a parametric representation of the characteristic. In contrast to the linear case, where the characteristics through two points having the same x and y coordinates can be obtained from each other by vertical displacement, the characteristics do not have any simple relationship in the quasi-linear case, and, in particular, the characteristic ground curves (i.e., the projections on the x,y plane of the characteristics) do not in general constitute a one-parameter family, so that a reduction to form (14) cannot be attained. Nevertheless, as will be seen, the characteristics and characteristic ground curves play an important role in the case of the quasi-linear equation also.

Let any solution of (1) be considered. Since the normal at any point of the surface representing the solution has direction numbers z_x, z_y, -1, (1) may be interpreted as requiring that the normal shall be orthogonal at each point of the surface to the characteristic through that point. This, in turn, is equivalent to requiring that the characteristic shall be tangent to the surface. It follows that if a (continuously differentiable) function $f(x,y)$ has the property that the surface $z = f(x,y)$ contains the characteristic through each of its points, then the given function satisfies (1). Suppose, in particular, that a curve C is given, and that by constructing the characteristic through each point of C there is obtained a smooth surface, no two of whose points have the same projection on the x,y plane. Then the surface evidently represents a function satisfying (1). Conversely, it will be shown that every solution of (1) can be obtained by such a construction.

The essential step in proving the last assertion consists in showing that a surface S representing a solution $f(x,y)$ of (1) must contain the characteristic through each one of its points. Let a point (x_0,y_0,z_0) be selected on S, and let a curve Γ be determined on S by means of the system

$$\frac{dx}{ds} = a(x,y,f(x,y)) \qquad \frac{dy}{ds} = b(x,y,f(x,y)) \qquad z = f(x,y) \qquad (18)$$

---

[1] In particular, it will suffice if these functions satisfy the Lipschitz condition (cf. Sec. 1-5) and if a and b do not both vanish at any point of G.

where the first pair of equations is to be solved with the initial conditions $x(0) = x_0$, $y(0) = y_0$. Then along this curve the identity

$$\frac{dz}{ds} = f_x \frac{dx}{ds} + f_y \frac{dy}{ds} = af_x + bf_y \tag{19}$$

must be satisfied. Since, by (1), $af_x + bf_y = c$, we conclude, from a comparison of (18) and (19) with (17), that the curve Γ must coincide with the characteristic through (x_0, y_0, z_0). [This argument shows that if two solutions of (1) possess a point P in common, then they actually have in common the characteristic through P. Hence, if the surfaces representing two different solutions intersect along a curve, this curve must be a characteristic. Conversely, it will be evident from the discussion in the following two paragraphs that any number of solutions can be constructed which will intersect along any prescribed characteristic, and also that any number of solutions can be constructed which will coincide on one side of a given characteristic but not on the other side. These remarks may be summed up in the statement that the characteristics are "branching lines" of solutions.]

It therefore follows that if a curve C is drawn on S and meets each characteristic on S at exactly one point, the surface can be reconstructed by drawing through each point of C the (unique) characteristic. While such a curve might be tangent at one or more of its points to the characteristic, we rule out this case. It then follows that the curve C [parametrized as in (4)] does not satisfy (6) at any point. [While this is obvious geometrically, it also follows analytically from the fact that if (6) were satisfied at any point of C, then (7) would also have to hold, and this would contradict the condition that has been imposed on C.]

Conversely, if a curve C is given which does not satisfy (6) at any point,[1] and if the characteristics are drawn through each point of C, it is geometrically evident that a surface is thus generated which represents a single-valued function satisfying (1), at least if attention is confined to a sufficiently small neighborhood of C. (Cf. Exercise 5.) Analytically, this follows from the following simple argument. The surface may be considered parametrized in the form

$$x = x(s,t) \qquad y = y(s,t) \qquad z = z(s,t) \tag{20}$$

where s and t are the parameters appearing in (17) and (4), respectively.[2] At any point of C the Jacobian $x_s y_t - x_t y_s$ reduces to $a\eta'(t) - b\xi'(t)$,

---

[1] It is understood, of course, that C is required to satisfy the conditions imposed in Sec. 1.

[2] Each characteristic has assigned to it the value of t corresponding to the point where the characteristic meets C, and each point on this characteristic also has a value of s associated with it by (17) and the initial conditions.

which does not vanish, by assumption. By continuity, the Jacobian remains distinct from zero in some neighborhood of C, and therefore s and t, and hence also z, can be expressed as single-valued functions of x and y.

We sum up the above discussion in the form of a theorem.

Theorem 1. If C is a curve which does not satisfy (6) at any point, then there exists a unique solution of (1) containing C, which can be constructed by passing the characteristic through each point of C. If C is itself a characteristic, there exist infinitely many solutions of (1) containing C, each of which can be obtained by selecting an arbitrary curve C' which intersects C and never satisfies (6), and then repeating the above construction. If C satisfies (6) but not (7) at some point, no solution of (1) exists which contains C.

It is now obvious that the same reasoning may be applied to the general quasi-linear equation (3) in n variables. We define the characteristics as the curves in $(n + 1)$-dimensional space satisfying the system

$$\frac{dx_1}{ds} = a_1, \quad \frac{dx_2}{ds} = a_2, \quad \ldots, \quad \frac{dx_n}{ds} = a_n, \quad \frac{dz}{ds} = c \tag{21}$$

Then every solution $z = f(x_1, x_2, \ldots, x_n)$ of (3) is generated by an $(n - 1)$-parameter family of characteristics, and conversely, given any $(n - 1)$-dimensional "surface" C in the $(n + 1)$-dimensional space of the variables $x_1, x_2, \ldots, x_n, z$ satisfying suitable restrictions analogous to those imposed in the above theorem (cf. Exercise 6), there exists a unique solution of (3) containing C, and this solution may be constructed by passing the characteristic through each point of C.

EXERCISES

5. Show analytically that if C satisfies (6) but not (7) at a certain point P, then the surface generated by the characteristics through C possesses a vertical tangent plane at P.

6. Determine the "suitable restrictions" mentioned in the last paragraph of the present section.

3. The General First-order Equation

We consider the equation

$$F(x, y, u, p, q) = 0 \qquad (p = z_x, q = z_y) \tag{22}$$

where F is defined and of class C^2 in some domain D of the x, y, z, p, q space. We assume that the equality (22) is satisfied at some point $(x_0, y_0, z_0, p_0, q_0)$

of D and that at this point at least one of the inequalities

$$F_p \neq 0 \qquad F_q \neq 0 \tag{23}$$

holds, so that (22) can be solved in some neighborhood of this point for p or q in terms of the four remaining variables. It is then evident that (22) may be interpreted geometrically, in the x,y,z space, as imposing on the surface $z(x,y)$ the condition of being tangent at each of its points to one of a one-parameter family of planes which pass through the point. More precisely, assuming (without loss of generality) that (22) can be solved for p in terms of x, y, z, and q, we permit q to vary, for each fixed set of values of x, y, z, over its range, and we consider all planes through (x,y,z) whose normals possess direction numbers $(p, q, -1)$. If (22) assumes the linear or quasi-linear form (1) at any point (x,y,z), the associated family of planes form a coaxial bundle, whose common line is known as the "Monge axis," but otherwise these planes envelop a conical surface with vertex at (x,y,z). This surface is known as the "Monge cone," and (22) may be interpreted as requiring that the surface $z(x,y)$ shall be tangent at each of its points to the Monge cone associated with that point.

The generators of the Monge cone associated with any point (x,y,z) can be determined by the following simple argument. Assuming, as above, that p is expressible in terms of x, y, z, and q, we choose distinct values of q, say q and $q + \Delta q$, and then determine from (22) the corresponding values of p, say p and $p + \Delta p$. The planes whose normals possess direction numbers p, q, -1 and $p + \Delta p$, $q + \Delta q$, -1 intersect in a line perpendicular to both normals and hence possessing direction numbers

$$1, -\frac{\Delta p}{\Delta q}, p - q\frac{\Delta p}{\Delta q}$$

Letting Δq approach zero and taking account of the relationship

$$F_p \frac{dp}{dq} + F_q = 0 \tag{24}$$

[which is obtained by differentiating (22), considered as an identity in p and q for fixed values of x, y, and z], we obtain for the direction numbers of the generators the quantities F_p, F_q, $pF_p + qF_q$.

It was shown in the two preceding sections that when (22) is linear or quasi-linear, each surface representing a solution of (22) can be generated by a one-parameter family of curves, the characteristics, and this simple observation served as the foundation of a very satisfactory theory. It will be shown here that a somewhat similar situation holds in the general case. However, the theory is slightly more difficult, the essential difference between the linear or quasi-linear case and the general case being that in the latter we must deal at each point with a family of directions, namely, those of the generators of the Monge cone, instead of with a single direction.

Assume for the moment that a solution $z(x,y)$ of (22) has been obtained, so that (22) reduces to an identity in x and y when z, p, and q are replaced by the functions $z(x,y)$, $z_x(x,y)$, and $z_y(x,y)$, respectively. At each point of the surface determined by the function $z(x,y)$ a unique direction is determined [cf. (23)] by the condition

$$dx:dy = F_p:F_q \tag{25}$$

Thus, introducing an auxiliary variable s, we find that the pair of differential equations

$$\frac{dx}{ds} = F_p \qquad \frac{dy}{ds} = F_q \tag{26}$$

determines in a unique manner, by Theorem 1-6, a one-parameter family of curves $x(s)$, $y(s)$, $z(s)$ $[= z(x(s), y(s))]$ which generate the surface $z(x,y)$. Furthermore, the functions $z(x,y)$, $p(x,y)$, $q(x,y)$ satisfy along each of these curves the differential equations

$$\frac{dz}{ds} = p\frac{dx}{ds} + q\frac{dy}{ds} \qquad \frac{dp}{ds} = p_x\frac{dx}{ds} + p_y\frac{dy}{ds} \qquad \frac{dq}{ds} = q_x\frac{dx}{ds} + q_y\frac{dy}{ds} \tag{27}$$

Taking account of (26) and the equality $p_y = q_x$, we obtain for (27) the alternative forms

$$\frac{dz}{ds} = pF_p + qF_q \qquad \frac{dp}{ds} = p_xF_p + q_xF_q \qquad \frac{dq}{ds} = p_yF_p + q_yF_q \tag{28}$$

Now, considering (22) as an identity in x and y, as explained above, and differentiating with respect to each variable, we obtain the equations

$$F_x + pF_z + p_xF_p + q_xF_q = 0 \qquad F_y + qF_z + p_yF_p + q_yF_q = 0 \tag{29}$$

We can therefore rewrite (28) in the following form:

$$\frac{dz}{ds} = pF_p + qF_q \qquad \frac{dp}{ds} = -(F_x + pF_z) \qquad \frac{dq}{ds} = -(F_y + qF_z) \tag{30}$$

Equations (26) and (30) constitute a system of five first-order ordinary differential equations in the five unknowns x, y, z, p, q. [Note that the reason for replacing (28) by (30) is that the former set of equations involves the additional quantities p_x, p_y, q_x, q_y.] Referring again to Theorem 1-6, we conclude that, given a point (x_0,y_0,z_0) on the surface and the values at this point of z_x and z_y [which must, of course, be consistent with (22)], the equations (26) and (30) determine a "strip" $(x(s), y(s), z(s), p(s), q(s))$ lying on the surface—i.e., the curve $(x(s),y(s),z(s))$ lies on the surface, and the normal to the surface at each point of this curve has the direction numbers p, q, -1.

Now, disregarding the manner in which equations (26) and (30) were obtained, we consider them as determining a system of curves in the x,y,z,p,q space. [Since s does not appear on the right side of any of the differential equations, it is not necessary to consider solutions of (26) and (30) as determining curves in the six-dimensional x,y,z,p,q,s space.] Taking account of the identity

$$\frac{dF}{ds} = F_x \frac{dx}{ds} + F_y \frac{dy}{ds} + F_z \frac{dz}{ds} + F_p \frac{dp}{ds} + F_q \frac{dq}{ds} \tag{31}$$

and replacing $\dfrac{dx}{ds}, \dfrac{dy}{ds}, \dfrac{dz}{ds}, \dfrac{dp}{ds}, \dfrac{dq}{ds}$ by the right sides of (26) and (30), we conclude that $\dfrac{dF}{ds}$ vanishes, and hence that F remains constant,[1] along each solution curve of the system of differential equations under consideration. Those solution curves on which the constant value of F reduces to zero—and such curves must exist, in accordance with the assumption made at the beginning of the section—are termed "characteristic strips" of equation (22).

We consider, as in Sec. 1, a curve C defined by equations (4) and satisfying the same conditions as those stated following (4). We assume that functions $P(t)$ and $Q(t)$ are given such that the equalities

$$\zeta'(t) = P(t)\xi'(t) + Q(t)\eta'(t) \tag{32a}$$
$$F(\xi(t),\eta(t),\zeta(t),P(t),Q(t)) = 0 \tag{32b}$$

and the inequality

$$F_p\eta'(t) - F_q\xi'(t) \neq 0 \tag{32c}$$

hold for each value of t. [It should be stressed that when (22) assumes the linear or quasi-linear form (1), the above inequality asserts that the projection of C does not possess the characteristic ground direction at any point; as explained in Sec. 1, it is then unnecessary to *prescribe* $P(t)$ and $Q(t)$, for these quantities are then uniquely determined by (32a) and (32b).] For each value of t we construct the characteristic strip which for $s = 0$ reduces to $(\xi(t),\eta(t),\zeta(t),P(t),Q(t))$; i.e., we determine the functions $x(s,t), y(s,t), z(s,t), p(s,t), q(s,t)$ which satisfy (26) and (30) and are such that $x(0,t) = \xi(t), y(0,t) = \eta(t), z(0,t) = \zeta(t), p(0,t) = P(t), q(0,t) = Q(t)$. Taking account of (26) and (32c), we observe that the Jacobian $x_s y_t - x_t y_s$ does not vanish on the segment $t_1 \leq t \leq t_2$, $s = 0$, and hence (by continuity) in some domain of the s,t plane containing this segment. It is therefore possible to express s and t as functions of x and y, and from this it follows, in turn, that the equations $x = x(s,t)$, $y = y(s,t)$, $z = z(s,t)$

---

[1] Given a system of differential equations, a function of the (independent and dependent) variables which remains constant along each solution curve is termed an "integral" of the system.

define a surface S, which can be expressed in the form

$$z = f(x,y) \tag{33}$$

such that
$$\zeta(t) = f(\xi(t),\eta(t)) \tag{34}$$

Thus, the surface S contains curve C. It is to be expected that when $p(s,t)$ and $q(s,t)$ are expressed as functions of x and y they reduce to z_x and z_y, and that (22) reduces to an identity in x and y. First, it follows from (32b) and the constancy of F along each characteristic strip that

$$F(x(s,t),y(s,t),z(s,t),p(s,t),q(s,t)) = 0 \tag{35}$$

Therefore, if we can show that

$$z_x = p \qquad z_y = q \tag{36}$$

the proof that S furnishes a solution to (22) will be complete. By referring to the identities

$$z_s = z_x x_s + z_y y_s \qquad z_t = z_x x_t + z_y y_t \tag{37}$$

and the inequality (32c), we observe that to prove (36) it will suffice to establish the identities

$$z_s - px_s - qy_s = 0 \qquad z_t - px_t - qy_t = 0 \tag{38}$$

The first of these identities is an immediate consequence of (26) and the first equation of (30). Referring to the left side of the second identity as V, we obtain, by differentiation with respect to s (and reversing the order of differentiation several times), the equality[1]

$$V_s = \frac{\partial z_s}{\partial t} - p_s x_t - p \frac{\partial x_s}{\partial t} - q_s y_t - q \frac{\partial y_s}{\partial t} \tag{39}$$

Replacing the quantities x_s, y_s, z_s, p_s, q_s by the right sides appearing in equations (26) and (30), we obtain, after an elementary computation,

$$V_s = (F_x x_t + F_y y_t + F_z z_t + F_p p_t + F_q q_t) - F_z V \tag{40}$$

The quantity in parentheses is equal to $\dfrac{\partial F}{\partial t}$, and it now follows from (35) that (40) simplifies to

$$V_s = -F_z V \tag{41}$$

Solving (41) with the initial condition $V(s,t)\Big]_{t=0} = 0$ [cf. (32a)], we conclude that $V(s,t) \equiv 0$, and the proof is complete.

A very close analogy has thus been established between the linear or quasi-linear equation (1) and the general first-order equation (22) in the

---

[1] The symbol $\dfrac{\partial}{\partial t}$ is used several times in (39) in preference to the subscript notation, for clarity of exposition.

noncharacteristic case (32c). The analogy extends, of course, to the characteristic case, in which (32c) is replaced by the equality

$$F_p \eta'(t) - F_q \xi'(t) = 0 \qquad (32d)$$

The extension to the case of three or more independent variables is also entirely routine. The reader should formulate a theorem analogous to (and including as a particular case) Theorem 1.

EXERCISES

7. Illustrate the theory developed in the present section by determining the two solutions of the equation $z_x z_y = 1$ passing through the straight line $x = 2t$, $y = 2t$, $z = 5t$.

8. At a given point P in the x, y, z space, let the characteristic strip corresponding to each generator of the Monge cone (which is assumed not to degenerate to a single line, as in the linear or quasi-linear case) be determined. Each characteristic strip determines, in an obvious manner, a characteristic curve tangent at P to one of the generators of the Monge cone. Show that these curves form a surface $z(x, y)$ containing P which, in general, has a conical vertex at P but which satisfies (22) elsewhere. This surface is termed the "characteristic conoid" associated with the point P.

3. THE CAUCHY PROBLEM

1. Classification of Equations with Linear Principal Parts

In this chapter we shall be concerned primarily with partial differential equations of the form

$$au_{xx} + 2bu_{xy} + cu_{yy} - f(x,y,u,p,q) = 0 \qquad (p = u_x,\ q = u_y) \quad (1)$$

where a, b, and c are given functions of the independent variables x and y, while f is a given function of the five indicated variables. A large fraction of the partial differential equations arising in mathematical physics are of this form. We list here three especially important examples:

Wave equation

$$u_{xx} - u_{yy} = 0 \qquad (a \equiv 1,\ b \equiv 0,\ c \equiv -1,\ f \equiv 0) \qquad (2a)$$

Heat equation

$$u_{xx} - u_y = 0 \qquad (a \equiv 1,\ b \equiv 0,\ c \equiv 0,\ f \equiv q) \qquad (2b)$$

Laplace equation

$$u_{xx} + u_{yy} = 0 \qquad (a \equiv 1,\ b \equiv 0,\ c \equiv 1,\ f \equiv 0) \qquad (2c)$$

Any equation of form (1) is termed a "second-order equation with linear principal part," the expression "second-order" referring, of course, to the fact that (1) contains derivatives of second order but not higher, while the "principal part" is the combination of terms containing second derivatives, namely,

$$L(u) = au_{xx} + 2bu_{xy} + cu_{yy} \qquad (3)$$

$L(u)$ is "linear" in the sense that the equality

$$L(\alpha_1 u_1 + \alpha_2 u_2) = \alpha_1 L(u_1) + \alpha_2 L(u_2) \qquad (4)$$

holds for arbitrary constants α_1, α_2 and arbitrary (sufficiently differentiable) functions u_1, u_2. By a "solution" of (1) is meant a function of x and y defined in a domain D such that (1) reduces in D to an identity in x and y when u is replaced by this function.

Equations of form (1) fall quite naturally into a three-way classification which can, perhaps, be introduced most simply by considering the effect of performing a change of independent variables. Let a pair of functions $g(x,y)$ and $h(x,y)$ be given, and suppose that in some neighborhood of the point (x_0,y_0) these functions are of class C^2. If the Jacobian $g_x h_y - g_y h_x$ does not vanish at (x_0,y_0) (and hence, by continuity, near this point), we are assured that, in some sufficiently small neighborhood of this point, the pair of equations

$$\xi = g(x,y) \qquad \eta = h(x,y) \tag{5}$$

can be solved for x and y as single-valued functions of ξ and η, say $x = \tilde{g}(\xi,\eta)$, $y = \tilde{h}(\xi,\eta)$. Furthermore, these new functions are of class C^2 in a neighborhood of the point (ξ_0,η_0) corresponding to (x_0,y_0), so that any C^2 function of x and y is also of class C^2 when considered as a function of ξ and η. The principal part of (1) may be expressed in terms of the new variables as follows (cf. Exercise 1):

$$au_{xx} + 2bu_{xy} + cu_{yy} = \alpha u_{\xi\xi} + 2\beta u_{\xi\eta} + \gamma u_{\eta\eta} + \cdots \tag{6}$$

where
$$\alpha = a\xi_x^2 + 2b\xi_x\xi_y + c\xi_y^2$$
$$\beta = a\xi_x\eta_x + b(\xi_x\eta_y + \xi_y\eta_x) + c\xi_y\eta_y$$
$$\gamma = a\eta_x^2 + 2b\eta_x\eta_y + c\eta_y^2 \tag{7}$$

and the quantities indicated in (6) by $\cdots$ involve only u and its first derivatives. Evidently (1) is transformed into a second-order equation whose principal part is precisely $\alpha u_{\xi\xi} + 2\beta u_{\xi\eta} + \gamma u_{\eta\eta}$, and an explicit calculation (cf. Exercise 2) shows that

$$\beta^2 - \alpha\gamma = (b^2 - ac)(\xi_x\eta_y - \xi_y\eta_x)^2 \tag{8}$$

Since the Jacobian $\xi_x\eta_y - \xi_y\eta_x$ does not vanish, it follows that the discriminants $b^2 - ac$ and $\beta^2 - \alpha\gamma$ are both positive, both zero, or both negative at any given pair of corresponding points. Thus, if equation (1) is classified at any given point according to the sign of its discriminant, its classification will not change under (sufficiently differentiable) change of variables. We now make this classification by defining equation (1) to be "hyperbolic," "parabolic," or "elliptic" at a given point, or throughout a given domain, according to whether the discriminant $b^2 - ac$ is positive, zero, or negative at the given point, or throughout the given domain.

We note that equations (2a), (2b), and (2c) are hyperbolic, parabolic, and elliptic, respectively, in any domain. It is of interest and importance (especially in the hyperbolic case) that (1) can be brought, by a suitable change of independent variables, into a form whose principal part coincides with that of one of the three equations appearing in (2). This fact will be established in Sec. 3.

EXERCISES

1. Work out (6) and (7) in detail.
2. Prove (8).

2. Characteristics

Reasoning by analogy with the considerations of Sec. 2-1, one is led to expect that a unique solution of (1) is determined by the requirements that the surface $u(x,y)$ shall contain a given curve C and be tangent at each point of C to a specified plane. These conditions ("Cauchy data") are formulated analytically as follows. Five functions, $\xi(t)$, $\eta(t)$, $\zeta(t)$, $h_1(t)$, $h_2(t)$, are given, and a function $u(x,y)$ is sought which satisfies (1) and the further equalities

$$u(\xi(t),\eta(t)) = \zeta(t) \qquad u_x(\xi(t),\eta(t)) = h_1(t) \qquad u_y(\xi(t),\eta(t)) = h_2(t) \quad (9)$$

[We assume, as in Sec. 2-1, that $\xi'(t)$ and $\eta'(t)$ do not vanish simultaneously, and also that all functions encountered possess whatever differentiability properties may be needed.]

We observe that the five given functions of t cannot be prescribed entirely independently, for the equality

$$\zeta'(t) = h_1(t)\xi'(t) + h_2(t)\eta'(t) \tag{10}$$

must be satisfied. [Cf. (2-5).] Assuming that (10) holds and that a function $u(x,y)$ satisfying (1) and (9) exists, we proceed to determine the values of its second derivatives, u_{xx}, $u_{xy} (= u_{yx})$, u_{yy}, at points of C. Just as (10) is obtained by differentiating the first equation appearing in (9), differentiation of the remaining two equations yields

$$h_1'(t) = u_{xx}\xi'(t) + u_{xy}\eta'(t) \qquad h_2'(t) = u_{xy}\xi'(t) + u_{yy}\eta'(t) \tag{11}$$

Replacing x, y, u, p, and q in a, b, c, and f by $\xi(t)$, $\eta(t)$, $\zeta(t)$, $h_1(t)$, and $h_2(t)$, respectively, we obtain in (1) and (11), for each fixed value of t, a system of three linear equations in the unknowns u_{xx}, u_{xy}, u_{yy}. Evaluating the determinant of this system, we find that the second derivatives are uniquely determined at any point of C unless, for the corresponding value of t, the "characteristic condition"

$$c\xi'(t)^2 - 2b\xi'(t)\eta'(t) + a\eta'(t)^2 = 0 \tag{12}$$

is satisfied.   When (12) holds, the system possesses either infinitely many solutions or none, according to whether the "compatibility condition"

$$\text{Rank} \begin{Vmatrix} a & 2b & c & f \\ \xi'(t) & \eta'(t) & 0 & h_1'(t) \\ 0 & \xi'(t) & \eta'(t) & h_2'(t) \end{Vmatrix} < 3 \tag{13}$$

is satisfied or not.

It is not difficult to show that the derivatives of u of all orders beyond the second are also uniquely determined at any point of C at which (12) does not hold.   (Cf. Exercise 3.)   It is therefore to be expected (since the Taylor expansion of the function $u(x,y)$ about a given point is determined by the values of all the derivatives at that point) that, in order to guarantee existence and uniqueness of a solution, it is necessary and sufficient that the characteristic condition (12) should not be satisfied at any point of C.   It will be seen later that this expectation is confirmed for hyperbolic equations.

It is to be noted that (12) involves only functions a, b, c, $\xi(t)$, and $\eta(t)$, not $\zeta(t)$, $h_1(t)$, $h_2(t)$, or f; geometrically, this means that two exceptional, or characteristic, directions are determined at each point of the x,y plane by (12), and that if the projection of C on the x,y plane (the "ground curve" of C) does not possess either of these directions at any point, the higher derivatives are uniquely determined by C and the prescribed values of the first derivatives.   A ground curve possessing a characteristic direction at each of its points is termed a characteristic ground curve, and the curve C in x,y,u space is termed a characteristic if it possesses a characteristic ground curve.   From (12) we find that the characteristic ground curves are obtained by solving the differential equations

$$\frac{dy}{dx} = \lambda_1(x,y) \qquad \frac{dy}{dx} = \lambda_2(x,y) \tag{14}$$

where $\lambda_1(x,y)$ and $\lambda_2(x,y)$ are the roots of the quadratic equation[1]

$$am^2 - 2bm + c = 0 \tag{15}$$

The assertion made in the preceding paragraph that condition (12) determines two characteristic directions at each point of the x,y plane is, of course, not strictly correct, for the quadratic equation (15) may possess complex conjugate roots, or real coincident roots, rather than real distinct roots.   Examining the discriminant of (15), we find that the number of (real) characteristic directions at a given point is zero, one, or two, according to whether equation (1) is elliptic, parabolic, or hyperbolic there.

[1] If a vanishes at any point (x,y), a trivial modification is necessary; however, it must be assumed that coefficients a, b, and c do not vanish simultaneously.

EXERCISES

3. Prove that if the ground curve of C does not possess a characteristic direction at any point, then all derivatives of u are determined at each point of C by the Cauchy data (9).

4. Prove that under a (sufficiently differentiable) change of independent variables characteristic directions map into characteristic directions (and hence characteristic curves map into characteristic curves).

3. Canonical Forms

In this section we consider the problem of reducing equation (1) to an especially simple, or "canonical," form by means of a suitable change of variables.

First we consider the case that (1) is hyperbolic at a given point (x_0, y_0). Then the quadratic equation (15) possesses distinct real roots $\lambda_1(x,y)$, $\lambda_2(x,y)$ in some neighborhood of this point. Let functions $\xi(x,y)$, $\eta(x,y)$ be determined which satisfy the homogeneous linear first-order equations

$$\xi_x + \lambda_1 \xi_y = 0 \qquad \eta_x + \lambda_2 \eta_y = 0 \tag{16}$$

and the inequalities

$$\xi_y \neq 0 \qquad \eta_y \neq 0 \tag{17}$$

in a neighborhood of (x_0, y_0). The existence of such functions is assured by the theory developed in Sec. 2-1. The Jacobian $\xi_x \eta_y - \xi_y \eta_x$ then assumes the form $(\lambda_2 - \lambda_1)\xi_y \eta_y$, and hence does not vanish, so that ξ and η may be used as independent variables in place of x and y; furthermore, by taking account of (7), (15), and (16), we see that the quantities α and γ defined in (7) must vanish. Therefore, (1) now assumes a form in which the principal part consists of the single term $2\beta u_{\xi\eta}$. Dividing[1] by 2β, we conclude that (1) may be reduced to the form

$$u_{\xi\eta} - \phi(\xi, \eta, u, u_\xi, u_\eta) = 0 \tag{18}$$

This is the canonical form of a hyperbolic equation, and will be used in establishing the basic theory of such equations. An alternative canonical form is obtained by performing the additional change of variables $\xi' = \xi + \eta$, $\eta' = \xi - \eta$; this leads to the form [cf. (2a)]

$$u_{\xi'\xi'} - u_{\eta'\eta'} - \psi(\xi', \eta', u, u_{\xi'}, u_{\eta'}) = 0 \tag{19}$$

We note that the two systems of characteristics of (18) are given by the equations $\xi = \text{constant}$, $\eta = \text{constant}$, while those of (19) are given by the equations $\xi' \pm \eta' = \text{constant}$.

---

[1] If β vanishes at any point, then we see from (7), by interchanging the roles of the two pairs of variables (x,y) and (ξ,η), that a, b, and c all vanish, a degenerate case which we exclude from consideration. (Cf. the preceding footnote.)

If (1) is parabolic throughout a neighborhood of (x_0, y_0) we may obtain, in exactly the same way as above, the function $\eta(x,y)$, but the function $\xi(x,y)$ would now be dependent on $\eta(x,y)$, since the Jacobian $\xi_x \eta_y - \xi_y \eta_x$ now vanishes. Instead, we choose for $\xi(x,y)$ any (sufficiently differentiable) function such that $\xi_x \eta_y - \xi_y \eta_x \neq 0$. Then, as before, the introduction of the variables ξ, η causes γ to vanish in a neighborhood of (x_0, y_0). Since, by (8), $\alpha\gamma - \beta^2 \equiv 0$, we conclude that β also vanishes. Thus, (1) assumes a form in which the principal part reduces to $\alpha u_{\xi\xi}$; division by α yields the canonical form

$$u_{\xi\xi} - \phi(\xi, \eta, u, u_\xi, u_\eta) = 0 \tag{20}$$

Finally, if (1) is elliptic at (x_0, y_0), the functions λ_1, λ_2 become complex conjugates, and a *formal* repetition of the reasoning used in the hyperbolic case would lead to (18), where ξ and η are complex conjugates. Introducing the real variables $\xi' = \frac{1}{2}(\xi + \eta)$, $\eta' = \frac{1}{2}i(\xi - \eta)$, we would then obtain the form

$$u_{\xi'\xi'} + u_{\eta'\eta'} - \phi(\xi', \eta', u, u_{\xi'}, u_{\eta'}) = 0 \tag{21}$$

The foregoing procedure can be justified with little difficulty if the definitions of the functions a, b, c can be extended to complex values of the independent variables—more precisely, if there exist analytic functions of the complex variables $x + ix'$, $y + iy'$ which reduce to $a(x,y)$, $b(x,y)$, $c(x,y)$ for $x' = y' = 0$. When such an extension is not possible, the proof that (1) can be transformed into form (21) is quite deep.

To sum up, we list the canonical forms which have been introduced in this section.[1]

Type of (1)	Discriminant condition	Canonical form
Hyperbolic.......	$b^2 - ac > 0$	$u_{xy} + \cdots = 0$ or $u_{xx} - u_{yy} + \cdots = 0$
Parabolic........	$b^2 - ac = 0$	$u_{xx} + \cdots = 0$
Elliptic.........	$b^2 - ac < 0$	$u_{xx} + u_{yy} + \cdots = 0$

EXERCISE

5. Reduce the following equations to canonical form:

(a) $(1 + x^2)^2 u_{xx} - (1 + y^2)^2 u_{yy} = 0.$

(b) $(1 + x^2)^2 u_{xx} - 2(1 + x^2)(1 + y^2)u_{xy} + (1 + y^2)^2 u_{yy} = 0.$

(c) $u_{xx} + (1 + x^2)^2 u_{yy} = 0.$

[1] For convenience we replace the letters ξ (or ξ') and η (or η') appearing in (18) to (21) by x and y.

4. The Cauchy Problem for Hyperbolic Equations

The "Cauchy problem" is the problem of obtaining a solution of (1) which satisfies conditions (9) along a given curve C. The heuristic argument presented in Sec. 2 suggests that, for any point (x_0, y_0) at which C has a noncharacteristic direction, there exists a neighborhood within which the Cauchy problem is solvable. Although the afore-mentioned argument seems to apply equally well to hyperbolic, parabolic, or elliptic equations (subject to the afore-mentioned restriction on the direction of C, which is always satisfied in the elliptic case), a satisfactory theorem can be obtained only in the hyperbolic case. We shall later discuss briefly the difficulties that arise when a Cauchy problem is formulated for a nonhyperbolic equation.

Taking account of the discussion in Sec. 3, we confine attention, when (1) is hyperbolic, to the case $a \equiv c \equiv 0$, $2b \equiv 1$, so that (1) assumes the form

$$u_{xy} = f(x,y,u,p,q) \qquad (p = u_x, \; q = u_y) \tag{22}$$

Since the two systems of characteristics are given in this case by $x = $ constant, $y = $ constant, the curve C can be expressed in the form $y = g(x)$, where $g'(x)$ is always positive (or always negative). The Cauchy data (9) can then be expressed in the form

$$u(x,g(x)) = h_1(x) \qquad u_x(x,g(x)) = h_2(x) \qquad u_y(x,g(x)) = h_3(x) \tag{23}$$

and the "consistency condition" (10) assumes the form

$$h_1'(x) = h_2(x) + h_3(x)g'(x) \tag{24}$$

Now, without loss of generality, we may assume that functions $h_1(x)$, $h_2(x)$, $h_3(x)$ vanish identically. This assertion is justified by introducing a new dependent variable,

$$v = u - h_1(x) - [y - g(x)]h_3(x) \tag{25}$$

Then equation (22) is converted into a similar equation, while the Cauchy data now vanish identically. (Cf. Exercise 6.) Therefore, it suffices to consider equation (22) in conjunction with the simplest possible form of Cauchy data, namely,

$$u \equiv p \equiv q \equiv 0 \qquad \text{on } C \tag{26}$$

We now state the principal result of this chapter.[1]

Theorem 1. Let $f(x,y,u,p,q)$ be a continuous function of the five indicated variables defined in a neighborhood N of the point $(x_0,y_0,0,0,0)$,

---

[1] The analogy, both in content and in method of proof, with Theorem 1-6 should be noted carefully.

where (x_0, y_0) lies on a smooth curve C whose tangent is never horizontal or vertical. Let f satisfy in N a Lipschitz condition (cf. Sec. 1-5) with respect to each of the variables u, p, and q. Then, in a sufficiently small neighborhood of (x_0, y_0), there exists a unique solution of (22) satisfying the Cauchy data (26).

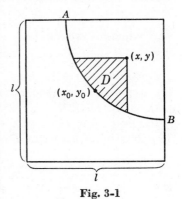

Fig. 3-1

Proof. Let $\delta > 0$ be chosen so small that the region S defined by the inequalities $|x - x_0| \leq \delta$, $|y - y_0| \leq \delta$, $|u| \leq \delta$, $|p| \leq \delta$, $|q| \leq \delta$ lies entirely inside N; let m denote the maximum of $|f|$ in S; let k_1, k_2, k_3 denote the Lipschitz constants associated with u, p, q respectively; and let $M = \max(m, k_1, k_2, k_3)$. Now choose a positive number l so small that all four of the following inequalities are satisfied:

$$Ml^2 < \delta \qquad Ml < \delta \qquad Ml(l + 2) < 1 \qquad l < 2\delta$$

Let A and B be the two points where C intersects the square formed by the lines $x = x_0 \pm l/2$, $y = y_0 \pm l/2$, and let R be the rectangle formed by the horizontal and vertical lines through A and B. (Cf. Fig. 3-1.) Then we define within R three sequences of functions $\{u_n\}$, $\{p_n\}$, $\{q_n\}$ as follows:

$$u_0(x,y) \equiv 0 \tag{27a}$$

$$u_{n+1}(x,y) = \iint_D f(\xi,\eta,u_n(\xi,\eta),p_n(\xi,\eta),q_n(\xi,\eta))\, d\xi\, d\eta \qquad (n \geq 0) \tag{27b}$$

$$p_n(x,y) = \frac{\partial u_n}{\partial x} \qquad q_n(x,y) = \frac{\partial u_n}{\partial y} \tag{27c}$$

In (27b) D denotes the region bounded by C and the horizontal and vertical lines through the point (x,y).

It is necessary, before proceeding further, to show that the definitions of the functions u_n, p_n, q_n are meaningful. We assume that for some value of n, say $n = k$, the functions u_n, p_n, q_n are defined and continuous in R, and satisfy there the inequalities $|u_n| \leq \delta$, $|p_n| \leq \delta$, $|q_n| \leq \delta$. Then the function $f(x,y,u_k(x,y),p_k(x,y),q_k(x,y))$ is well defined in R, and thus the definitions of $u_{k+1}(x,y)$, $p_{k+1}(x,y)$, $q_{k+1}(x,y)$ are meaningful. Furthermore, from (27b), we obtain the inequality

$$|u_{k+1}(x,y)| \leq M \cdot \text{area of } D \leq M \cdot \text{area of } R \leq Ml^2 < \delta \tag{28}$$

Similarly, by differentiating (27b) with respect to x, we obtain

$$p_{k+1}(x,y) = \int f(x,\eta,u_k(x,\eta),p_k(x,\eta),q_k(x,\eta))\, d\eta \tag{29}$$

where the integration is performed over the vertical boundary of D. We immediately obtain from (29) the inequality

$$|p_{k+1}(x,y)| \leq Ml < \delta \tag{30a}$$

and, similarly, we obtain

$$|q_{k+1}(x,y)| \leq Ml < \delta \tag{30b}$$

Inequalities (28), (30a), and (30b), together with the fact that for $n = 0$ the hypotheses imposed above for $n = k$ are obviously satisfied, show that definitions (27b) and (27c) are meaningful for all n.

Replacing n in (27b) by $n + 1$, subtracting, and taking account of the Lipschitz condition, we obtain, for any point in R, the inequality

$$|u_{n+2}(x,y) - u_{n+1}(x,y)| \leq Ml^2 S_n \tag{31}$$

where $S_n = \max_R (|u_{n+1} - u_n| + |p_{n+1} - p_n| + |q_{n+1} - q_n|)$. Similarly, from (29) we obtain

$$|p_{n+2}(x,y) - p_{n+1}(x,y)| \leq MlS_n \tag{32a}$$

and analogously we obtain

$$|q_{n+2}(x,y) - q_{n+1}(x,y)| \leq MlS_n \tag{32b}$$

Adding these three inequalities and taking the maximum of the left-hand side, we obtain

$$S_{n+1} \leq Ml(l + 2)S_n = \alpha S_n \qquad \alpha = Ml(l + 2) < 1 \tag{33}$$

By a trivial induction we obtain

$$S_n \leq \alpha^n S_0 \tag{34}$$

and we conclude that each of the three series

$$\sum_{n=0}^{\infty} (u_{n+1} - u_n) \qquad \sum_{n=0}^{\infty} (p_{n+1} - p_n) \qquad \sum_{n=0}^{\infty} (q_{n+1} - q_n) \tag{35}$$

is dominated by the convergent geometric series $\sum_{n=0}^{\infty} S_0 \alpha^n$. Hence, taking account of the telescoping character of the series (35) and the fact that $u_0 \equiv p_0 \equiv q_0 \equiv 0$, we conclude that the sequences $\{u_n\}$, $\{p_n\}$, $\{q_n\}$ are uniformly convergent throughout R to continuous functions, which we denote by u, p, q, respectively. Passing to the limit under the integral sign in (27b) and (29) (as in Sec. 1-5), we conclude that

$$u(x,y) = \iint_D f(\xi,\eta,u,p,q) \, d\xi \, d\eta \tag{36}$$

and

$$p(x,y) = \int f(x,\eta,u,p,q) \, d\eta \tag{37}$$

Differentiating (36) with respect to x and (37) with respect to y, we obtain

$$u_x = p \qquad p_y = f(x,y,u,p,q) \tag{38}$$

Similarly,
$$u_y = q \tag{39}$$

From (38) and (39) we conclude that

$$u_{xy} = f(x,y,u,p,q) = f(x,y,u,u_x,u_y) \tag{40}$$

Furthermore, from (36) and (37) we conclude that u and p (and similarly q) vanish if (x,y) is taken on C, for then D shrinks to the single point (x,y). Thus, u satisfies equation (22) and the Cauchy data (26).

The uniqueness portion of the theorem is established very much as was done in the proof of Theorem 1-6. Suppose that in the aforementioned rectangle R a second solution, say $U(x,y)$, of the same Cauchy problem exists. Then U and $P = U_x$ are expressible in exactly the same forms as (36) and (37). Subtracting (36) and (37), together with an analogous equation for q, from the corresponding equations for U, P, and Q $(= U_y)$, adding, and exploiting the Lipschitz condition, we obtain

$$\max_R (|u - U| + |p - P| + |q - Q|)$$
$$\leq Ml(l + 2) \max_R (|u - U| + |p - P| + |q - Q|) \tag{41}$$

Since $Ml(l + 2) < 1$, we conclude that each side of (41) must vanish, and hence the uniqueness is established.

We conclude this section with a few remarks concerning various aspects of the Cauchy problem.

1. The proof of Theorem 1 shows that the value of the solution u of (22) at a point P does not depend on *all* the Cauchy data (23), but only on that portion pertaining to the segment of C (the segment of determination) which is intercepted by the horizontal and vertical lines through P. Conversely, the Cauchy data at a point Q on C influences the values of the solution only in a portion of the region in which the solution is defined; this portion (the region of effect of Q) is readily seen to consist of two of the four quadrants determined by the horizontal and vertical lines through Q, namely, those quadrants which do not contain the curve C. Analogous statements hold, of course, for hyperbolic equations in noncanonical form, with horizontal and vertical lines replaced by characteristics.

2. A slight extension of the analysis presented in the proof of Theorem 1 shows that [under very mild hypotheses on the function $f(x,y,u,p,q)$] the Cauchy problem for equation (22) is a "stable" problem, in the sense that slight changes in the Cauchy data (23) result in slight changes in

the solution. This fact is of great physical significance, for a mathematical problem which is not stable cannot be expected to correspond accurately to a physical problem.

3. As indicated at the beginning of the present section, it is not possible to extend Theorem 1 in any straightforward manner to elliptic or parabolic equations, despite the heuristic argument presented in Sec. 2. This is demonstrated conclusively by considering the Laplace equation (2c) with the following Cauchy data, prescribed on the x axis:

$$u(x,0) = 0 \qquad u_x(x,0) = 0 \qquad u_y(x,0) = g(x) \tag{42}$$

Anticipating from Chap. 6 (or from the theory of analytic functions of a complex variable) the fact that a function satisfying (2c)—even on only one side of the x axis—and the first condition of (42) must possess derivatives of all orders on the x axis, we see, in particular, that the function $g(x)$ appearing in (42) must be indefinitely differentiable. Thus, for example, there can be no solution if $g(x) = |x|^3$, for $g'''(x)$ fails to exist at $x = 0$.

Continuing further, we may note that even when $g(x)$ is so chosen that the Cauchy problem *is* solvable, the solution will fail to possess the property of stability discussed in aspect 2. The classical illustration of this fact is furnished by the following example: If $g(x) \equiv 0$, the solution is given by $u \equiv 0$, but if $g(x) = n^{-1} \sin nx$ (which can be made uniformly small by choosing the parameter n sufficiently large), the solution is given by $u = n^{-2} \sinh ny \sin nx$, which does *not* become small as n increases.

A counterpart of the foregoing remarks is furnished by the fact that the Dirichlet problem (cf. Sec. 6-6) is not, in general, solvable for a hyperbolic equation. Referring to the especially simple hyperbolic equation $u_{xy} = 0$, we observe that any solution must be of the form $u = f(x) + g(y)$, and from this fact it follows that if P_1, P_2 and Q_1, Q_2 are pairs of opposite vertices of any rectangle with sides parallel to the axes, then $u(P_1) + u(P_2) = u(Q_1) + u(Q_2)$. Thus, if values of u are prescribed on the boundary of the rectangle which are not consistent with the above equality, no solution can be found.

4. In complete analogy with the concluding remarks of Sec. 1-5, we note that the restriction in Theorem 1 to a sufficiently small neighborhood is unavoidable, but that a "global" or "in-the-large" theorem holds when equation (22) is linear. More precisely, consider the equation

$$u_{xy} + au_x + bu_y + cu = f \tag{43}$$

where a, b, c, and f are given functions of x and y defined in a rectangle R. Then, for any sufficiently smooth monotone curve C lying entirely in R and any set of Cauchy data prescribed on C, there exists a unique solution of (43) defined throughout the rectangle determined by the horizontal and vertical lines through the end points of C.

EXERCISE

6. Prove that substitution (25) reduces the Cauchy problem consisting of equations (22) and (23) to another such problem in which the Cauchy data vanish identically.

5. The One-dimensional Wave Equation

Some of the material presented in the preceding sections, in particular the discussion of the role played by the characteristics, can be illustrated by the especially simple example of equation (2a), which permits all the equations involved in the analysis to be written out quite explicitly. As explained in Sec. 3, equation (2a) is equivalent to the equation $u_{xy} = 0$, but we prefer form (2a) on account of its immediate physical significance; we shall, however, replace the variable y by t, to represent the time variable.

First, we consider the problem with Cauchy data

$$u(x,0) = f(x) \qquad u_x(x,0) = f'(x) \qquad u_t(x,0) = 0 \qquad (44)$$

where $f(x)$ is defined for all x. Then the solution of the Cauchy problem defined by (2a) and (44) is given by (cf. Exercise 7)

$$u = \tfrac{1}{2}[f(x + t) + f(x - t)] \qquad (45)$$

Thus, a solution exists and is uniquely determined for all values of x and t. In order that (45) should actually constitute a solution of (2a), it is necessary that $f''(x)$ exist for all values of x. Suppose, however, that $f''(x)$ fails to exist for a certain value of x; for the sake of dealing with a specific example, suppose that $f(x) = x^2$ for positive x and $f(x) = -x^2$ for negative x, so that $f''(0)$ fails to exist. Then u assumes the following form[1] (cf. Exercise 9):

$$u = \begin{cases} x^2 + t^2 & x > t \\ 2tx & |x| \leq t \\ -(x^2 + t^2) & x < -t \end{cases} \qquad (46)$$

It is immediately evident from (46) that u, u_x, and u_t are continuous without exception, but that each of the second derivatives, u_{xx}, u_{xt}, u_{tt}, fails to exist along the lines $x - t = 0$, $x + t = 0$, undergoing jump discontinuities as point (x,t) crosses either of these lines. It is to be noted that these lines are the characteristics through the point $(0,0)$ at which a

---

[1] We confine attention, here and subsequently, to positive values of t (as is done in most physical problems), but (46) holds, with trivial modifications, for $t < 0$.

breakdown in the differentiability properties of the prescribed Cauchy data occurs, so that this discontinuity may be said to propagate along the characteristics. The precise nature of the propagated discontinuity can be seen most conveniently by introducing the change of variables $x + t = \xi$, $x - t = \eta$, so that (46) assumes the form (when account is taken of the restriction of t to positive values)

$$u = \begin{cases} \tfrac{1}{2}(\xi^2 + \eta^2) & 0 < \eta < \xi \\ \tfrac{1}{2}(\xi^2 - \eta^2) & \xi > 0,\, \eta < 0 \\ -\tfrac{1}{2}(\xi^2 + \eta^2) & \eta < \xi < 0 \end{cases} \tag{46'}$$

From (46') it is seen that $u_{\xi\xi}$ and $u_{\xi\eta}$ remain continuous across the line $\eta = 0$, $\xi > 0$, while $u_{\eta\eta}$ is discontinuous across this line; similarly, $u_{\eta\eta}$ and $u_{\xi\eta}$ remain continuous across the line $\xi = 0$, $\eta < 0$, while $u_{\xi\xi}$ is discontinuous across this line. Returning now to (46), we may assert that the second-order normal derivative is discontinuous across each point of the afore-mentioned characteristics $x \pm t = 0$, while the second derivatives involving at least one differentiation in a characteristic direction remain continuous across a characteristic. Essentially the same phenomenon occurs for the most general hyperbolic (or parabolic) equation.

If the function $f(x)$ appearing in (44) is defined only on some finite segment of the x axis, say the interval $0 \leq x \leq 1$, then the Cauchy data determine a solution only in the triangle bounded by the x axis and the lines $x = t$, $x = 1 - t$. Physically, this corresponds to the fact that the vibrations of a string occupying in its rest position the specified portion of the x axis are influenced by the constraints which are imposed on the ends of the string, the time at which the point P with abscissa x is first affected by the constraints being equal to the distance of P from the nearer end of the string. Thus, it may be said that the influence of the constraints on the ends (boundary conditions) is propagated along the string with unit velocity. (Similarly, in the case of the equation $u_{xx} - c^{-2}u_{tt}$ the influence is propagated with velocity c.) Confining attention to the simplest boundary conditions, namely,

$$u(0,t) \equiv 0 \qquad u(1,t) \equiv 0 \qquad (t \geq 0) \tag{47}$$

we readily confirm that the solution to the problem consisting of (2a), (44), and (47) is still given by (45), provided that the function $f(x)$, originally defined only on the interval $0 \leq x \leq 1$, is now defined for all values of x by extending its definition in the following manner:

$$f(x) = -f(-x) \qquad f(x + 2) = f(x) \tag{48}$$

(Cf. Exercise 10.)

7. Derive (45).

8. Show that when the third condition of (44) is replaced by $u_t(x,0) = g(x)$, the solution is given by

$$u = \tfrac{1}{2}\left[f(x + t) + f(x - t) + \int_{x-t}^{x+t} g(\xi)\,d\xi \right] \qquad (45')$$

Show that in this case, as well as in the case of (45), the displacement $u(x,t)$ can be interpreted as the superposition of two disturbances moving with unit velocity to the right and left, respectively, without change of shape.

9. Derive (46).

10. Confirm the solution given for the last problem discussed in the text, and show that the displacement $u(x,t)$ can be interpreted as the superposition of two disturbances which move with unit velocity and are reflected, with change of sign, at the end points.

6. The Riemann Function

For a linear hyperbolic equation (43), the solution of the Cauchy problem can be expressed in a form which brings out clearly the features of uniqueness and stability which are mentioned in Sec. 4. Let the left side of (43) be denoted by Lu, and let the "adjoint" operator L^* of L be defined by the identity

$$L^*v = v_{xy} - (av)_x - (bv)_y + cv \qquad (49)$$

It is easily confirmed (cf. Exercise 11) that the identity

$$vLu - uL^*v = (\tfrac{1}{2}vu_y - \tfrac{1}{2}uv_y + auv)_x + (\tfrac{1}{2}vu_x - \tfrac{1}{2}uv_x + buv)_y \qquad (50)$$

holds for arbitrary functions u and v, and so, with the aid of the divergence

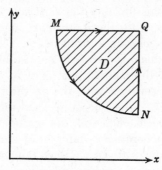

Fig. 3-2

theorem (cf. Sec. 6-3), we obtain, referring to Fig. 3-2,

$$\iint_D (vLu - uL^*v)\,dx\,dy = \int_M^N (\tfrac{1}{2}vu_y - \tfrac{1}{2}uv_y + auv)\,dy$$

$$- (\tfrac{1}{2}vu_x - \tfrac{1}{2}uv_x + buv)\,dx + \int_N^Q (\tfrac{1}{2}vu_y - \tfrac{1}{2}uv_y + auv)\,dy$$

$$+ \int_M^Q (\tfrac{1}{2}vu_x - \tfrac{1}{2}uv_x + buv)\,dx \qquad (51)$$

Employing integration by parts, we obtain

$$\int_N^Q \cdots dy + \int_M^Q \cdots dx = u(Q)v(Q) - \tfrac{1}{2}u(M)v(M) - \tfrac{1}{2}u(N)v(N)$$
$$+ \int_N^Q u(av - v_y)\, dy + \int_M^Q u(bv - v_x)\, dx \quad (52)$$

Now suppose that it is possible to determine a function v, depending on the pair of points $P\ [= (x,y)]$ and $Q\ [= (\xi,\eta)]$, such that the following conditions are all satisfied:

$$L^*v = 0 \qquad\qquad\qquad\qquad\qquad (53a)$$
$$bv - v_x = 0 \qquad \text{on the line } y = \eta \qquad (53b)$$
$$av - v_y = 0 \qquad \text{on the line } x = \xi \qquad (53c)$$
$$v(Q) = 1 \qquad\qquad\qquad\qquad\qquad (53d)$$

(It is understood that the independent variables in the above equations are the coordinates x, y of P, while the coordinates ξ, η of Q are to be considered as parameters.) Then, taking account of (51) and (52), we obtain for any function u satisfying (43) the integral representation

$$u(Q) = \tfrac{1}{2}u(M)v(M) + \tfrac{1}{2}u(N)v(N) + \iint_D fv\, dx\, dy$$
$$+ \int_M^N \left(\tfrac{1}{2}vu_x - \tfrac{1}{2}uv_x + buv\right) dx - \left(\tfrac{1}{2}uv_y - \tfrac{1}{2}uv_y + auv\right) dy \quad (54)$$

This formula, which can be written down once the appropriate function v is known and the Cauchy data are prescribed, demonstrates, quite independently of Theorem 1, the *unique* determination of the solution of the Cauchy problem by the Cauchy data, and it is easy to prove, on the other hand, from this formula the *existence* part of the theorem as well. (Cf. Exercise 13.) Furthermore, it is immediately evident that small changes in the Cauchy data result in small changes in the solution, so that the Cauchy problem is shown to be a stable one.

The function v defined by equations (53a) to (53d) is known as the "Riemann function" associated with the operator L, and is denoted as $R_L(P;Q)$, or $R_L(x,y;\xi,\eta)$. The question of existence of this function is readily settled. For each choice of the parameter point Q equations (53b) and (53c) become *ordinary* differential equations of first order, so that in combination with (53d) they uniquely determine v along the characteristics through Q; in fact, v is given along these lines by the formulas

$$v(x,\eta) = \exp\left[\int_\xi^x b(t,\eta)\, dt\right] \qquad (55a)$$
$$v(\xi,y) = \exp\left[\int_\eta^y a(\xi,t)\, dt\right] \qquad (55b)$$

We are thus led to a so-called "characteristic Cauchy problem," in which a solution of a hyperbolic equation [(53a) in the present case] is sought whose values are prescribed on a pair of intersecting characteristics. (Cf. Exercise 16.) A minor modification of the proof of Theorem 1 shows that this problem is solvable, in essentially the same manner as the noncharacteristic Cauchy problem. Thus the existence of the Riemann function is established.

In two simple cases the Riemann function can be expressed in an explicit form. If $Lu\ (= L^*u) = u_{xy}$, then it is seen that the Riemann function is given by

$$R_L(x,y;\xi,\eta) \equiv 1 \tag{56}$$

If $L(u)(= L^*u) = u_{xy} + cu$, c being any constant, the Riemann function is given by

$$R_L(x,y;\xi,\eta) = J_0(2[c(x - \xi)(y - \eta)]^{1/2}) \tag{57}$$

[Since the Taylor expansion of the Bessel function $J_0(t)$, namely,

$$J_0(t) = 1 - \frac{t^2}{2^2} + \frac{t^4}{2^2 4^2} - \cdots \tag{58}$$

contains only even powers of t, the ambiguity in the square root appearing in (57) is of no importance.]

EXERCISES

11. Confirm (50), and show that L^* is completely determined by the requirement that the left side of (50) shall be expressible as a "divergence expression" $(\cdot\cdot\cdot)_x + (\cdot\cdot\cdot)_y$.

12. Prove that any linear second-order differential operator L,

$$Lu = au_{xx} + 2bu_{xy} + cu_{yy} + du_x + eu_y + fu$$

has a unique adjoint L^* such that the requirement imposed in the preceding exercise is satisfied, and then show that $(L^*)^* = L$.

13. Prove that the function u defined by (54) satisfies (43) and the Cauchy data that are inserted on the right side.

14. Show that the Riemann functions R_L and R_{L^*} are related by the equality

$$R_L(x,y;\xi,\eta) = R_{L^*}(\xi,\eta;x,y)$$

so that if L is self-adjoint (i.e., $L = L^*$), R_L is symmetric in the pair of points (x,y), (ξ,η).

15. Show that it is possible to determine from the values of a solution of (22) on a pair of intersecting characteristics the values of u_x and u_y on both characteristics.

16. Show that a characteristic Cauchy problem can be reduced, by means of a substitution similar to (25), to one in which the unknown function is required to vanish on both characteristics. Then show that the iteration procedure employed in the proof of Theorem 1 applies in the present case.

7. Classification of Second-order Equations in Three or More Independent Variables

We consider briefly in this section a generalization of equation (1) to the case of three or more independent variables: $x_1, x_2, \ldots, x_n$. We begin by considering the effect of a (real) nonsingular homogeneous linear transformation

$$\xi_i = \sum_{j=1}^{n} c_{ij} x_j \tag{59}$$

on an expression of the form

$$L(u) = \sum_{j,k=1}^{n} a_{jk} u_{x_j x_k} \tag{60}$$

where the coefficients a_{jk} are given constants. Restricting attention to the case that u is of class C^2, so that $u_{x_j x_k} = u_{x_k x_j}$, we may assume that

$$a_{jk} = a_{kj} \tag{61}$$

An elementary computation shows that under the substitution (59) the expression $L(u)$ assumes the form

$$L(u) = \sum_{j,k=1}^{n} \alpha_{jk} u_{\xi_j \xi_k} \tag{62}$$

where
$$\alpha_{jk} = \sum_{i,m=1}^{n} c_{ki} c_{jm} a_{mi} \tag{63}$$

Now, by the theory of quadratic forms, it is possible to select the coefficients c_{ij} appearing in (59) in such a manner that $\alpha_{im} = 0$ if $i \neq m$ and $\alpha_{ii} = 1, -1,$ or 0. While there is wide freedom in the choice of the coefficients c_{ij}, it is of importance that for every possible choice the afore-mentioned numbers $1, -1,$ and 0 appear the same number of times, which we denote by n_1, n_{-1}, n_0, respectively. Thus, the numbers n_1, n_{-1}, n_0 are *invariantly* associated with the differential operator L appearing in (60), so that $L(u)$ can be reduced to a uniquely determined canonical form,

$$L(u) = \sum_{k=1}^{n_1} u_{\xi_k \xi_k} - \sum_{k=n_1+1}^{n_1+n_{-1}} u_{\xi_k \xi_k} \tag{64}$$

We now classify the differential operator L as follows: If $n_1 = n$ or $n_{-1} = n$, L is termed elliptic; if $n_0 > 0$, L is termed parabolic; if $n_0 = 0$ and $0 < n_1 < n$, L is termed hyperbolic. If $n_1 = n - 1$ and $n_{-1} = 1$, or vice versa, L is termed "properly hyperbolic"; otherwise "improperly hyperbolic," or "ultrahyperbolic."

The same classification can be formulated in terms of the quadratic form $\sum\limits_{i,j=1}^{n} a_{ij}\lambda_i\lambda_j$. If this form is positive- or negative-definite, L is elliptic; if the form is degenerate (i.e., reducible by a suitable linear transformation to a form in fewer than n variables), L is parabolic; if the form is not degenerate and not definite, L is hyperbolic. (Cf. Exercise 18.)

Now we consider a differential equation of the form

$$\sum_{j,k=1}^{n} a_{jk}u_{x_jx_k} - f(x_1,x_2, \ldots ,x_n,u,p_1,p_2, \ldots ,p_n) = 0 \qquad (p_i = u_{x_i}) \qquad (65)$$

where the a_{jk} may be given functions of the x's, all defined in some domain D of the $(x_1,x_2, \ldots ,x_n)$ space. It is now no longer true, in general, that the right sides of (60) and (62) are equal under the substitution (59). However, the right sides of the afore-mentioned equations will differ only by an expression involving the first derivatives of u; this will hold true even if the change of variables is no longer linear, as in (59), provided the quantities c_{ij} now are taken as $\dfrac{\partial \xi_i}{\partial x_j}$, so that

$$\alpha_{jk} = \sum_{i,m=1}^{n} a_{im} \frac{\partial \xi_j}{\partial x_i} \frac{\partial \xi_k}{\partial x_m} \qquad (66)$$

If (65) is now classified at each point P according to the classification assigned above to the "principal part" $L(u)$ (the coefficients a_{jk} being assigned their values at P), it follows from the above discussion that (65) will become, under a sufficiently differentiable change of variables, an equation of the same type (elliptic, parabolic, or hyperbolic) at the point P' corresponding to P.

In contrast to the case of two independent variables (and of an arbitrary number of independent variables with *constant* coefficients a_{jk}), it is not possible, in general, to effect a reduction of the principal part of (65) to a canonical form (64). This fact is suggested by the observation that (66), taken together with the symmetry condition (61) (which implies the symmetry of the α_{jk}) and the requirement that the new principal part shall be proportional to the right side of (64), constitutes a system of $\tfrac{1}{2}n(n + 1) - 1$ equations in the n unknown functions $\xi_k(x_1,x_2, \ldots ,x_n)$. Since $\tfrac{1}{2}n(n + 1) - 1$ exceeds n for $n \geq 3$, it is to be expected that this system will be unsolvable. Although this counting argument does not constitute a proof, the impossibility of a reduction to canonical form can be exhibited by means of a specific example.

Finally, we consider briefly the formulation of the concept of characteristics and of the Cauchy problem for equations of the form (65). By the Cauchy problem we now mean the problem of determining a solution u of (65) defined in a domain containing a given $(n - 1)$-dimensional surface S, such that u and all its first partial derivatives assume specified values (Cauchy data) on S. In analogy with Sec. 2, we shall define a characteristic surface as one on which the Cauchy data fail to determine [together with (65)] the values of all the second derivatives.

To obtain a criterion that a given surface S be a characteristic, let us first consider the case that S consists of part or all of the plane $x_1 = 0$. From the given values[1] of u, u_{x_1}, u_{x_2}, . . . , u_{x_n} it is evidently possible to determine at each point of S the values of all the second derivatives except $u_{x_1 x_1}$, which, in turn, can then be uniquely determined from (65) unless $a_{11} = 0$, in which case either no value or infinitely many values of $u_{x_1 x_1}$ can be determined. Now, turning to the general case of a surface defined by an equation of the form $\xi(x_1, x_2, . . . , x_n) = 0$, we determine additional functions $\xi_2, \xi_3, . . . , \xi_n$ of the x's such that the Jacobian $\dfrac{\partial(\xi, \xi_2, . . . , \xi_n)}{\partial(x_1, x_2, . . . , x_n)}$ does not vanish. Changing variables in (65), we conclude, by the argument presented above, that the given surface is a characteristic if and only if the coefficient α_{11} vanishes. Referring to (66), we thus conclude that the surface $\xi = 0$ (or, more generally, $\xi = $ constant) is a characteristic if and only if

$$\sum_{i,j=1}^{n} a_{ij} \frac{\partial \xi}{\partial x_i} \frac{\partial \xi}{\partial x_j} \equiv 0 \tag{67}$$

EXERCISES

17. Work out the details leading to (62) and (63).
18. Prove the equivalence of the two definitions following (64).

8. The Wave Equation in Two and Three Dimensions

In this section we shall consider briefly a specific Cauchy problem for the simplest of all hyperbolic equations in three and in four independent variables, namely, the wave equations

$$u_{xx} + u_{yy} - u_{tt} = 0 \tag{68a}$$
$$u_{xx} + u_{yy} + u_{zz} - u_{tt} = 0 \tag{68b}$$

---

[1] Actually only u and u_{x_1} need be given; in analogy with (10), the given values of $u_{x_2}, u_{x_3}, . . . , u_{x_n}$ would have to agree with those obtained by differentiating the values of u given on S. Similarly, in the general case only u and its derivative in the direction normal to the surface can be prescribed freely.

[In keeping with the physical significance of the independent variables, we denote them as x, y, z, t (time) rather than as in the preceding section.] Since equation (67) assumes the forms

$$\xi_x^2 + \xi_y^2 - \xi_t^2 = 0 \tag{69a}$$

and
$$\xi_x^2 + \xi_y^2 + \xi_z^2 - \xi_t^2 = 0 \tag{69b}$$

it follows that the surface $t = 0$ is not a characteristic of either (68a) or (68b). It is therefore plausible, although Theorem 1 does not carry over in any straightforward way to higher dimensions, that the Cauchy problem is solvable for Cauchy data prescribed on the afore-mentioned surface.

Somewhat surprisingly, the most effective procedure for obtaining solutions of this pair of problems involves solving first the one in the greater number of variables, and then solving the other by treating it as a particular case of the problem that has been solved. We therefore begin by considering (68b) in conjunction with the Cauchy data:

$$u(x,y,z,0) = \psi(x,y,z) \qquad u_t(x,y,z,0) = \phi(x,y,z) \tag{70}$$

It suffices to show how to obtain a solution in the particular case that $\psi \equiv 0$; for if we momentarily denote by $u^{(\phi)}$ a solution of (68b) satisfying (70) with $\psi \equiv 0$, then a simple calculation (cf. Exercise 19) shows that $u = u_t^{(\psi)}$ satisfies (68b) and (70) with $\phi \equiv 0$. Thus, by superposition, we find that $u^{(\phi)} + u_t^{(\psi)}$ solves the given Cauchy problem.

If $\phi(x,y,z)$ depends only on the distance r from a fixed point Q, then a solution depending only on r and t can easily be obtained, for the assumption $u = u(r,t)$ and the substitution $v = ru$ reduce (68b) to the form

$$v_{rr} - v_{tt} = 0 \qquad (r > 0) \tag{71}$$

and the conditions (70) (with $\psi \equiv 0$) assume the form

$$v(r,0) = 0 \qquad v_t(r,0) = r\phi(r) \tag{72}$$

Equations (71) and (72) constitute a linear noncharacteristic Cauchy problem, and hence determine a unique solution in the region $0 < |t| < r$, $r > 0$. However, the original problem requires a solution defined for $t > 0$, $r > 0$, and this can be obtained by the following device. The definition of v in terms of u shows that v must vanish for $r = 0$; if the definition of $\phi(r)$ is extended to negative values of r as an even function $[\phi(r) = \phi(-r)]$, then (71) and (72) determine the solution

$$v = \frac{1}{2} \int_{r-t}^{r+t} \xi\phi(\xi)\, d\xi \tag{73}$$

which vanishes for $t = 0$. Since the integrand in (73) is odd, the lower limit may be replaced by $|r - t|$. Reverting to the variable u, we thus obtain the solution

$$u(x,y,z,t) = u(r,t) = \frac{1}{2r} \int_{|r-t|}^{r+t} \xi\phi(\xi)\, d\xi \qquad (r > 0, t > 0) \qquad (74)$$

If the function $\phi(x,y,z)$ prescribed in (70) is expressible as the sum of a (finite) number of functions ϕ_1, ϕ_2, . . . , each depending only on the distance from points Q_1, Q_2, . . . , then the solution is given by a sum of the corresponding solutions each having the form (74), namely,

$$u(x,y,z,t) = \sum \frac{1}{2r_i} \int_{|r_i-t|}^{r_i+t} \xi\phi_i(\xi)\, d\xi \qquad (r_i = \overline{PQ_i}) \qquad (75)$$

It may then be expected that the solution for an arbitrary function

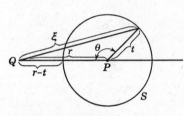

Fig. 3-3

$\phi(x,y,z)$ can be obtained by carrying out a suitable limiting process on (75). It is more convenient, however, to return to (74) and rewrite it in a manner which immediately suggests the solution for an arbitrary $\phi(x,y,z)$.

Referring to Fig. 3-3, in which S denotes the sphere of radius t with center at P and Q denotes the center of symmetry of $\phi(x,y,z)$, we find by elementary geometrical considerations (cf. Exercise 20) that (74) can be rewritten in the form[1]

$$u(x,y,z,t) = \frac{1}{4\pi t} \iint_S \phi\, dS \qquad (76)$$

Since (76) involves integration over a surface determined entirely by $P\ [= (x,y,z)]$ and t, it follows that equation (75) can also be written in form (76), and it is natural to expect that (76) furnishes the solution in the general case of a more or less arbitrary function $\phi(x,y,z)$. This expectation will now be confirmed.

Noting that the right side of (76) can be rewritten in the form $t\bar{\phi}(S)$, where $\bar{\phi}(S)$ denotes the mean value of ϕ over the afore-mentioned sphere S, so that $\lim_{t\to 0} \bar{\phi}(S) = \phi(P)$, it follows that

$$\lim_{t\to 0} u(P,t) = 0 \qquad u_t(P,0) = \lim_{t\to 0} \frac{u(P,t)}{t} = \lim_{t\to 0} \bar{\phi}(S) = \phi(P) \qquad (77)$$

The prescribed Cauchy data are therefore satisfied. It remains to show that (68b) is satisfied. For this purpose, it is convenient to rewrite (76) in the form

$$4\pi u(x,y,z,t) = t\iint\phi(x_1 + \alpha_1 t,\ x_2 + \alpha_2 t,\ x_3 + \alpha_3 t)\, d\Omega \qquad (76')$$

---

[1] Although Q is shown outside S in Fig. 3-3 (corresponding to the inequality $t < r$), the equivalence of (74) and (76) holds also for $t \geq r$.

where $x_1 = x$, $x_2 = y$, $x_3 = z$, α_1, α_2, α_3 denote direction cosines of the radius vector connecting the center P to the point of integration, and $d\Omega$ represents the element of solid angle subtended at P by dS ($d\Omega = t^{-2}\,dS$). Performing the indicated differentiations, we obtain from (76′)

$$4\pi(u_{xx} + u_{yy} + u_{zz}) = t \iint \Big(\sum_{i=1}^{3} \phi_{x_i x_i} \Big)\,d\Omega \tag{78}$$

and

$$4\pi u_{tt} = 2 \iint \Big(\sum_{j=1}^{3} \alpha_j \phi_{x_j} \Big)\,d\Omega + t \iint \Big(\sum_{i,j=1}^{3} \alpha_i \alpha_j \phi_{x_i x_j} \Big)\,d\Omega \tag{79}$$

Next we rewrite these integrals by setting $t\alpha_i = \xi_i$; using in (78) the fact that $\sum_{j=1}^{3} \alpha_j \xi_j = t$, and in each integral the fact that $\phi_{x_i} = \phi_{\xi_i}$, we obtain

$$4\pi t^2 (u_{xx} + u_{yy} + u_{zz} - u_{tt})$$

$$= \iint_S \sum_{j=1}^{3} \alpha_j \Big[-2\phi_{\xi_j} + \sum_{i=1}^{3} (\xi_j \phi_{\xi_i \xi_i} - \xi_i \phi_{\xi_i \xi_j}) \Big]\,dS \tag{80}$$

The integrand is seen to be the (outward) normal component of the vector whose jth component is the quantity appearing in brackets. Applying the divergence theorem,[1] we conclude that the right side of (80) is equal to the integral over the interior of S of the quantity

$$\sum_{j=1}^{3} \Big(\sum_{i=1}^{3} \xi_j \phi_{\xi_i \xi_i} - 2\phi_{\xi_j} - \sum_{i=1}^{3} \xi_i \phi_{\xi_i \xi_j} \Big)_{\xi_j} \tag{81}$$

and a simple computation (cf. Exercise 21) shows that this quantity vanishes identically. Thus, u satisfies (68b), as was to be proved.[2]

Turning to equation (68a), one might attempt, as above, to obtain first a solution for radially symmetric Cauchy data by reducing the problem to one in only two independent variables (t and the distance from the center of symmetry). However, it is easily shown that such an attempt must fail. Instead, as indicated above, we employ a very simple but effective method (Hadamard's method of descent) which furnishes the solution to the present problem by considering it as a particular case of the problem which has been solved. We simply consider the given function $\phi(x,y)$ as a function of the three variables x, y, z by employing the obvious definition

$$\phi(x,y,z) = \phi(x,y) \tag{82}$$

---

[1] The integral of the outward normal component of a vector field over a closed surface equals the integral over the enclosed volume of the divergence; cf. Sec. 6-3.

[2] We omit the proof of the fact that (76) furnishes the *unique* solution to the Cauchy problem under consideration.

The integral (76) is clearly independent in this case of z, for changing z merely involves averaging ϕ over a sphere S' obtained by translating S parallel to the z axis. The function $u(x,y,z,t)$ thus obtained may be considered as a function of the three variables x, y, t, and since u_{zz} vanishes identically, (68b) reduces to (68a).

Although the foregoing remarks may be considered as furnishing the solution to the problem under consideration, it is instructive to develop for the solution an explicit formula, which will demonstrate a remarkable contrast between the wave equations in three variables (68a) and in four variables (68b). To obtain such a formula, we merely observe that, since ϕ now depends only on x and y, the integral (76) can be converted into an integral over the disc D_t in the x,y plane with center at (x,y) and with radius t. A simple geometric consideration shows that the element of surface area dS located at (ξ,η,ζ) on S and its projection dA located at $(\xi,\eta,0)$ are related by the simple equation

$$dS = |\alpha_3|^{-1}\, dA \tag{83}$$

where α_3 is defined following (76′). Furthermore,

$$|\alpha_3|^{-1} = t[t^2 - (\xi - x)^2 - (\eta - y)^2]^{-\frac{1}{2}}$$

and so (76) assumes the form

$$u(x,y,t) = \frac{1}{2\pi} \iint_{D_t} \frac{\phi(\xi,\eta)}{[t^2 - (\xi - x)^2 - (\eta - y)^2]^{\frac{1}{2}}}\, d\xi\, d\eta \tag{84}$$

(The appearance of the factor $1/2\pi$, instead of $1/4\pi$, is explained by the fact that two surface elements on S project onto each element of D_t.)

We now point out the striking contrast between (76) and (84) which has been referred to previously. If, for three dimensions (four independent variables), the function $\phi(x,y,z)$ differs from 0 only in a bounded region R [so that the function $u(x,y,z,t)$ may be considered as describing the propagation of a disturbance in a medium which was originally $(t = 0)$ at rest except in R], then u remains zero at a point P located outside R until a time t_0 equal to the minimum distance from P to R. This follows from the observation that for $0 < t < t_0$ the right side of (76) is equal (aside from the factor t) to the mean value of the initial disturbance ϕ over a spherical surface which does not meet R. Beginning at $t = t_0$, the influence of the initial disturbance is experienced at P. However, for $t > t_1$, where t_1 denotes the maximum distance from P to R, the surface S again consists entirely of points which were originally undisturbed, so that the value of u at P returns permanently to the value zero. Thus, we may assert that the influence of an initial disturbance of limited extent is also limited in its duration at each point. In contrast to this, we note from (84) that, while a disturbance concentrated in a bounded region R at time $t = 0$ does not exert an influence on a point P outside R until some subsequent time t_0, it is not true that the effect of the disturbance disappears when t exceeds the value t_1 defined above. Instead, (84) shows that for $t > t_1$ the value of u at P may be expressed in the form

$$u(x,y,t) = \frac{1}{2\pi} \iint_R \frac{\phi(\xi,\eta)}{[t^2 - (\xi - x)^2 - (\eta - y)^2]^{1/2}} \, d\xi \, d\eta \qquad (85)$$

In general this quantity will not vanish; in particular, if $\phi(\xi,\eta)$ is positive throughout R, u will become and remain positive at point P after the effect of the initial disturbance begins to be experienced.

By taking account of the remarks which follow (70), we can see that the contrast described above between the behavior of the solutions in two and in three dimensions persists when the restriction that ψ shall vanish identically is removed.

The one-dimensional wave equation (2a) shows a behavior intermediate between that of (68a) and (68b). As can be seen by referring to (45'), if the initial values of u and u_t differ from zero only on a bounded portion of the x axis, the effect of the initial *displacement* $f(x)$ disappears at any given point after a finite time, but the influence of the initial *velocity* $g(x)$ is felt permanently, for, at each point, u reaches (after a finite time, and keeps permanently thereafter) the value $\int g(x) \, dx$, this integral being extended over the entire x axis, or, equivalently, over the portion of the x axis to which a nonzero velocity was initially imparted.

For a greater number of variables, it is found that solutions of the wave equation in an even number of dimensions (*odd* number of independent variables) exhibit a behavior similar to that found above in the two-dimensional case (68a), while in an odd number of dimensions the behavior of solutions is similar to that obtained in the three-dimensional case (68b). This striking differentiation in the nature of wave propagation in even- and odd-dimensional spaces is known as "Huygens' principle."

EXERCISES

19. Prove the assertion which follows (70).
20. Derive (76) from (74).
21. Prove that expression (81) reduces to zero.
22. Solve (68b) with the Cauchy data $u(x,y,z,0) \equiv 0$, $u_t(x,y,z,0) = x^2 + 2xy + 3z^3$.

9. The Legendre Transformation

In Sec. 3 it was shown that equation (1) can be reduced to one of several simple (canonical) forms by means of a suitable change of independent variables. In the present section we present a brief discussion of the Legendre transformation, which often proves effective in the study of partial differential equations possessing a more complicated structure than (1). On the other hand, the Legendre transformation, in contrast to the transformations considered in Sec. 3, involves the (unknown) solution to the differential equation under consideration, and the resulting difficulties connected with reformulating the conditions which are imposed on the solution (aside from the differential equation itself) may outweigh the advantages furnished by the simplification of the differential equation.

For simplicity, we consider a differential equation of first or second order, but it will be apparent how the method extends to equations of higher order. Let the given differential equation be expressed in the form

$$F(x,y,u,u_x,u_y,u_{xx},u_{xy},u_{yy}) = 0 \tag{86}$$

If a solution $u(x,y)$ exists in a neighborhood of a given point (x_0,y_0) and if $u_{xx}u_{yy} - u_{xy}{}^2$, or $(u_x)_x(u_y)_y - (u_x)_y(u_y)_x$, does not vanish in this neighborhood, then the pair of equations

$$\xi = u_x \qquad \eta = u_y \tag{87}$$

can be solved in a neighborhood (usually smaller than the afore-mentioned one) of (x_0,y_0) for x and y in terms of ξ and η, so that (87) defines a valid transformation of independent variables. If we introduce a new dependent variable, $\omega(\xi,\eta)$, in place of $u(x,y)$ by the substitution

$$\omega = \xi x + \eta y - u \tag{88}$$

we readily obtain from (86) an analogous equation which must be satisfied by ω. Differentiating both sides of (88) with respect to ξ and taking account of (87), we obtain

$$\omega_\xi = x + \xi x_\xi + \eta y_\xi - u_x x_\xi - u_y y_\xi = x \tag{89}$$

Similarly, $\omega_\eta = y$. Thus, if (86) does not contain any second derivatives, the new dependent variable ω must satisfy the differential equation obtained from (86) by replacing $x, y,$ $u,$ $u_x,$ and u_y with $\omega_\xi,$ $\omega_\eta,$ $\xi\omega_\xi + \eta\omega_\eta - \omega,$ $\xi,$ and $\eta,$ respectively. If (86) actually contains second derivatives, we proceed as follows: Differentiating the first equation of (87) with respect to ξ and η, we obtain

$$\begin{aligned}
1 &= u_{xx}x_\xi + u_{xy}y_\xi = u_{xx}\omega_{\xi\xi} + u_{xy}\omega_{\xi\eta} \\
0 &= u_{xx}x_\eta + u_{xy}y_\eta = u_{xx}\omega_{\xi\eta} + u_{xy}\omega_{\eta\eta}
\end{aligned} \tag{90}$$

Solving this pair of equations, we obtain (cf. Exercise 23)

$$u_{xx} = \frac{\omega_{\eta\eta}}{\omega_{\xi\xi}\omega_{\eta\eta} - \omega_{\xi\eta}{}^2} \qquad u_{xy} = -\frac{\omega_{\xi\eta}}{\omega_{\xi\xi}\omega_{\eta\eta} - \omega_{\xi\eta}{}^2} \tag{91}$$

Similarly,

$$u_{yy} = \frac{\omega_{\xi\xi}}{\omega_{\xi\xi}\omega_{\eta\eta} - \omega_{\xi\eta}{}^2} \tag{92}$$

These equations evidently enable us to convert (86) into a differential equation of second order for the dependent variable ω.

We repeat that the Legendre transformation is applicable only if $u_{xx}u_{yy} - u_{xy}^2$ does not vanish. The vanishing of this expression has a very simple geometrical interpretation, namely, that the surface $u(x,y)$ is developable—i.e., a sufficiently small neighborhood of any point on the surface can be mapped onto a plane region without changing the length of any curve. Thus, when the Legendre transformation is to be employed, separate consideration must be given to determining whether (86) possesses developable solutions. (Cf. Exercises 24 and 25.)

We conclude this section with a very brief discussion of the application of the Legendre transformation to a problem of great physical significance. The plane steady (i.e., time-independent) flow of a perfect fluid can be described by means of either of two functions: the "velocity potential" $\phi(x,y)$ or the "stream function" $\psi(x,y)$. The rectangular components U, V of the velocity vector at each point of the flow are related to ϕ and ψ by the equations

$$\rho U = \rho\phi_x = \psi_y \qquad \rho V = \rho\phi_y = -\psi_x \tag{93}$$

where ρ denotes the density of the fluid. Differentiating the first equation with respect to x and the second with respect to y and adding, we obtain

$$(\rho\phi_x)_x + (\rho\phi_y)_y = 0 \tag{94}$$

If ρ is constant throughout the fluid, then (94) reduces to the Laplace equation $\phi_{xx} + \phi_{yy} = 0$. However, if the fluid is compressible, then the variability of ρ must be taken into account. We assume that an "equation of state" relating the density ρ and the pressure p is known, and that "Bernoulli's law" relating p, ρ, and the speed $q[= (U^2 + V^2)^{1/2}]$ is satisfied, namely,

$$\tfrac{1}{2}q^2 + \int \frac{dp}{\rho} = \text{constant} \tag{95}$$

(The integral may be expressed in terms of either p or ρ with the aid of the equation of state.) Since ρ may now be expressed as a function of q, say $\rho = f(q) = f([\phi_x^2 + \phi_y^2]^{1/2})$, we obtain from (94) the more explicit form

$$\left(1 + \frac{\phi_x^2 f'}{qf}\right)\phi_{xx} + \frac{2\phi_x\phi_y f'}{qf}\,\phi_{xy} + \left(1 + \frac{\phi_y^2 f'}{qf}\right)\phi_{yy} = 0 \tag{96}$$

$\left[\text{Here } f' = f'(q) = \dfrac{df}{dq}.\right]$ This equation is not linear, for the function ϕ appears (through its first derivatives) in the coefficients of the second derivatives; however, since the derivatives of *highest* order appear linearly, (96) is said to be "quasi-linear." If we now apply the Legendre transformation we immediately see that (96) goes over into a *linear* equation, namely,

$$\left(1 + \frac{\eta^2 f'}{qf}\right) \omega_{\xi\xi} - \frac{2\xi\eta f'}{qf} \omega_{\xi\eta} + \left(1 + \frac{\xi^2 f'}{qf}\right) \omega_{\eta\eta} = 0 \qquad (97)$$

[Here $q = (\xi^2 + \eta^2)^{\frac{1}{2}}$.] Thus, instead of considering the flow in the "physical" plane of the variables x, y, one may study it in the "hodograph" plane of the variables $\xi(= U = \phi_x)$, $\eta(= V = \phi_y)$.

The discriminant of (97) works out to $-1 - qf'/f$, or $-1 - \frac{q}{\rho}\frac{d\rho}{dq}$. From (95) we obtain $\frac{dp}{dq} = -\rho q$, and from this we find that the discriminant can be rewritten in the form $(q^2 a^{-2} - 1)$, where a, the velocity of sound or "acoustic velocity," is equal to $\left(\frac{dp}{d\rho}\right)^{\frac{1}{2}}$. (It is assumed that $\frac{dp}{d\rho}$ is positive.) Thus, equation (97) is elliptic or hyperbolic according to whether the "Mach number" $M = q/a$ is below or above one, respectively. If the flow is "subsonic" ($M < 1$) in a portion of the region under consideration and "supersonic" ($M > 1$) elsewhere, equation (97) must be treated in a region of the ξ,η plane within which the type does not remain fixed.

EXERCISES

23. Assuming that the Legendre transformation is applicable (i.e., that $u_{xx}u_{yy} - u_{xy}^2$ does not vanish), prove that $(u_{xx}u_{yy} - u_{xy}^2)(\omega_{\xi\xi}\omega_{\eta\eta} - \omega_{\xi\eta}^2) = 1$, so that the determinant of system (90) does not vanish.

24. Prove that the only nondevelopable solution of the equation

$$xu_x + yu_y - u = u_x^2 + u_y^2$$

is given by $u = \frac{1}{4}x^2 + \frac{1}{4}y^2$.

25. Prove that the Legendre transformation cannot be applied to the equation $xu_x + yu_y = u$.

4. THE FREDHOLM ALTERNATIVE IN BANACH SPACES

The principal objective of the present chapter and the one following it is to develop the theory of linear integral equations (of Fredholm type). There is a very close connection between integral and differential equations, for it is often possible to formulate an integral equation which is equivalent to a given differential equation *plus* additional (initial or boundary) conditions. This has already been seen in Chap. 1, where the differential equation (1-33b) and the condition $g(x_0) = y_0$, taken together, are equivalent to the integral equation (1-45). Further examples of this equivalence will be presented in Chaps. 7 and 9.

Instead of dealing explicitly with integral equations, we shall present a more abstract development, in terms of "normed linear spaces," which includes, as will be shown, the theory of integral equations as a particular "concrete" interpretation. Thus, in addition to providing an exposition of the theory of integral equations, these two chapters may serve as an elementary introduction to the important subject of abstract linear analysis.

1. Linear Spaces

This section is devoted to a brief discussion of the most important purely algebraic concepts relating to linear spaces.

DEFINITION 1. A "linear space" (or vector space) V is a set of objects (elements or vectors) in which two operations, + (addition) and · (multiplication), are defined, where these two operations satisfy the following requirements.

1. For every pair of elements f, g of V, the element $f + g$ is defined, and $f + g = g + f$.

2. For all elements f, g, h, the equality $f + (g + h) = (f + g) + h$ holds. From this it follows, exactly as in the development of the real-number system, that any finite sum $f_1 + f_2 + \cdots + f_n$ of elements has an unambiguous meaning.

3. There exists a zero element o such that, for every element f, $f + o = f$.

4. For every scalar[1] α and every element f, the element $\alpha \cdot f$ is defined.

5. For every element f, $0 \cdot f$† $= o$ and $1 \cdot f = f$.

6. For all scalars α, β and elements f, g, $\alpha \cdot (f + g) = \alpha \cdot f + \alpha \cdot g$, $(\alpha + \beta) \cdot f = \alpha \cdot f + \beta \cdot f$,‡ $\alpha \cdot (\beta \cdot f) = (\alpha\beta) \cdot f$.

It is scarcely necessary to point out that the real- and complex-number systems constitute linear spaces. (Cf. Exercise 1.) Aside from these trivial examples, the most familiar linear spaces are furnished by classical vector analysis in euclidean space of two and three dimensions.

In order to help illustrate some of the subsequent definitions, we introduce here two rather simple examples of linear spaces.

1. The set V_n of all sequences of scalars, $\{\alpha_1, \alpha_2, \ldots, \alpha_n\}$, where n is a fixed positive integer, the operations $+$ and $\cdot$ being defined in the obvious manner:

$$\{\alpha_1, \alpha_2, \ldots, \alpha_n\} + \{\beta_1, \beta_2, \ldots, \beta_n\} = \{\alpha_1 + \beta_1, \alpha_2 + \beta_2, \ldots, \alpha_n + \beta_n\}$$
$$\lambda \cdot \{\alpha_1, \alpha_2, \ldots, \alpha_n\} = \{\lambda\alpha_1, \lambda\alpha_2, \ldots, \lambda\alpha_n\}$$

2. The set V_∞ of all infinite sequences $\{\alpha_1, \alpha_2, \ldots\}$, the operations $+$ and $\cdot$ being defined as above

DEFINITION 2. A finite set of elements f_1, f_2, $\ldots$, f_n is said to be (linearly) independent if the equality[2] $\alpha_1 f_1 + \alpha_2 f_2 + \cdots + \alpha_n f_n = o$ implies that each of the scalars α_1, α_2, $\ldots$, α_n must vanish. An infinite set of elements is said to be independent if every finite subset is independent.

DEFINITION 3. A (nonvacuous) subset M of a linear space V is said to be a (linear) manifold if, for every pair of elements f, g of M and every pair of scalars α, β, the linear combination $\alpha f + \beta g$ also belongs to M. (Clearly, V is itself a manifold, and any manifold is itself a linear space.)

---

[1] The term "scalar" is to be interpreted consistently as real number or complex number, V being accordingly termed a "real" or "complex" linear space. More generally, the scalars may be chosen from an arbitrary field F, in which case V is termed a "linear space over F," but we shall not be concerned with this generalization.

† Often the same symbol is used for the zero scalar 0 and the zero element o, since they cannot be confused.

‡ The symbol $+$ is used here with two different meanings, signifying addition between *scalars* on the left and addition between *vectors* on the right.

[2] Henceforth we omit the symbol $\cdot$ denoting the multiplication of an element by a scalar.

DEFINITION 4. A linear space is said to be "finite-dimensional" if there exists a finite set of elements $f_1, f_2, \ldots, f_n$ such that every element f can be expressed as a linear combination

$$f = \alpha_1 f_1 + \alpha_2 f_2 + \cdots + \alpha_n f_n \tag{1}$$

of these elements. If the afore-mentioned set of elements is independent, it is termed a "basis" of the space.

The proof of the following elementary but important theorem is left as Exercise 2.

Theorem 1. Every finite-dimensional linear space admits a basis. Any two of its bases contain the same number of elements, so that it is meaningful to *define* the "dimension" of the space as the number of elements in any basis. Given an n-dimensional linear space, a set of elements is a basis if and only if it consists of exactly n independent elements. The scalars $\alpha_1, \alpha_2, \ldots, \alpha_n$ appearing in representation (1) of any given element f are uniquely determined by f and the basis.

Referring to the spaces V_n and V_∞ defined above, we note that the elements $\{1,0, \ldots ,0\}, \{0,1, \ldots ,0\}, \ldots, \{0,0, \ldots ,1\}$ form a basis of V_n, so that this space is n-dimensional, while V_∞, on the other hand, is evidently not finite-dimensional.

As a simple example of a linear manifold (other than the given space itself), we may take the set of all elements (in either V_n or V_∞) whose first entry, α_1, is zero.

EXERCISES

1. Prove that the real-number system forms a real linear space of dimension one, while the complex-number system may be considered as either a complex space of dimension one or a real space of dimension two.

2. Prove Theorem 1.

3. Prove that a set of elements $\{\alpha_1{}^{(k)}, \alpha_2{}^{(k)}, \ldots, \alpha_n{}^{(k)}\}$, $1 \leq k \leq n$, forms a basis of V_n if and only if the determinant $|\alpha_j{}^{(k)}|$ does not vanish.

4. Prove that every manifold M of a linear space V of dimension n is of dimension $\leq n$, equality holding only if $M = V$.

5. Prove that the zero element o is unique.

6. Prove that if $f_1, f_2, \ldots, f_n$ are any finite set of elements belonging to a manifold, then every linear combination of them also belongs to the manifold. *Hint:* Use induction.

2. Normed Linear Spaces

In this section we introduce the concept of a norm, which is a generalization of the absolute value of a real or complex number and of length in elementary vector analysis.

DEFINITION 5.   A normed linear space is a linear space in which there is associated with each element f a real number, termed the "norm" of f and denoted $\|f\|$. The norm is required to satisfy the following conditions:

$$\|f\| \geq 0 \qquad \text{and} \qquad \|f\| = 0 \qquad \text{if and only if } f = o \qquad (2a)$$

For all elements f and g

$$\|f + g\| \leq \|f\| + \|g\| \qquad \text{(triangle inequality)} \qquad (2b)$$

For each scalar λ and each element f

$$\|\lambda f\| = |\lambda| \cdot \|f\| \qquad (2c)$$

We now present a number of simple examples of normed linear spaces.

1. Consider the linear space V_n defined in Sec. 1, and let the norm of any element be defined as follows:

$$\|\{\alpha_1, \alpha_2, \ldots, \alpha_n\}\| = |\alpha_1| + |\alpha_2| + \cdots + |\alpha_n| \qquad (3)$$

2. Consider V_n again, but let the norm now be defined as follows:

$$\|\{\alpha_1, \alpha_2, \ldots, \alpha_n\}\| = \max(|\alpha_1|, |\alpha_2|, \ldots, |\alpha_n|) \qquad (4)$$

3. Consider the class of all continuous complex-valued functions $f(x)$ defined on the finite closed interval $a \leq x \leq b$, addition of functions and multiplication by scalars being defined in the obvious manner. Let the norm be defined as follows:

$$\|f\| = \max_{a \leq x \leq b} |f(x)| \qquad (5)$$

We denote this normed space as $C(a,b)$.

In each of these three examples it is quite trivial that $(2b)$ is satisfied; we now present two examples in which the proof is not quite obvious. That $(2b)$ actually holds in these cases will follow from the Schwarz inequality, which is established in the following chapter.

4. Consider again the space V_n, as in examples 1 and 2, but with norm defined as follows:

$$\|\{\alpha_1, \alpha_2, \ldots, \alpha_n\}\| = (|\alpha_1|^2 + |\alpha_2|^2 + \cdots + |\alpha_n|^2)^{\frac{1}{2}} \qquad (6)$$

5. Consider the class of functions employed in example 3, but with norm defined as follows:

$$\|f\| = \left(\int_a^b |f(x)|^2 \, dx \right)^{\frac{1}{2}} \qquad (7)$$

We denote this normed space as $H(a,b)$.

The norm makes it possible to introduce the concept of convergence, in close analogy with the real- and complex-number systems.

DEFINITION 6. A sequence of elements $\{f_n\}$ is said to be "convergent in norm," or to be a Cauchy sequence, if $\lim\limits_{m,n \to \infty} \|f_n - f_m\| = 0$; i.e., if for every $\epsilon > 0$ there exists an integer N such that $\|f_n - f_m\| < \epsilon$ whenever m and n both exceed N.

DEFINITION 7. A sequence of elements $\{f_n\}$ is said to be "convergent" if there exists an element f such that $\lim\limits_{n \to \infty} \|f - f_n\| = 0$. Since the element f, if it exists, is unique (cf. Exercise 8), f may properly be termed the "limit" of the given sequence.

DEFINITION 8. An element f is said to be a "limit point" of a set S of elements if there exists a sequence $\{f_n\}$ of elements of S, each distinct from f, which converges to f.

DEFINITION 9. A set S of elements is said to be "closed" if every limit point of S belongs to S.

Returning now to linear manifolds, we introduce the following definition.

DEFINITION 10. A closed linear manifold is termed a "subspace."

We conclude this section by noting that, as follows immediately from Definition 8, any normed space is closed, but that, on the other hand, as shown by Exercise 12, a linear manifold in this space may fail to be closed (although, of course, when considered as a normed space by itself, without regard to the larger space, it is indeed closed). Thus, Definition 10 is not devoid of content.

EXERCISES

7. A metric space is defined as a nonvacuous set of elements such that, for every pair of elements f, g, a "distance" $\rho(f,g)$ is defined which satisfies the following conditions:

   (a) $\rho(f,f) = 0$, $\rho(f,g) > 0$ if $f \neq g$.

   (b) $\rho(f,g) = \rho(g,f)$.

   (c) $\rho(f,h) \leq \rho(f,g) + \rho(g,h)$.

Prove that any normed linear space becomes a metric space with the obvious definition of distance: $\rho(f,g) = \|f - g\|$.

8. Prove the assertion made in Definition 7. *Hint:* Use (2b).

9. Prove that f is a limit point of a set S if and only if, for each $\epsilon > 0$, there exist at least two elements f_1, f_2 of S such that $\|f_k - f\| < \epsilon$, $k = 1, 2$. (Why cannot "at least two" be replaced by "at least one"?) Prove that "at least two" can be replaced by "infinitely many."

10. Prove that the norm is a continuous function; more precisely, if $\|f - g\| < \epsilon$, then $|\|f\| - \|g\|| < \epsilon$.

11. Show that any set consisting of a finite number of elements is closed.

12. Let V denote the subset of V_∞ consisting of all bounded sequences, the norm $\|\{\alpha_1,\alpha_2, \ldots\}\|$ being defined as sup $|\alpha_k|$. Let M denote the subset of V consisting

of those sequences containing only a finite number of nonzero entries.   Prove that M is a manifold, but not a subspace, of V.

3. Banach Spaces

Among the five normed spaces introduced in Sec. 2, it will be noted that the space $H(a,b)$ differs in an important respect from each of the other four examples.   Consider the sequence of functions $\{f_n(x)\}$ defined as follows:[1]

$$f_n(x) = \begin{cases} 0 & 0 \leq x \leq \dfrac{1}{2} - \dfrac{1}{10^n} \\[2mm] \dfrac{1}{2} + \dfrac{1}{2} \cdot 10^n \left(x - \dfrac{1}{2} \right) & \dfrac{1}{2} - \dfrac{1}{10^n} \leq x \leq \dfrac{1}{2} + \dfrac{1}{10^n} \\[2mm] 1 & \dfrac{1}{2} + \dfrac{1}{10^n} \leq x \leq 1 \end{cases}$$
$$(n = 1, 2, 3, \ldots) \quad (8)$$

Each of these functions clearly belongs to $H(0,1)$, and the sequence is readily seen to be convergent in norm.   (This sequence is also convergent in the ordinary sense for each value of x under consideration, but this fact does not concern us here.)   However, there does not exist a function $f(x)$ of the space $H(0,1)$ such that

$$\lim_{n \to \infty} \|f - f_n\| = 0$$

It is perhaps worthwhile to prove this statement in detail.   Suppose there were such a function $f(x)$.   Then, from the relationship

$$\int_0^1 |f - f_n|^2 \, dx \to 0$$

there would follow, a fortiori,

$$\int_0^c |f - f_n|^2 \, dx \to 0$$

where c is any positive constant less than $\frac{1}{2}$.   Now from (8) we see that, except perhaps for a finite number of values of n, $f_n \equiv 0$ in the interval $0 \leq x \leq c$, so that the condition

$$\int_0^c |f|^2 \, dx = 0$$

must be satisfied; since f, being a member of $H(0,1)$, is continuous, this last equality implies that $f \equiv 0$ for $0 \leq x \leq c$, and hence for $0 \leq x < \frac{1}{2}$. Similarly, we conclude that $f \equiv 1$ for $\frac{1}{2} < x \leq 1$.   Thus, $f(x)$ cannot

---

[1] For convenience we choose $a = 0$, $b = 1$.

be continuous at $x = \frac{1}{2}$, contradicting the assumption that it belongs to $H(0,1)$.

Thus, it has been shown that a Cauchy sequence in $H(a,b)$ may fail to be convergent. This is in contrast, as stated at the beginning of this section, to the other normed spaces defined in Sec. 2. In particular, the space $C(a,b)$ consists of the same functions as $H(a,b)$, but the norm in $C(a,b)$ is such that convergence in norm is synonymous with uniform convergence, so that, by a fundamental theorem of analysis, a Cauchy sequence in $C(a,b)$ is convergent. Similarly, the same assertion is easily proved for the other three normed spaces. (Cf. Exercise 16.)

The afore-mentioned contrast between $H(a,b)$ and the other normed spaces serves to introduce the concept of completeness, which we now define.

DEFINITION 11. A normed linear space V is said to be "complete" if for every Cauchy sequence $\{f_n\}$ of elements of V there exists an element f of V which is the limit of the sequence. A complete normed linear space is termed a "Banach space."

EXERCISES

13. Construct a sequence of functions $\{f_n(x)\}$ belonging to $H(a,b)$ such that

(a) $\{f_n(x)\}$ converges at every point, but not in norm, to a function $f(x)$ which also belongs to $H(a,b)$.

(b) $\{f_n(x)\}$ converges in norm, but not pointwise for any value of x, to a function $f(x)$ which also belongs to $H(a,b)$.

14. Show that if a sequence of functions $\{f_n(x)\}$ belonging to $H(a,b)$ converges in norm to a function $f(x)$, and pointwise to a function $g(x)$ also belonging to $H(a,b)$, then $f(x) \equiv g(x)$.

15. *Continuation of Exercise* 14: Let $a = 0$, $b = 1$. Show that the sequence of functions $\{x^n\}$ converges in norm to the function $f \equiv 0$, while it converges pointwise to the function $g(x)$ which vanishes identically except that $g(1) = 1$. (Note that this does not contradict the assertion of Exercise 14.)

16. Show that the spaces 1, 2, and 4 defined in Sec. 2 are complete.

17. Let A denote the set of all functions $f(x)$ defined on the interval $0 \leq x \leq 1$ such that $f(x)$ vanishes everywhere with the possible exception of a countable number of points, the values of $f(x)$ being subject to the condition that $\Sigma |f(x)|$ converges, the summation being extended over those values of x where $f(x) \neq 0$. Let $\|f\|$ be defined as the above sum. Show that, with the obvious definition of $+$ and $\cdot$, A is a Banach space.

18. Let $f_1, f_2, \ldots, f_n$ be a fixed linearly independent set of elements of a normed space. Show that there exists a positive constant c such that, for every choice of the scalars $\lambda_1, \lambda_2, \ldots, \lambda_n$, the inequality $\|\lambda_1 f_1 + \lambda_2 f_2 + \cdots + \lambda_n f_n\| \geq c(|\lambda_1| + |\lambda_2| + \cdots + |\lambda_n|)$ holds. *Hint:* Investigate the minimum of $\|\lambda_1 f_1 + \lambda_2 f_2 + \cdots + \lambda_n f_n\|$ under the restriction $|\lambda_1| + |\lambda_2| + \cdots + |\lambda_n| = 1$.

Use this result to prove that a finite-dimensional normed space must be not only complete, but also "locally compact"—i.e., every bounded sequence $\{f_n\}$ possesses a convergent subsequence.

19. A normed space B is said to be "separable" if it possesses a denumerable dense subset—i.e., a sequence of elements $\{f_n\}$ such that for every element g of B and every $\epsilon > 0$ there exists an element f_n of the sequence such that $\|g - f_n\| < \epsilon$. Prove:

(a) Every finite-dimensional Banach space is separable.

(b) The space $C(a,b)$, which is obviously not finite-dimensional, is separable. *Hint:* Either construct a dense sequence of "polygonal" functions, or employ the Weierstrass approximation theorem to construct a dense sequence of polynomials.

(c) The space A defined in Exercise 17 is not separable.

20. Prove that every finite-dimensional linear manifold of a Banach space is closed (i.e., it is a subspace).

21. Prove that every subspace S of a Banach space B is complete, and hence is itself a Banach space. Furthermore, if B is separable, so is S.

4. Linear Functionals and Linear Operators

Two types of functions are of especial importance in the study of Banach spaces. First, there are functions which associate with each element f a scalar, which may be denoted as $l(f)$, and, second, there are functions which associate with each element f of the space an element, which may be denoted as Tf. Following common usage, functions of these two types will be denoted as "functionals" and "operators," respectively. The only functionals and operators that we shall consider are those which are *additive* and *continuous;* we proceed to define these two terms.

DEFINITION 12. A functional l (or an operator T) is said to be additive if for every pair of elements f, g and every pair of scalars α, β the equality $l(\alpha f + \beta g) = \alpha l(f) + \beta l(g)$ [or $T(\alpha f + \beta g) = \alpha Tf + \beta Tg$] holds. [Setting $\alpha = \beta = 0$, we find that $l(o) = 0$, $T(o) = o$.]

DEFINITION 13. A functional l (or an operator T) is said to be "continuous at the element f" if for every $\epsilon > 0$ there exists a $\delta > 0$ such that the inequality $\|f - g\| < \delta$ implies the inequality $|l(f) - l(g)| < \epsilon$ (or $\|Tf - Tg\| < \epsilon$). A functional (or operator) is said to be "continuous" if it is continuous at every element of the space. A "linear" functional (or operator) is one which is both additive and continuous.

Closely connected with the concept of continuity is the concept of boundedness, which we now formulate. The connection will then be established in Theorem 2.

DEFINITION 14. A functional l (or an operator T) is said to be "bounded" if there exists a constant C such that, for each element f, the inequality $|l(f)| \leq C\|f\|$ (or $\|Tf\| \leq C\|f\|$) holds. The bound, or norm, of l (or T) is the minimum value of C for which the above inequalities are valid for all f. (Cf. Exercise 22.) The norm is denoted by $\|l\|$ (or $\|T\|$).

We are now in position to prove the following simple but important theorem.

Theorem 2. An additive functional (or operator) is continuous, and hence linear, if and only if it is continuous at o, and also if and only if it is bounded. If l is a linear functional and the sequence $\{f_n\}$ converges to f, then $\{l(f_n)\}$ converges to $l(f)$; similarly for a linear operator.

Proof. It clearly suffices to give the proof for a functional. If l is continuous at o, we can find a $\delta > 0$ such that $|l(f) - l(o)| = |l(f)| \leq 1$ whenever $\|f\| < \delta$; by additivity, for any f we obtain $|l(f)| \leq \delta^{-1} \|f\|$; this shows that l is bounded. Given any element f and any $\epsilon > 0$, the inequality $\|g - f\| < \delta\epsilon$ implies, again by additivity, that

$$|l(g) - l(f)| = |l(g - f)| \leq \delta^{-1}\|g - f\| < \delta^{-1} \cdot \delta\epsilon = \epsilon$$

Thus, l is continuous at f, and hence at all points of the space. Thus l is a continuous functional, and, in particular, is continuous at o. We have thus shown that any one of the three conditions (continuity, continuity at o, boundedness) implies the other two, and the proof of the first part is therefore complete. The latter part of the theorem follows immediately from the inequality $|l(f - f_n)| \leq \|l\| \cdot \|f - f_n\|$.

Henceforth it will be understood that all functionals and operators under consideration are linear. Let us explain addition and multiplication of operators, and scalar multiplication of operators, by the following rather obvious definitions:

$$(T_1 + T_2)f \equiv T_1 f + T_2 f \qquad (\lambda T)f = \lambda(Tf) \qquad (T_1 T_2)f = T_1(T_2 f) \qquad (9)$$

By elementary arguments (cf. Exercise 23) we obtain, from the first two of these definitions,

$$\|T_1 + T_2\| \leq \|T_1\| + \|T_2\| \qquad \|\lambda T\| = |\lambda| \cdot \|T\| \qquad (10)$$

Furthermore, if the operators $\{T_n\}$ constitute a Cauchy sequence, in the sense that $\lim\limits_{m,n \to \infty} \|T_m - T_n\| = 0$, we can conclude that this sequence is convergent to an operator T, in the sense that $\lim\limits_{n \to \infty} \|T - T_n\| = 0$. For, given any element f, the sequence $\{T_n f\}$ is a Cauchy sequence (since $\|T_m f - T_n f\| \leq \|T_m - T_n\| \cdot \|f\| \to 0$) and hence, *on account of the completeness* of the space, convergent to an element which we *define* to be Tf. T is then readily shown to be additive, bounded, and the limit of the sequence $\{T_n\}$ in the sense explained above. (Cf. Exercise 24.)

Let $R(B)$ denote the set of all operators defined on the Banach space B. Then the foregoing remarks, taken together with the obvious fact that the only operator having zero norm is the zero operator O defined in the obvious manner

$$Of = o \tag{11}$$

show that $R(B)$ is itself a Banach space. Furthermore, returning to (9), we see that $R(B)$ admits an operation of multiplication between any two of its elements, and the following inequality is readily proved (cf. Exercise 23):

$$\|T_1 T_2\| \le \|T_1\| \cdot \|T_2\| \tag{12}$$

A Banach space admitting, aside from the operations of addition and scalar multiplication, an operation of multiplication between its elements satisfying (12) (where T_1 and T_2 are now understood to denote any two elements of the space, not necessarily operators on some Banach space) is known as a Banach algebra; thus, the set of all operators on any Banach space constitutes a Banach algebra.

A particular class of linear operators will prove to be of especial importance; these are the operators having the property of "complete continuity," which we now proceed to formulate.

DEFINITION 15.   An operator T is said to be completely continuous (or compact) if for every sequence $\{f_n\}$ which is bounded (i.e., $\|f_n\| < C$ for some sufficiently large constant C) the sequence $\{Tf_n\}$ contains a convergent subsequence.

In a finite-dimensional Banach space this concept is of no significance. (Cf. Exercise 29.)   On the other hand, if B is any infinite-dimensional Banach space, the identity operator I, which is, of course, defined by the identity

$$If \equiv f \tag{13}$$

is not completely continuous.   This follows from the fact, easily shown with the aid of Lemma 1 of Theorem 6, that such a space possesses a sequence of unit elements $\{f_n\}$ such that $\|f_i - f_j\| > \frac{1}{2}$, $i \ne j$.   The operator O is, trivially, completely continuous, but a more interesting example is furnished by integral transforms.   Let $K(x,y)$ be any continuous function defined on the square $0 \le x$, $y \le 1$, and let an operator K be defined on the space $C(0,1)$ (cf. Sec. 2) as follows:

$$g(x) = Kf = \int_0^1 K(x,y)f(y)\, dy \qquad [f \in C(0,1)] \tag{14}$$

Clearly $g \in C(0,1)$ and K is additive, while its boundedness follows from the trivial estimate $\|g\| \le \max |K(x,y)| \cdot \|f\|$, which leads to the following inequality for $\|K\|$:

$$\|K\| \le \max |K(x,y)| \tag{15}$$

Now consider any bounded sequence $\{f_n\}$, and let M be an upper bound on the norms; given any $\epsilon > 0$, we can, by uniform continuity, choose $\delta > 0$ such that whenever $|x_1 - x_2| < \delta$ the inequality

$$|K(x_1,y) - K(x_2,y)| < \frac{\epsilon}{M} \tag{16}$$

is satisfied. From this it follows that the inequality

$$|g(x_1) - g(x_2)| < \int_0^1 \frac{\epsilon}{M} \cdot M \, dy = \epsilon \tag{17}$$

also holds. From (15) and (17) we observe that the functions of the sequence $\{Kf_n\}$ are uniformly bounded and equicontinuous; invoking Theorem 1-1, we conclude that the latter sequence contains a uniformly convergent subsequence—i.e., a sequence convergent in the norm defined on the space $C(0,1)$. This argument establishes the complete continuity of K. The interval $0 \leq x \leq 1$ employed in this example may, of course, be replaced by any other finite interval. (Cf. Exercise 37.)

As was shown earlier in this section, a sequence of operators $\{T_n\}$ which converges in the sense that $\lim\limits_{m,n \to \infty} \|T_m - T_n\| = 0$ also converges in the sense that there exists an operator T such that $\lim\limits_{n \to \infty} \|T - T_n\| = 0$; the analogous assertion for completely continuous operators is also true. While the proof of this fact is similar to that of Theorem 1-1, we present it in detail.

Theorem 3. Let the sequence $\{T_n\}$ of completely continuous operators converge to the operator T; then T is also completely continuous.

Proof. Let $\{f_n\}$ be any bounded sequence of elements, $\|f_n\| < M$. Since T_1 is completely continuous, we can select a subsequence $\{f_{k1}\}$ such that the sequence $\{T_1 f_{k1}\}$ is convergent. From the sequence $\{f_{k1}\}$ we select a subsequence $\{f_{k2}\}$ such that $\{T_2 f_{k2}\}$ converges. Continuing this procedure and then "diagonalizing," we obtain a single sequence $\{f_{kk}\}$ such that for each n the sequence $\{T_n f_{kk}\}$ converges. Given $\epsilon > 0$, we choose any m such that the inequality

$$\|T - T_m\| < \frac{\epsilon}{4M} \tag{18}$$

holds. Keeping m fixed, we then select N so large that

$$\|T_m(f_{jj} - f_{kk})\| < \frac{\epsilon}{2} \tag{19}$$

holds whenever $j,k > N$; but then,

$$\|T(f_{jj} - f_{kk})\| = \|(T - T_m)(f_{jj} - f_{kk}) + T_m(f_{jj} - f_{kk})\|$$
$$\leq \|(T - T_m)(f_{jj} - f_{kk})\| + \|T_m(f_{jj} - f_{kk})\|$$
$$< \|T - T_m\| \cdot (\|f_{jj}\| + \|f_{kk}\|) + \frac{\epsilon}{2} < \frac{\epsilon}{4M} \cdot 2M + \frac{\epsilon}{2} = \epsilon \quad (20)$$

The complete continuity of T is thus established.

We now turn to the important concept of the inverse of an operator. Suppose that T is an operator which maps the Banach space B onto itself in a one-to-one manner, so that for each element g of B there exists a unique element f such that

$$Tf = g \quad (21)$$

Then f may be considered as a (single-valued) function of g, and we denote this functional relationship by T^{-1}. Thus, (21) is equivalent to

$$T^{-1}g = f \quad (22)$$

Clearly T^{-1}, like T, is additive, and it can be shown that T^{-1} is bounded, and hence linear. However, we shall not prove this assertion. T^{-1} is known as the "inverse" of T. It is evident that T is the inverse of T^{-1}, and that both of the equalities

$$TT^{-1} = I \qquad T^{-1}T = I \quad (23)$$

hold. Conversely, these equalities characterize the inverse operator, as is shown by the following theorem.

Theorem 4. Suppose that the operators S and T, defined on a Banach space B, satisfy the equalities

$$ST = I \qquad TS = I \quad (24)$$

Then S and T are the inverses of each other.

Proof. Suppose that S carries elements f and g into the same element—that is, $Sf = Sg$. Multiplying both sides of this equality by T and taking account of the second equality of (24), we conclude that $f = g$. Thus, S carries distinct elements into distinct elements. Now, according to the first equality of (24), S maps the range of T (i.e., the set of all elements of B which are expressible in the form $Tf, f \in B$) onto all of B; a fortiori, the range of S must coincide with B. Thus, we have proved that S provides a one-to-one mapping of B onto itself, and the second equality of (24) now shows that T is the inverse of S. As pointed out above, S must then be the inverse of T; this is also evident from the symmetry of (24). [It is noteworthy that both equalities of (24) are used in the proof. As shown by Exercise 35, neither one of them implies the other.]

Suppose, in particular, that T can be expressed in the form

$$T = I - K \qquad \|K\| < 1 \tag{25}$$

Since an obvious induction based on (12) shows that

$$\|K^n\| \leq \|K\|^n \qquad (n = 1, 2, 3, \ldots) \tag{26}$$

it follows that the series

$$\|I\| + \|K\| + \|K^2\| + \cdots$$

is dominated by the geometric series $1 + \|K\| + \|K\|^2 + \cdots$, and is therefore convergent. According to Exercise 27, it is meaningful to define an operator S by the "Neumann series"

$$S = I + K + K^2 + \cdots \tag{27}$$

From Exercise 28 it now follows that both equalities (24) are satisfied. We have therefore proved the following simple result, which, on account of its importance, is stated as a theorem.

Theorem 5. Let the operator T be expressible in the form (25). Then T possesses an inverse, which is given by (27).

EXERCISES

22. Show that the set of values of C which can be used in Definition 14 is closed, so that it is proper to speak of the minimum value (rather than of the infimum). Prove that for a linear functional l the norm is given by

$$\|l\| = \sup_{\|f\|=1} |l(f)|$$

and similarly for an operator T. Prove by a simple example that "sup" may not be replaced by "max."

23. Prove the assertions made in (10) and (12).

24. Prove the assertions made in the text concerning the operator T which was defined as the limit of the Cauchy sequence $\{T_n\}$.

25. Prove that if T is the limit of the Cauchy sequence $\{T_n\}$, then $\lim\limits_{n \to \infty} \|T_n\|$ exists and equals $\|T\|$; in particular, it follows that the quantities $\|T_n\|$ are bounded.

26. Prove that if the series $f_1 + f_2 + \cdots$ converges to f (in the sense that the partial sums converge to f), then the series $Tf_1 + Tf_2 + \cdots$ converges to Tf.

27. Show that if $\|T_n\| \leq c_n$, where the series $c_1 + c_2 + \cdots$ is convergent, then the series $T_1 + T_2 + \cdots$ is convergent in the sense that there exists an operator T such that $\lim\limits_{n \to \infty} \left\| T - \sum\limits_{k=1}^{n} T_k \right\| = 0$; furthermore, $\|T\| \leq \sum\limits_{k=1}^{\infty} c_k$.

28. Show that if $\lim\limits_{n \to \infty} \|T - T_n\| = 0$ and if S is any operator, then $\lim\limits_{n \to \infty} \|TS - T_nS\| = \lim\limits_{n \to \infty} \|ST - ST_n\| = 0$.

29. Prove that every additive operator defined on a finite-dimensional Banach space is bounded and completely continuous. *Hint:* Use Exercise 18.

30. Prove that any (finite) linear combination of completely continuous operators is also completely continuous.

31. Prove that if S is any operator and T is completely continuous, then so are ST and TS.

32. Prove that if $K_1, K_2, \ldots, K_n$ are each completely continuous, then $(I - K_1)$ $(I - K_2) \cdots (I - K_n)$ may be expressed as $I - T$, where T is also completely continuous.

33. Accepting Lemma 1 of Theorem 6, prove the assertion following (13), and then show that the set of all elements f such that $(I - K)f = o$, where K is completely continuous, is a finite-dimensional subspace.

34. Prove that if operators S_1 and S_2 exist such that $S_1 T = T S_2 = I$, then $S_1 = S_2$; hence, by Theorem 4, T^{-1} exists.

35. Show that neither equality of (24) implies the other. *Hint:* Consider V, as defined in Exercise 12, and let the operators S and T be defined as follows:

$$T\{\alpha_1, \alpha_2, \alpha_3, \ldots\} = \{0, \alpha_1, \alpha_2, \ldots\}$$
$$S\{\alpha_1, \alpha_2, \alpha_3, \ldots\} = \{\alpha_2, \alpha_3, \alpha_4, \ldots\}$$

[If S and T satisfy the first equality of (24) we call T a "right-inverse" of S, and S a "left-inverse" of T.]

36. In contrast to the preceding exercise, show that either equality of (24) implies the other if B is finite-dimensional.

37. Prove that (15) may be replaced by the stronger inequality $\|K\| \leq \max\limits_{0 \leq x \leq 1}$ $\int_0^1 |K(x,y)| \, dy$. Generalize both of these inequalities from $C(0,1)$ to $C(a,b)$.

38. Suppose that $\|K^m\| < 1$ for some positive integer m. Then prove that $I - K$ is invertible and that its inverse is given by (27), even though K need not satisfy the condition imposed in (25).

39. Obtain (27) *formally* by writing the equation $f = (I - K)^{-1}g$ in the form $f = g + Kf$ and replacing f on the right side of the latter equation by the entire right side. (This is an abstract formulation of the "method of successive approximation"—cf. Theorem 1-6.)

5. The Fredholm Alternative

The theory of linear integral equations shows a strong resemblance to a much more elementary theory, namely, that of a finite number of simultaneous linear (algebraic) equations in the same number of unknowns. Indeed, the first successful approach to the theory of integral equations, due to Fredholm, was based on considering an integral equation as a limiting case of such an algebraic system. However, we shall develop in this section a theorem concerning linear equations in Banach spaces which includes as particular cases the basic theorems concerning *both* linear algebraic systems and linear integral equations. These two particular cases are stated as corollaries of Theorem 6. It is of interest that the proof of the latter is not based, as one might expect, on the first corollary, which is the simplest of all particular cases. Indeed, it will be found instructive to recast the proof of the theorem into forms specifically pertaining to each corollary.

Theorem 6.   Fredholm Alternative.   Let K be a completely continuous operator defined on a Banach space B.   If the equation

$$f - Kf = o \tag{28}$$

admits only the trivial solution $f = o$, then the equation

$$f - Kf = g \tag{29}$$

has one and only one solution for each element g.   Otherwise, the solutions of (28) form a finite-dimensional manifold, there exist elements g for which (29) is not solvable, and whenever (29) *is* solvable the solution is not unique, since the sum of a nontrivial solution of (28) and a solution of (29) constitutes a new solution of the latter equation.

Proof.   The proof will be carried out by establishing a number of lemmas, some of which are of interest in themselves.

LEMMA 1.   If B_0 is a proper subspace of B and α is any positive number less than one, there exists an element x in B of unit norm such that $\|x - y\| \geq \alpha$ for every element y of B_0.   If B_0 is finite-dimensional, α may even be chosen *equal* to one.

Proof.   Let x_0 be any element of B not belonging to B_0, and let

$$\delta = \inf \|x_0 - y\| \quad \text{for } y \in B_0$$

If δ were zero, x_0 would be a limit point of B_0, and hence an element of B_0 (since B_0, being a subspace, is closed).   Therefore, δ is positive, and, by its definition, there exists an element y_0 in B_0 such that

$$\|x_0 - y_0\| < \alpha^{-1}\delta$$

For any element y of B_0 the element $y_0 + \|x_0 - y_0\|y$ also belongs to B_0, and hence

$$\delta < \|x_0 - (y_0 + \|x_0 - y_0\|y)\| = \|x_0 - y_0\| \cdot \|x - y\| < \alpha^{-1}\delta \cdot \|x - y\|$$

where

$$x = \|x_0 - y_0\|^{-1}(x_0 - y_0)$$

Thus, $\|x - y\| > \alpha$, and since $\|x\| = 1$, the proof of the first part of the lemma is complete.   If B_0 is finite-dimensional, we select a sequence $y_1, y_2, \ldots$ of elements of B_0 such that $\|y_k - x_0\| \to \delta$, and express each of these elements in terms of a basis $\{f_1, f_2, \ldots, f_n\}$ of B_0:

$$y_k = \lambda_1^{(k)}f_1 + \lambda_2^{(k)}f_2 + \cdots + \lambda_n^{(k)}f_n \tag{30}$$

The quantities $\|y_k\|$ are bounded, for, by the triangle inequality,

$$\|y_k\| = \|x_0 + (y_k - x_0)\| \le \|x_0\| + \|y_k - x_0\|$$

It then follows from Exercise 18 that the λ's appearing in (30) are bounded. Hence, a subsequence of the sequence $\{y_k\}$ may be chosen for which each of the sequences $\{\lambda_1^{(k)}\}$, $\{\lambda_2^{(k)}\}$, . . . , $\{\lambda_n^{(k)}\}$ is convergent. Consider the element

$$\tilde{y} = \lambda_1 f_1 + \lambda_2 f_2 + \cdots + \lambda_n f_n$$

where each λ_m is the limit of the sequence $\{\lambda_m^{(k)}\}$. Clearly $\tilde{y}$ belongs to B_0, and from the inequalities

$$\|x_0 - \tilde{y}\| \le \|x_0 - y_k\| + \|y_k - \tilde{y}\|$$

$$\le \|x_0 - y_k\| + \sum_{m=1}^{n} |\lambda_m^{(k)} - \lambda_m| \cdot \|f_m\|$$

we conclude, by letting k become infinite, that $\|x_0 - \tilde{y}\|$ cannot exceed δ, and hence must equal δ. We may now repeat the argument used previously, but with the inequality $\|x_0 - y_0\| < \alpha^{-1}\delta$ replaced by the equality $\|x_0 - \tilde{y}\| = \delta$. We then conclude that the unit vector

$$x = \|x_0 - \tilde{y}\|^{-1}(x_0 - \tilde{y})$$

satisfies the inequality $\|x - y\| \ge 1$ for each element y of B_0.

LEMMA 2. Let N_k ($k = 0, 1, 2, \ldots$) denote the null space of $(I - K)^k$ —i.e., the set of all elements f such that $(I - K)^k f = o$. (Note that N_0 consists exclusively of the element o.) Then: (1) Each N_k is a finite-dimensional subspace of B. (2) $N_0 \subset N_1 \subset N_2 \subset \cdots$. (3) There exists an integer ν such that $N_k = N_{k+1}$ if and only if $k \ge \nu$.

Proof. Part 1 follows directly from Exercises 32 and 33. Part 2 follows from the fact that

$$(I - K)^{k+1}f = (I - K)(I - K)^k f$$

so that, if $f \in N_k$,

$$(I - K)^{k+1}f = (I - K)(I - K)^k f = (I - K)o = o$$

To prove part 3, we proceed in two steps. First, if $N_j = N_{j+1}$, then

$$(I - K)^{j+1}f = o$$

implies that

$$(I - K)^j f = o$$

Replacing f by $(I - K)g$, we conclude that

$$(I - K)^{j+2}g = o$$

implies that

$$(I - K)^{j+1}g = o$$

so that $N_{j+2} = N_{j+1} = N_j$. By induction, $N_j = N_{j+1}$ implies that $N_j = N_{j+r}$, $r = 1, 2, 3, \ldots$. Secondly, suppose that for each index j the spaces N_j and N_{j+1} are different. Then N_j is a proper subspace of N_{j+1}, so that, by Lemma 1, we can choose unit elements x_1 in N_1, x_2 in N_2, $\ldots$ such that $\|x_j - y\| \geq 1$ whenever $y \in N_{j-1}$. Now, on the one hand, the sequence $\{Kx_1, Kx_2, \ldots\}$ must admit a convergent subsequence, while, on the other hand, the equality

$$Kx_i - Kx_j = x_i - f_{ij} \qquad f_{ij} = x_j + (I - K)x_i - (I - K)x_j \quad (31)$$

shows that, for $i > j$, $f_{ij} \in N_{i-1}$, so that $\|x_i - f_{ij}\| \geq 1$. Thus,

$$\|Kx_i - Kx_j\| \geq 1$$

so that, in contradiction with an earlier statement, we conclude that the sequence $\{Kx_1, Kx_2, \ldots\}$ does *not* admit a convergent subsequence. Thus the proof is complete.

LEMMA 3. Let M_k ($k = 0, 1, 2, \ldots$) denote the range of $(I - K)^k$ —i.e., the set of all elements expressible in the form $(I - K)^k g$, $g \in B$. (It is understood that $M_0 = B$.) Then for each k there exists a positive number σ_k such that, for every element $z \in M_k$, the equation $z = (I - K)^k x$ possesses a solution whose norm does not exceed $\sigma_k \|z\|$.

Proof. For $k = 0$ the lemma is trivial, with $\sigma_0 = 1$. Turning to the case $k = 1$, we first show that for each $z \in M_1$ the equation

$$z = (I - K)x$$

or

$$x = z + Kx \qquad (32)$$

possesses a solution whose norm is a minimum. For, if we select a sequence $\{x_n\}$ of solutions of (32) whose norms approach the minimum, we can, since K is completely continuous, select a subsequence such that $\{Kx_n\}$ is convergent. Referring to (32), we conclude that the subsequence $\{x_n\}$ converges, and its limit satisfies (32) and has the minimum norm. (The solution of minimum norm is not necessarily unique.) Now assume the lemma were false. Then we could find a sequence $\{z_n\}$ of elements of M_1 such that $\{z_n\}$ converges to o and the corresponding sequence $\{x_n\}$ of solutions of $(I - K)x = z_n$, or

$$x = z_n + Kx \qquad (33)$$

having minimum norm would consist entirely of unit elements. As above, we could find a subsequence of $\{x_n\}$ such that $\{Kx_n\}$ converges. Referring to (33), we see that the indicated subsequence of $\{x_n\}$ would converge and that its limit, x, would satisfy the homogeneous equation

$$x = Kx \qquad (34)$$

Now, since $(x_n - x) = z_n + K(x - x_n)$, we find that $\|x_n - x\| \geq 1$, for no solution of (33) can have norm less than one. This inequality cannot hold, however, since a subsequence of $\{x_n\}$ converges to x. This contradiction proves the lemma for $k = 1$. For $k > 1$ we merely note that since, according to Exercise 32, $(I - K)^k$ can be written in the form $I - K_k$, where K_k is also completely continuous, the above argument applies equally well to this case.

LEMMA 4. (1) $M_0 \supset M_1 \supset M_2 \supset \cdots$. (2) Each M_k is a subspace of B. (3) There exists an integer μ such that $M_k = M_{k+1}$ if and only if $k \geq \mu$.

Proof. If $f = (I - K)^{k+1}g$, then $f = (I - K)^k h$, where $h = (I - K)g$. Thus, any element of M_{k+1} is also an element of M_k. This proves part 1. If $f_1 = (I - K)^k g_1$ and $f_2 = (I - K)^k g_2$, then

$$\alpha_1 f_1 + \alpha_2 f_2 = (I - K)^k (\alpha_1 g_1 + \alpha_2 g_2)$$

This shows that M_k is a linear manifold. To show that M_k is closed, we select any limit point z of M_k and a sequence $\{z_n\}$ of elements of M_k converging to z. According to the preceding lemma, we can find a sequence $\{x_n\}$ such that $\|x_n\| \leq \sigma_k \|z_n\|$ and $(I - K)^k x_n = z_n$, or (cf. preceding proof)

$$x_n = z_n + K_k x_n \qquad (35)$$

Once again we conclude that a suitably selected subsequence of $\{x_n\}$ converges, and that the limit of this subsequence satisfies the equality $x = z + K_k x$, or

$$(I - K)^k x = z \qquad (36)$$

Thus, $z \in M_k$, and so M_k is a subspace. This completes the proof of part 2. To prove part 3, we first show that if $M_k = M_{k+1}$, then $M_{k+1} = M_{k+2}$, from which it follows (by induction) that $M_{k+n} = M_k$, $n = 1, 2, 3, \ldots$. Consider any element $f \in M_{k+1}$. Then

$$f = (I - K)^{k+1}g$$

for some element g. But $(I - K)^k g$ can be expressed as $(I - K)^{k+1}h$, and so $f = (I - K)(I - K)^{k+1}h = (I - K)^{k+2}h$. Thus, $M_{k+1} = M_{k+2}$. It remains to prove that for some value of k the equality $M_k = M_{k+1}$ must hold. In the contrary case, M_{k+1} would be a proper subspace of M_k for each value of k, and then, according to Lemma 1, we could choose unit elements $x_0 \in M_0 \, (= B)$, $x_1 \in M_1$, $\ldots$, such that $\|x_i - y\| > \frac{1}{2}$ for every element y belonging to M_{i+1}. (Note that, according to Exercise

21, each M_i is a Banach space, so that Lemma 1 applies.) We now refer to (31), where we assign to x_i and x_j the meanings developed in the present proof. Expressing x_i and x_j as $(I - K)^i g$ and $(I - K)^j h$, respectively, and confining our attention to the case $i < j$, we obtain for f_{ij} the expression $(I - K)^{i+1}[(I - K)^{j-i-1}h + g - (I - K)^{j-i}h]$, from which it follows that $f_{ij} \in M_{i+1}$. Therefore, $\|Kx_i - Kx_j\| > \frac{1}{2}$, and so it is not possible to extract a convergent subsequence from $\{Kx_n\}$. Since K is completely continuous, we have obtained a contradiction, and so the proof is complete.

LEMMA 5. *If $g \in M_\mu$ and k is any positive integer, the equation $(I - K)^k f = g$ has one and only one solution belonging to M_μ.*

Proof. Since $M_\mu = M_{\mu+k}$, g can be expressed as $(I - K)^{\mu+k}y$ for some element y, and $(I - K)^\mu y$ belongs to M_μ and is a solution of the given equation. To establish uniqueness, suppose that there were two different solutions, f_1 and f_2, such that $f_1 \in M_\mu$, $f_2 \in M_\mu$. Let $h_1 = f_1 - f_2$. Then

$$(I - K)^k h_1 = (I - K)^k f_1 - (I - K)^k f_2 = g_2 - g_2 = o$$

Also, $h_1 \in M_\mu$, since M_μ is a manifold. By the existence part of the present lemma, which has already been proved, we can find elements $h_2, h_3, \ldots$, which belong to M_μ and satisfy the equations

$$(I - K)^k h_{n+1} = h_n \qquad n = 1, 2, 3, \ldots$$

Then

$$(I - K)^{nk} h_{n+1} = h_1 \neq o \qquad (I - K)^{(n+1)k} h_{n+1} = (I - K)^k h_1 = o$$

Thus, h_{n+1} would belong to $N_{(n+1)k}$ but not to N_{nk}. This would contradict Lemma 2, and the proof is complete.

LEMMA 6. $\nu = \mu$.

Proof. Let f belong to $N_{\mu+1}$. Then $(I - K)^{\mu+1}f = o$, and so the equation $(I - K)g = o$ possesses the solution $g = (I - K)^\mu f$ belonging to M_μ. According to Lemma 5, $g = o$, and therefore f belongs to N_μ. Thus, $N_\mu = N_{\mu+1}$, and so $\mu \geq \nu$. Suppose now that $\mu > \nu$. Then $\mu \geq 1$, and so we can choose an element f in $M_{\mu-1}$ not belonging to M_μ. Then f may be expressed as $(I - K)^{\mu-1}g$. Let $h = (I - K)^\mu g$, so that $h \in M_\mu$. By Lemma 5, the equation $(I - K)^\mu x = h$ possesses a *unique* solution x in M_μ. Then

$$(I - K)^\mu (g - x) = (I - K)^\mu g - (I - K)^\mu x = h - h = o$$

and so $g - x \in N_\mu$. On the other hand,

$$(I - K)^{\mu-1}(g - x) = (I - K)^{\mu-1}g - (I - K)^{\mu-1}x$$
$$= f - (I - K)^{\mu-1}x \neq o$$

[since $f \notin M_\mu$, while the equality

$$(I - K)^{\mu-1}x = (I - K)^{\mu-1}(I - K)^\mu y = (I - K)^\mu (I - K)^{\mu-1}y$$

valid for some element y, shows that $(I - K)^{\mu-1}x \in M_\mu$]. Thus, $g - x \notin N_{\mu-1}$. Hence, $N_{\mu-1} \neq N_\mu$, and so $\mu \leq \nu$. The assumption that $\mu > \nu$ is therefore incorrect, and it follows that μ and ν are equal.

Turning back now to the theorem, we observe that if ν and μ both vanish, $M_1 = B$ and N_1 consists exclusively of the element o. Therefore, as f ranges over all of B, $(I - K)f$ does the same, so that (29) is always solvable and (28) possesses only the trivial solution. That the solution of (29) is unique follows from the observation that the difference between two solutions would be a solution of (28), and hence equal to the element o. If, on the other hand, ν and μ are both positive, M_1 is a proper subset of B, so that there exist elements g for which (29) is not solvable. In this case, N_1 contains elements other than o, and so whenever (29) is solvable the solution is not unique, for to any solution we may add any element of N_1. Furthermore, Lemma 3 shows that when uniqueness holds the solution of (29) satisfies the inequality $\|f\| \leq \sigma_1 \|g\|$, whereas, when uniqueness does not hold, it is still true that (29), *if solvable*, possesses a solution satisfying the above inequality.

It may be remarked that the full strength of Lemma 6 has not been used, for we have exploited only the fact that μ and ν are either both positive or both zero.

Now, consider the space V_n defined in Sec. 1, and let it be normed in any manner, such as in examples 1, 2, or 4 of Sec. 2. Consider the operator K defined as follows:

$$K\{\alpha_1, \alpha_2, \ldots, \alpha_n\} = \{\beta_1, \beta_2, \ldots, \beta_n\} \tag{37}$$

where
$$\beta_i = \sum_{k=1}^{n} c_{ik}\alpha_k \tag{38}$$

the c_{ik} being n^2 given scalars. Since V_n is finite-dimensional, K is trivially bounded and completely continuous, and so Theorem 6 assumes in this case the following form, if we set $d_{ik} = \delta_{ik} - c_{ik}$.

COROLLARY. The system

$$\gamma_i = \sum_{k=1}^{n} d_{ik}\alpha_k \tag{39}$$

admits a unique solution for each choice of the scalars γ_i if and only if the homogeneous system obtained by setting each γ_i equal to zero admits only the trivial solution.

This is, of course, the familiar basic theorem on systems of linear algebraic equations, but its formulation differs from the customary one in that no reference is made to determinants, which may therefore be looked on simply as a computational aid for solving (39) by Cramer's rule.

Finally, referring to the integral operator K defined on $C(0,1)$ by (14), and taking account of the obvious fact that the interval $0 \leq x \leq 1$ may be replaced by any other finite (closed) interval, we obtain the following result, which constitutes the basic theorem of the Fredholm theory of integral equations.

COROLLARY. Let $K(x,y)$ be defined and continuous in the closed square $a \leq x, y \leq b$. Then the integral equation[1]

$$g(x) = f(x) - \int_a^b K(x,y)f(y) \, dy \tag{40}$$

possesses a unique solution for each continuous function $g(x)$ if and only if the homogeneous equation obtained by setting $g \equiv 0$ in (40) possesses only the trivial solution.

EXERCISES

40. Show that the second corollary is, in some plausible sense, a generalization of the first.

41. It might appear that a more natural generalization of the first corollary than the second would be the following: The equation

$$\int_a^b K(x,y)f(y) \, dy = g(x) \tag{40'}$$

admits a continuous solution for each continuous function g if and only if the equation $\int_a^b K(x,y)f(y) \, dy \equiv 0$ admits no continuous solution other than $f \equiv 0$. However, show that this is not true.

42. Let $K(x,y)$ be defined and continuous in the closed triangle $0 \leq y \leq x \leq a$, where a is any positive constant. Let the operator K be defined on $C(0,a)$ as follows:

$$Kf = \int_0^x K(x,y)f(y) \, dy \tag{41}$$

Prove that $\|K^n\| \leq (aM)^n/n!$, where $M = \max |K(x,y)|$. Therefore, according to Exercise 38, the "Volterra integral equation"

$$f(x) = g(x) + \int_0^x K(x,y)f(y) \, dy \tag{42}$$

possesses a unique solution for each continuous function $g(x)$.

---

[1] (40) is known as a "Fredholm integral equation of the second kind." The function $K(x,y)$ is known as the "kernel" of the equation.

5. THE FREDHOLM ALTERNATIVE IN HILBERT SPACES

In this chapter an important class of Banach spaces, known as Hilbert spaces, will be studied. In particular, it will be shown that the Fredholm alternative (Theorem 4-6) admits in these spaces a more precise formulation, especially when the (completely continuous) operator K appearing in (4-29) possesses a certain additional property, namely, that of being hermitian. In this case a very complete theory of equations of the form (4-29) can be obtained, which can be applied to integral equations whose kernels satisfy very mild conditions.

1. Inner-product Spaces

DEFINITION 1. An "inner-product space" is a linear space in which to every ordered pair of elements f, g there is associated a scalar, denoted (f,g) and termed the inner product of f and g, such that, for all elements f, g, h and all scalars λ,

$$(\lambda f, g) = \lambda(f,g) \tag{1a}$$

$$(f,f) > 0 \qquad \text{unless } f = o\dagger \tag{1b}$$

$$(f + g, h) = (f,h) + (g,h) \tag{1c}$$

$$(f,g) = \begin{cases} (g,f) & \text{in a real linear space[1]} \\ \overline{(g,f)} & \text{in a complex linear space} \end{cases} \tag{1d}$$

We shall find it convenient henceforth to restrict our attention to complex spaces, but there will be no difficulty in seeing what modifica-

---

† Setting $\lambda = 0$ and $g = o$ in (1a), we obtain $(o,o) = 0(f,o) = 0$.

[1] In this case every inner product must be real, for by (1c) and the first line of (1d) we obtain $2(f,g) = (f + g, f + g) - (f,f) - (g,g)$, and the right side of this equality is real by (1b).

tions, if any, are needed for real spaces. Two very simple, but important, examples of inner-product spaces are now introduced:

1. Consider the linear space V_n defined in Sec. 4-1, and let the inner product of any two elements

$$a = \{\alpha_1, \alpha_2, \ldots, \alpha_n\}$$

and

$$b = \{\beta_1, \beta_2, \ldots, \beta_n\}$$

be given by

$$(a,b) = \sum_{k=1}^{n} \alpha_k \bar{\beta}_k \tag{2}$$

The inner-product space obtained from V_n in this manner will be denoted U_n, "unitary n-dimensional space."

2. Consider the class of functions appearing in examples 3 and 5 of Sec. 4-2, and let the inner product of any two elements be given by

$$(f,g) = \int_a^b f(x)\overline{g(x)}\, dx \tag{3}$$

The resemblance between these two inner-product spaces and the normed spaces defined in examples 4 and 5 of Sec. 4-2 is quite apparent, and will be discussed later in this section. First we return to Definition 1 and observe that (1b) suggests that any inner-product space may be considered as a normed space by defining the norm of any element in the obvious manner

$$\|f\| = (f,f)^{\frac{1}{2}} \tag{4}$$

Indeed, comparison of Definition 1 with Definition 4-5 shows that it is necessary only to establish the "triangle inequality" (4-2b). For this purpose we need the following theorem, which is of major importance.

Theorem 1. Schwarz Inequality. For any elements f and g,

$$|(f,g)| \leq \|f\| \cdot \|g\| \tag{5}$$

The equality holds if and only if f and g are linearly dependent.

Proof. If $f = o$, (5) holds trivially, for then $(f,g) = (0f,g) = 0(f,g) = 0$. If $f \neq o$, let $h = \|f\|^{-1}f$ (so that $\|h\| = 1$) and let $\alpha = (g,h)$. Using Definition 1 repeatedly, we obtain

$$0 \leq (g - \alpha h, g - \alpha h) = (g,g) - \alpha(h,g) - \bar{\alpha}(g,h) + |\alpha|^2(h,h)$$
$$= \|g\|^2 - |(h,g)|^2 \tag{6}$$

and hence

$$|(h,g)| \leq \|g\| \tag{7}$$

which is equivalent to (5). Equality holds in (7), and hence in (5), only if $g - \alpha h = o$, and this is equivalent to the assertion that f and g are linearly dependent.

The triangle inequality now follows readily.

Theorem 2. For any elements f and g,

$$\|f + g\| \leq \|f\| + \|g\| \tag{8}$$

The equality holds if and only if f and g are positive multiples of each other.[1]

Proof. Using Definition 1 and Theorem 1, we obtain

$$\|f + g\|^2 = (f + g, f + g) = \|f\|^2 + 2 \operatorname{Re}(f,g) + \|g\|^2 \leq \|f\|^2 + 2\|f\| \cdot \|g\| + \|g\|^2 = (\|f\| + \|g\|)^2 \tag{9}$$

Comparing the end expressions of (9), we obtain (8). Equality holds only if $\operatorname{Re}(f,g) = |(f,g)| = \|f\| \cdot \|g\|$. The latter equality implies that $f = \lambda g$, while the former equality then implies that $\operatorname{Re} \lambda = |\lambda|$; that is, λ is real and positive.

Thus, we have proved the fact suggested previously, which we now state as a theorem.

Theorem 3. Any inner-product space becomes a normed linear space if the norm is defined by (4).

By comparing illustrative examples 1 and 2 of this section with examples 4 and 5, respectively, of Sec. 4-2, we find that we have justified the assertion made there that the triangle inequality does indeed hold. The essential point in the argument is, of course, that it is possible to introduce into each of the two spaces under consideration an inner product which yields a norm coinciding with that originally defined. The question arises as to whether this can always be done. That the answer is negative follows readily from a simple example based on the "parallelogram law,"

$$\|f + g\|^2 + \|f - g\|^2 = 2\|f\|^2 + 2\|g\|^2 \tag{10}$$

which holds in any inner-product space. (Cf. Exercise 2.) Now, consider the space V_n with norm as in example 1 of Sec. 4-2. For convenience, we take $n = 2$ and let f and g denote the elements $\{1,0\}$ and $\{0,1\}$, respectively. Clearly $\|f\| = \|g\| = 1$, $\|f + g\| = \|f - g\| = 2$, so that (10) is violated. It is of interest that (10) is known to be sufficient, as well as necessary, for the existence of an inner product consistent with the specified norm.

An important role is played by the concept of orthogonality, which we now define.

---

[1] We set aside, for convenience, the trivial case that either of the given elements is o.

DEFINITION 2. Two elements are said to be "orthogonal" if their inner product vanishes.[1] An "orthogonal set" of elements is a set any two of whose elements are orthogonal; if each element possesses unit norm, the set is said to be "orthonormal."

We note that any orthonormal set is linearly independent, for we obtain from the relationship

$$\alpha_1 f_1 + \alpha_2 f_2 + \cdots + \alpha_n f_n = o \tag{11}$$

by taking the inner product of each side with any one of the elements f_k, the equality $\alpha_k = 0$.

From the relationship $(\alpha_1 f_1 + \alpha_2 f_2, g) = \alpha_1(f_1, g) + \alpha_2(f_2, g)$, it follows that the set of all elements orthogonal to a given element is a linear manifold. With the aid of the Schwarz inequality we easily show that this manifold is a subspace, for if the sequence $f_1, f_2, \ldots$ converges to f, and if each element of the sequence is orthogonal to g, then

$$|(f,g)| = |(f - f_n, g) + (f_n, g)| = |(f - f_n, g)| \leq \|f - f_n\| \cdot \|g\| \to 0$$

and hence $(f,g) = 0$. More generally, it is evident from the above argument that if S denotes any (nonvacuous) set of elements, the set $S^\perp$ (the orthogonal complement of S) of elements which are orthogonal to every element of S is a subspace.

Theorem 4. Every finite-dimensional inner-product space possesses an orthonormal basis.

Proof. Select an arbitrary basis $f_1, f_2, \ldots, f_n$, and let $e_1 = f_1/\|f_1\|$, so that $\|e_1\| = 1$. Then the scalar α is so chosen that $(f_2 - \alpha e_1, e_1) = 0$; i.e., $\alpha = (f_2, e_1)$. We note that $f_2 - \alpha e_1 \neq o$, for otherwise the original set of n elements would not be independent. We may therefore divide $f_2 - \alpha e_1$ by its norm, thus obtaining a unit element e_2 orthogonal to e_1. Similarly, we can choose scalars β and γ in a unique manner such that both e_1 and e_2 are orthogonal to $f_3 - \beta e_1 - \gamma e_2$. Since the latter vector is distinct from o, we may divide it by its norm, thus obtaining a unit vector e_3 orthogonal to e_1 and e_2. Continuing this procedure, we obtain an orthonormal set of elements $e_1, e_2, \ldots, e_n$ which are seen, either directly or by combining Theorem 4-1 with the remark immediately following Definition 2, to form a basis.

It is clear that the "Gram-Schmidt" orthogonalization procedure used in the above proof may be applied to replace any denumerable set of independent elements in an infinite-dimensional inner-product space by an equivalent orthonormal set. By "equivalent" is meant that any finite linear combination of elements of either set is also expressible as a finite linear combination of elements of the other.

---

[1] It follows from (1d) that the relationship of orthogonality is symmetric.

Given an element f of an inner-product space and an orthonormal set $e_1, e_2, \ldots, e_n$, we consider the problem of choosing scalars $\alpha_1, \alpha_2, \ldots, \alpha_n$ so as to minimize $\|f - \alpha_1 e_1 - \alpha_2 e_2 - \cdots - \alpha_n e_n\|$. This problem is solved by the following simple argument. Let the "Fourier coefficients" f_i ($i = 1, 2, \ldots, n$) be defined as follows:

$$f_i = (f, e_i) \tag{12}$$

Then an elementary calculation furnishes the following result:

$$\left\| f - \sum_{i=1}^{n} \alpha_i e_i \right\|^2 = \|f\|^2 - \sum_{i=1}^{n} |f_i|^2 + \sum_{i=1}^{n} |f_i - \alpha_i|^2 \tag{13}$$

From (13) it follows that the left side is minimized by choosing each of the coefficients α_i equal to f_i, and that the minimum value of the left side is $\|f\|^2 - \sum_{i=1}^{n} |f_i|^2$. Since the latter quantity cannot be negative, we obtain "Bessel's inequality,"

$$\sum_{i=1}^{n} |f_i|^2 \leq \|f\|^2 \tag{14}$$

with equality holding if and only if f is a linear combination of the elements $\{e_i\}$. It follows that the "Parseval equality"

$$\sum_{i=1}^{n} |f_i|^2 = \|f\|^2 \tag{15}$$

holds for every element f if and only if the elements $\{e_i\}$ form a basis of (span) the space, which would therefore have dimension n. Finally, we note that if there exists a denumerable set of elements $e_1, e_2, e_3, \ldots$ in the space, then (14) holds for each n, and hence, by a passage to the limit, we obtain

$$\sum_{i=1}^{\infty} |f_i|^2 \leq \|f\|^2 \tag{16}$$

This inequality, which in particular assures the convergence of the series appearing on the left, is also known as Bessel's inequality.

EXERCISES

1. Prove the Schwarz inequality by exploiting the fact that $(f - \lambda g, f - \lambda g)$ is nonnegative for all values of λ.

2. Prove (10), and explain the term "parallelogram law."

3. Show that (13) may be interpreted as a generalization of the theorem that the shortest distance from a point to a line is the perpendicular.

4. Prove that any n-dimensional inner-product space is isomorphic to U_n.

5. Given n elements $f_1, f_2, \ldots, f_n$ in any inner-product space, prove that set $|(f_i,f_j)|$ is positive if the given elements are independent and zero otherwise.

6. Work out (13) in detail.

2. Hilbert Spaces

Just as we concentrated in Chap. 4 on *complete* normed linear spaces, here we shall concentrate on *complete* inner-product spaces, or "Hilbert spaces."[1]   Clearly, any Hilbert space is a Banach space, but not conversely.

According to Exercise 4-18, any finite-dimensional inner-product space, in particular U_n, is a Hilbert space. A less elementary example is furnished by the infinite-dimensional generalization of U_n; by this we mean the set of all infinite sequences $\{\alpha_1,\alpha_2, \ldots\}$ of complex numbers such that $\sum_{k=1}^{\infty} |\alpha_k|^2$ converges, addition of elements and multiplication by scalars being defined in the obvious manner, while the inner product is defined by the obvious generalization of (2), namely,

$$(a,b) = \sum_{k=1}^{\infty} \alpha_k \bar{\beta}_k \qquad (17)$$

We denote this space as l_2. Because of the infinite summation involved, it is necessary to show that (17) is meaningful. This is readily seen by taking account of the elementary inequality[2]

$$2|\alpha_k \bar{\beta}_k| \leq |\alpha_k|^2 + |\beta_k|^2$$

which shows that series (17) converges whenever a and b belong to l_2. It is also necessary to show that the addition of elements is meaningful; this follows from summing over all values of k the inequality[3]

$$|\alpha_k + \beta_k|^2 \leq 2|\alpha_k|^2 + 2|\beta_k|^2$$

These considerations show that l_2 is well defined as an inner-product space. The proof of completeness, which is somewhat more subtle, proceeds as follows. Let the sequence[4] $\{a_1,a_2, \ldots\}$ be convergent in norm (i.e., in the Cauchy sense), and let $\alpha_k{}^{(n)}$ denote the kth component of a_n. Since

$$|\alpha_k{}^{(m)} - \alpha_k{}^{(n)}| \leq \|a_m - a_n\|$$

[1] Sometimes the condition of separability is imposed, and sometimes also the additional requirement of infinite dimensionality.

[2] $0 \leq (|\alpha_k| - |\beta_k|)^2 = |\alpha_k|^2 + |\beta_k|^2 - 2|\alpha_k| \cdot |\beta_k| = |\alpha_k|^2 + |\beta_k|^2 - 2|\alpha_k \bar{\beta}_k|$.

[3] $|\alpha_k + \beta_k|^2 + |\alpha_k - \beta_k|^2 = 2|\alpha_k|^2 + 2|\beta_k|^2$.   [Cf. (10).]

[4] It is necessary to distinguish carefully between the sequence of scalars $\{\alpha_1,\alpha_2, \ldots\}$, which constitutes an element of l_2, and the sequence $\{a_1,a_2, \ldots\}$ of elements of l_2.

it follows that, for each fixed k, the scalars $\alpha_k{}^{(1)}$, $\alpha_k{}^{(2)}$, . . . constitute a Cauchy sequence, and therefore converge[1] to a limit, which we denote as α_k. It will now be shown that the sequence $\{\alpha_1, \alpha_2, \ldots\}$, which we denote as a, belongs to l_2 and is the limit of the given sequence of elements $\{a_n\}$. From Exercise 4-10 we conclude that the quantities $\|a_1\|$, $\|a_2\|$, . . . constitute a convergent, and hence a bounded, sequence of real numbers. Letting M denote an upper bound on these numbers, and N any positive integer, temporarily fixed, we conclude that, for all n,

$$\sum_{k=1}^{N} |\alpha_k{}^{(n)}|^2 \leq M^2$$

Letting *first* n and *then* N become infinite, we conclude that

$$\sum_{k=1}^{\infty} |\alpha_k|^2 \leq M^2$$

Hence, a does in fact belong to l_2. Now let $\epsilon > 0$ be given. Then, since the sequence a_1, a_2, . . . is convergent in norm, we can find an integer μ such that $\|a_n - a_m\| < \epsilon$ whenever n, $m \geq \mu$. Furthermore, letting $b^{(\nu)}$ denote the element whose first ν components agree with those of b and whose remaining components are all zero, we choose ν so large that

$$\|a - a^{(\nu)}\| < \epsilon \qquad \|a_\mu - a_\mu{}^{(\nu)}\| < \epsilon$$

Then, by applying the triangle inequality to the right side of the equality

$$a - a_n = (a - a^{(\nu)}) + (a^{(\nu)} - a_n{}^{(\nu)}) + (a_\mu{}^{(\nu)} - a_n) + (a_\mu{}^{(\nu)} - a_\mu) \\ + (a_n{}^{(\nu)} - a_\mu{}^{(\nu)}) \quad (18)$$

we obtain

$$\|a - a_n\| \leq \|a - a^{(\nu)}\| + \|a^{(\nu)} - a_n{}^{(\nu)}\| \\ + \|a_\mu - a_n\| + \|a_\mu{}^{(\nu)} - a_\mu\| + \|a_n{}^{(\nu)} - a_\mu{}^{(\nu)}\| \quad (19)$$

It follows from the manner in which the integers μ and ν were chosen that the first, third, fourth, and fifth[2] terms on the right-hand side of (19) are each less than ϵ (for $n \geq \mu$). Since

$$\|a^{(\nu)} - a_n{}^{(\nu)}\|^2$$

is given by the *finite* sum

$$\sum_{k=1}^{\nu} |\alpha_k - \alpha_k{}^{(n)}|^2$$

---

[1] It should be stressed that we exploit here and elsewhere the completeness of the complex-number system.

[2] For the fifth term we use the trivial inequality $\|a_n{}^{(\nu)} - a_\mu{}^{(\nu)}\| \leq \|a_n - a_\mu\|$.

we may employ a termwise passage to the limit to conclude that, with increasing n, $\|a^{(\nu)} - a_n^{(\nu)}\|$ approaches zero, and hence is less than ϵ for all sufficiently large n. Therefore, for n sufficiently large,

$$\|a - a_n\| < 5\epsilon$$

The completeness of l_2 is thus established.

A second example of an infinite-dimensional Hilbert space, which plays an important role later in this chapter, is obtained by suitably extending the inner-product space defined in example 2 of Sec. 1. As shown in Sec. 4-3, this space is not complete, but from Theorem 1-7 it is evident that completeness may be attained by dropping the requirement of continuity and admitting into the space all quadratically integrable functions (in the Lebesgue sense). However, a new difficulty then arises, for it is evident that any "null function"—i.e., a function which vanishes almost everywhere—possesses zero norm, not only the identically vanishing function. This difficulty is overcome by "identifying" any two functions whose difference is a null function. By this we mean that we form "equivalence classes" of functions, two functions being in the same class if and only if their difference is a null function. (Of course, we use here the trivial fact that if $f - g$ and $g - h$ are null functions, then so is their sum $f - h$.) Thus, we form a Hilbert space whose elements are certain *classes* of functions, not individual functions. Nevertheless, one may speak of "the element $f(x)$" instead of "the class of all functions differing from $f(x)$ by a null function." The inner product between any two elements is given, of course, by (3), where $f(x)$ and $g(x)$ are any functions chosen from the two elements. This Hilbert space is denoted $L_2(a,b)$. Its infinite dimensionality is obvious, while its separability follows from the fact, immediately evident from the definition of the Lebesgue integral, that the step functions with (complex) rational values and rational points of discontinuity form a denumerable dense set. Closely related to $L_2(a,b)$ is the Hilbert space $\mathbf{L}_2(a,b)$ of all functions $f(x,y)$ quadratically integrable on the square $a < x,\ y < b$, the identification of functions and definition of inner product being entirely analogous to the case of $L_2(a,b)$.

We turn now to the concept of a basis of an infinite-dimensional linear space. (Definition 4-4 applies only to a finite-dimensional space.) For any *normed* space, which is the only case with which we are concerned, we introduce the following definition, which is easily seen to be consistent with the afore-mentioned one.

DEFINITION 3. A collection C of elements of a normed linear space is said to form a basis if the collection is linearly independent, and if every element of the space can be approximated arbitrarily closely by a finite linear combination of elements of C; i.e., given $\epsilon > 0$ and an element f of the space, there exist elements ϕ_1, ϕ_2, . . . , ϕ_n of C and scalars α_1, α_2, . . . , α_n such that $\|f - \alpha_1\phi_1 - \alpha_2\phi_2 - \cdots - \alpha_n\phi_n\| < \epsilon$.

We shall now show that every separable[1] space contains a finite or denumerable basis. Let the sequence of elements $\{\phi_1, \phi_2, . . .\}$ be dense in the space. We may assume that $\phi_1 \neq o$, for otherwise we could merely change the numbering of the elements. We now seek the first element in this sequence which is a linear combination of its predecessors; if no such element exists, the given sequence clearly constitutes a basis, but if there *is* such an element, we delete it from the sequence and then repeat the above procedure with the diminished sequence. The elements that are never deleted by this procedure are then readily seen to form a basis.

If we are dealing with a separable inner-product space, we may apply the Gram-Schmidt procedure described in the preceding section to the basis obtained by the foregoing method, and in this manner we obtain an orthonormal basis, finite or denumerable, according to whether the space is finite- or infinite-dimensional.

Finally, suppose that we are dealing with a separable infinite-dimensional Hilbert space H. (It may be stressed that completeness does not play any role in the two preceding paragraphs.) Let the elements e_1, e_2, . . . constitute an orthonormal basis; let f be any element of H; and let the coefficients f_i be defined by (12). Letting

$$f^{(n)} = \sum_{i=1}^{n} f_i e_i$$

we obtain, for $m > n$, the equality

$$\|f^{(m)} - f^{(n)}\|^2 = \sum_{i=n+1}^{m} |f_i|^2$$

and, by taking account of (16), we conclude that the sequence $\{f^{(1)}, f^{(2)}, . . .\}$ is convergent in norm. On account of the completeness of the space, the sequence converges to an element g of H. We shall now show that $g = f$. Letting $h = g - f$, we have, for any n,

$$(h, e_n) = (g - f^{(m)}, e_n) + (f^{(m)} - f, e_n)$$

For $m > n$, the last of these inner products obviously vanishes, and so, with the aid of the Schwarz inequality, we obtain

$$|(h, e_n)| \leq \|g - f^{(m)}\| \cdot \|e_n\| = \|g - f^{(m)}\|$$

---

[1] Even without the hypothesis of separability, it is possible to demonstrate the existence of a basis by using transfinite induction, but a basis of a nonseparable space cannot be denumerable.

Letting $m \to \infty$, we obtain $(h, e_n) = 0$. Thus, h is orthogonal to every element of the orthonormal basis e_1, e_2, ... Given $\epsilon > 0$, it is possible, by the definition of a basis, to select a finite number of scalars $\alpha_1, \alpha_2, \ldots, \alpha_n$ such that

$$\|h - \alpha_1 e_1 - \alpha_2 e_2 - \cdots - \alpha_n e_n\| < \epsilon$$

On the other hand, by an argument employed in Sec. 1, the left side of this inequality is minimized by choosing each α_i equal to (h, e_i), which is equal to zero. Thus, we conclude that $\|h\| < \epsilon$, and hence $h = o$. Therefore $g = f$, and so we may write

$$f = \sum_{i=1}^{\infty} f_i e_i \tag{20}$$

This equation is to be interpreted as meaning that $f = \lim_{n \to \infty} \sum_{i=1}^{n} f_i e_i$.

Without going into detail, we remark that the above reasoning can be modified to give the following result: Given any orthonormal basis e_1, e_2, ... and any set of scalars f_1, f_2, ... such that $\sum_{i=1}^{\infty} |f_i|^2$ converges, then the series $\sum_{i=1}^{\infty} f_i e_i$ converges, in the sense explained above, to an element f which is related to the scalars f_i by (12), and the infinite-dimensional analogue of (15), namely,

$$\sum_{i=1}^{\infty} |f_i|^2 = \|f\|^2 \tag{21}$$

holds.

EXERCISES

7. Prove that l_2 is separable, and that the elements $\{1,0,0, \ldots\}$, $\{0,1,0, \ldots\}$, $\{0,0,1, \ldots\}$, ... form an orthonormal basis.

8. Prove the statements made in the last paragraph of the present section.

9. Let H denote the set of all complex-valued functions defined on the interval $0 \leq x \leq 1$, vanishing everywhere except on a finite or denumerable set (which may differ from function to function) and such that $\Sigma |f(x)|^2$ converges, the summation being extended over the points where $f \neq 0$. Prove that, if $(f,g) = \Sigma f(x) \overline{g(x)}$, then H is a nonseparable Hilbert space. Determine an orthonormal basis of H.

3. Projections, Linear Functionals, Adjoint Operators

Given any vector in three-dimensional euclidean space and any plane (or line), it is possible to express the vector as the sum of two vectors, one of them parallel and the other perpendicular to the given plane (or line); furthermore, the decomposition is unique. Our first objective in the present section is to establish the corresponding fact (Theorem 6) in any Hilbert space.

DEFINITION 4. A subset M of a linear space is said to be "convex" if, for every pair of elements f, g in M, and every real number t between zero and one, the element $tf + (1 - t)g$ also is in M.

This is an obvious generalization of the definition of convexity for figures in two- and three-dimensional euclidean space. A simple example of convexity is furnished by the set of all elements of V_n (cf. Sec. 4-1) whose first component α_1 has a specified value α; note that this set is a manifold if and only if $\alpha = 0$.

Theorem 5. Let M be any nonvacuous closed convex subset of a Hilbert space H. Then there exists a unique element h of M having minimum norm; i.e., if $g \in M$ and $g \neq h$, then $\|g\| > \|h\|$.

Proof. Let the infimum of $\|f\|$ for all $f \in M$ be denoted by δ. Then we can select a sequence $\{h_n\}$ of elements of M such that $\|h_n\| \to \delta$. From the equality

$$\tfrac{1}{4}\|h_n - h_m\|^2 = \tfrac{1}{2}\|h_n\|^2 + \tfrac{1}{2}\|h_m\|^2 - \|\tfrac{1}{2}(h_n + h_m)\|^2 \qquad (22)$$

[cf. (10)] and the fact (which follows from the convexity of M and the definition of δ) that $\|\tfrac{1}{2}(h_n + h_m)\| \geq \delta$, we conclude that, given any $\epsilon > 0$, we can choose N so large that the inequality

$$\tfrac{1}{4}\|h_n - h_m\|^2 \leq \tfrac{1}{2}\delta^2 + \tfrac{1}{2}\delta^2 + \epsilon - \delta^2 = \epsilon \qquad (23)$$

holds for all m and n exceeding N. Hence, the sequence $\{h_n\}$ is convergent in norm; since H is complete, there exists an element $h \in H$ such that $\|h - h_n\| \to 0$, and, since M is closed, $h \in M$. The continuity of the norm then assures that $\|h\| = \delta$. It remains only to prove the uniqueness of h. Suppose that there are two distinct elements h and $\bar{h}$ in M each having norm δ. Then, by convexity, $\tfrac{1}{2}(h + \bar{h})$ also belongs to M, and from the equality

$$\|\tfrac{1}{2}(h + \bar{h})\|^2 = \tfrac{1}{2}\|h\|^2 + \tfrac{1}{2}\|\bar{h}\|^2 - \|\tfrac{1}{2}(h - \bar{h})\|^2 = \delta^2 - \|\tfrac{1}{2}(h - \bar{h})\|^2 \qquad (24)$$

we conclude that M contains an element of norm less than δ, contrary to the definition of δ. Thus the uniqueness of h is proved.

COROLLARY. Let M be any nonvacuous closed convex subset of a Hilbert space H, and let f be any element of H. Then there exists a unique element h of M such that $\|f - h\| < \|f - \bar{h}\|$ whenever $\bar{h}$ is an element of M distinct from h.

Proof. Let M_f denote the set of all elements of H of the form $f - g$, $g \in M$. Then M_f is, like M, nonvacuous, closed, and convex, so that the above theorem is applicable.

Theorem 6. Projection Theorem. Let M be any subspace of a Hilbert space H. Then every element f of H can be expressed in a unique manner in the form

$$f = g + h \qquad (g \in M, h \in M^\perp) \qquad (25)$$

Proof. Let g be that element of M, whose existence is assured by the above corollary, such that $\|f - \tilde{g}\| > \|f - g\|$ for all elements $\tilde{g}$ of M other than g, and let $h = f - g$. Then, for any element e in M and any real number λ,

$$\|f - (g + \lambda e)\|^2 \geq \|h\|^2 \qquad (26)$$

Since $\|f - (g + \lambda e)\|^2 = \|h - \lambda e\|^2 = \|h\|^2 + \lambda^2\|e\|^2 - 2\lambda \operatorname{Re}(h,e)$, we obtain from (26) the inequality

$$\lambda[\lambda\|e\|^2 - 2 \operatorname{Re}(h,e)] \geq 0 \qquad (27)$$

Now, if $\operatorname{Re}(h,e) < 0$, λ can be chosen negative and so close to zero that the second factor of the left side of (27) is positive, and thus a contradiction with (27) would be obtained. Similarly, the possibility that $\operatorname{Re}(h,e) > 0$ is ruled out, so that $\operatorname{Re}(h,e)$ is shown to be zero. Taking λ as pure imaginary instead of real, we can show similarly that $\operatorname{Im}(h,e) = 0$. Thus, $(h,e) = 0$; i.e., h is orthogonal to every element of M, as was to be proved. It now remains only to establish the uniqueness of the decomposition (25). Let another such decomposition be given by $f = \tilde{g} + \tilde{h}$. Then by subtraction we obtain $(g - \tilde{g}) + (h - \tilde{h}) = o$, or $(g - \tilde{g}) = -(h - \tilde{h})$, and hence

$$\|g - \tilde{g}\|^2 = (g - \tilde{g}, \tilde{h} - h) = (g,\tilde{h}) - (\tilde{g},\tilde{h}) - (g,h) + (\tilde{g},h) = 0 \quad (28)$$

Thus $g - \tilde{g} = o$, $h - \tilde{h} = o$, and the uniqueness is established.

The element g which is uniquely associated with each element f of H by (25) is termed the (orthogonal) projection of f on M; we denote this relationship as follows:

$$g = P_M f \qquad (29)$$

From (25) and the resulting equality $\|f\|^2 = \|g\|^2 + \|h\|^2$, we easily conclude that the operator P_M is additive and bounded—hence linear—with $\|P_M\| = 1$ except in the trivial case that M consists of the single element o; in this case, P_M is the zero operator and $\|P_M\| = 0$. Furthermore, whenever $f \in M$ the decomposition (25) evidently assumes the form $f = f + o$, and so we conclude that $P_M^2 f = P_M f$ for every element f. Thus, $P_M^2 = P_M$; an operator having this property is said to be "idempotent." We also note that for every pair of elements f_1, f_2 the equality

$$(P_M f_1, f_2) = (f_1, P_M f_2) \tag{30}$$

holds; the proof follows trivially from (25).

With the aid of the concept of projection which has just been introduced we can now give a strikingly simple description of all linear functionals in a Hilbert space.

Theorem 7. Riesz Representation Theorem. Let l be any linear functional defined on a Hilbert space H. Then there exists a unique element e of H such that, for every element f,

$$l(f) = (f, e) \tag{31}$$

Conversely, for each element e of H, (31) defines a linear functional.

Proof. If $l(f) \equiv 0$, l can be defined by (31) with $e = o$. In any other case, the set of elements f such that $l(f) = 0$ is a proper subspace M of H. (That M is a manifold follows from the additivity of l, while the closure follows from the boundedness of l.) Next we show that any two elements of $M^\perp$ must be linearly dependent.[1] To show this, we select two such elements, h and h'. If either is the element o, the assertion is trivial. Otherwise, the quantities $l(h)$ and $l(h')$ must both be different from zero, and so the element $h'' = h - [l(h)/l(h')]h'$ is well defined and, like h and h', orthogonal to M. On the other hand, $l(h'') = l(h) - [l(h)/l(h')]l(h') = 0$, and so $h'' \in M$. Therefore $h'' = o$; the linear dependence of h and h' is thus established. Now, let h_0 be any fixed element of $M^\perp$ other than o. Then, in view of the foregoing discussion, each element f of H can be expressed in the form

$$f = P_M f + \lambda h_0 \tag{32}$$

From (32) we get

$$(f, h_0) = (P_M f, h_0) + (\lambda h_0, h_0) = \lambda (h_0, h_0) \tag{33}$$

and

$$l(f) = l(P_M f) + l(\lambda h_0) = \lambda l(h_0) \tag{34}$$

Eliminating λ from (33) and (34), we obtain

$$l(f) = \frac{l(h_0)(f, h_0)}{(h_0, h_0)} = \left(f, \frac{\overline{l(h_0)}}{(h_0, h_0)} h_0 \right) \tag{35}$$

Thus (31) holds with $e = [\overline{l(h_0)}/(h_0, h_0)]h_0$.

The uniqueness of e is trivial, for if $(f, e) \equiv (f, e')$, then, in particular, the equalities $(e, e - e') = 0$ and $(e', e - e') = 0$ must hold. Subtracting, we obtain $(e - e', e - e') = 0$, and hence $e - e' = o$, or

[1] The reader may find it helpful to visualize the following argument in three-dimensional euclidean space.

$e = e'$. Finally, it is easily established that each element e of H determines a linear functional. Additivity follows from the properties of the inner product, and boundedness from the Schwarz inequality: $|(f,e)| \leq \|f\| \cdot \|e\|$, so that the norm of the functional (f,e) is at most $\|e\|$. By setting $f = e$, we conclude that the norm of the functional is precisely $\|e\|$.

Now we turn to the concept of the "adjoint" of a given operator T. For each fixed element g we define a linear functional, which we denote as l_g, as follows:

$$l_g(f) = (Tf, g) \tag{36}$$

The additivity of l_g is an immediate consequence of the additivity of T and of the inner product (as a function of the first element); the boundedness follows from the Schwarz inequality and the boundedness of T, for

$$|l_g(f)| \leq \|Tf\| \cdot \|g\| \leq C\|f\| \qquad C = \|T\| \cdot \|g\| \tag{37}$$

By the preceding theorem, we can associate with g a uniquely determined element T^*g such that

$$l_g(f) = (f, T^*g) \tag{38}$$

From (36) and (38) we obtain

$$(Tf, g) = (f, T^*g) \tag{39}$$

for all elements f, g. The additivity of T^* is immediately established, for, from the identities

$$
\begin{aligned}
(f, T^*(g_1 + g_2)) &= (Tf, g_1 + g_2) = (Tf, g_1) + (Tf, g_2) \\
&= (f, T^*g_1) + (f, T^*g_2) = (f, T^*g_1 + T^*g_2)
\end{aligned} \tag{40}
$$

and

$$(f, T^*\lambda g) = (Tf, \lambda g) = \bar{\lambda}(Tf, g) = \bar{\lambda}(f, T^*g) = (f, \lambda T^*g) \tag{41}$$

we conclude, as in the proof of the uniqueness of e in representation (31), that $T^*(g_1 + g_2) = T^*g_1 + T^*g_2$, $T^*\lambda g = \lambda T^*g$. The boundedness of T^* is established by the following consideration. For any $g \in H$,

$$
\begin{aligned}
\|T^*g\|^2 = (T^*g, T^*g) &= (T(T^*g), g) = |(T(T^*g), g)| \\
&\leq \|T(T^*g)\| \cdot \|g\| \leq \|T\| \cdot \|g\| \cdot \|T^*g\|
\end{aligned} \tag{42}
$$

If $T^*g \neq o$, we divide by $\|T^*g\|$ and obtain

$$\|T^*g\| \leq \|T\| \cdot \|g\| \tag{43}$$

whereas if $T^*g = o$ the inequality (43) holds trivially. Hence T^* is bounded, and $\|T^*\| \leq \|T\|$. We are therefore assured of the existence of an adjoint $(T^*)^*$ of T^*; from the identities

$$(Tf, g) = (f, T^*g) = \overline{(T^*g, f)} = \overline{(g, (T^*)^*f)} = ((T^*)^*f, g) \tag{44}$$

we conclude that $(T^*)^* = T$, and hence $\|T\| = \|(T^*)^*\| \leq \|T^*\| \leq \|T\|$. Summing up, we have the following theorem.

Theorem 8. Every operator T possesses a uniquely determined adjoint T^*, and $\|T\| = \|T^*\|$.

Finally, we state a number of elementary facts concerning adjoint operators, leaving the proofs as Exercise 11.

Theorem 9

(a) $(T_1 + T_2)^* = T_1^* + T_2^*$ (b) $(\lambda T)^* = \bar{\lambda} T^*$

(c) $(T_1 T_2)^* = T_2^* T_1^*$ (d)† $(T^{-1})^* = (T^*)^{-1}$

EXERCISES

10. Prove that every nonvacuous subset M of a linear space contains a unique "convex hull"—i.e., a convex set $\tilde{M}$ such that $M \subset \tilde{M}$ and $\tilde{M} \subset N$ for every convex set N containing M. *Hint:* Consider the class of all convex sets containing M, and show that convexity is preserved under intersection.

11. Prove Theorem 9.

12. Prove that the adjoint of a completely continuous operator is also completely continuous. *Hint:* Use the identity $(TT^*f,f) = \|T^*f\|^2$.

13. A sequence of operators $\{T_n\}$ is said to converge "strongly" to an operator T if, for each element f, the sequence $\{T_n f\}$ converges to Tf. Show, by means of an example, that, in contrast to Theorem 4-3, strong convergence of a sequence of completely continuous operators does *not* imply the complete continuity of the limit.

14. Prove that, in a complex Hilbert space, the identical vanishing of (Tf,f) implies that $T = 0$. *Hint:* Replace f in the expression (Tf,f) by $f + g$, $f - g$, $f + ig$, and $f - ig$.

15. Prove that the assertion of the preceding exercise is not true in a real Hilbert space. *Hint:* Consider the rotations of the plane about the origin.

16. Give an alternate proof of Theorem 6, in the case that H is separable, by constructing orthonormal bases for M and $M^\perp$.

4. Hermitian and Completely Continuous Operators

DEFINITION 5. An operator T is said to be "hermitian" (or self-adjoint[1]) if $T = T^*$.

By referring to (30), we see that any projection operator is hermitian.

Theorem 10. An operator T is hermitian if and only if (Tf,f) is real for every element f.

Proof. If T is hermitian, then $(Tf,f) = (f,T^*f) = (f,Tf) = \overline{(Tf,f)}$, and so (Tf,f) is real. Conversely, if (Tf,f) is always real, then

---

† Assertion (d) is to be understood as follows: If T^{-1} exists, then $(T^*)^{-1}$ also exists, and (d) holds.

[1] A distinction is made between these two terms in the theory of *unbounded* operators.

$(Tf,f) \equiv \overline{(Tf,f)} \equiv \overline{(f,T^*f)} \equiv (T^*f,f)$, and hence $((T - T^*)f,f) \equiv 0$, from which we conclude, by Exercise 14, that $T = T^*$.

Theorem 11. Let $N_T = \sup\limits_{\|f\|=1} |(Tf,f)|$. If T is hermitian, then $N_T = \|T\|$.

Proof. Whether or not T is hermitian, we obtain, from the Schwarz inequality,

$$|(Tf,f)| \leq \|Tf\| \cdot \|f\| \leq \|T\| \cdot \|f\|^2 \tag{45}$$

and hence

$$N_T \leq \|T\| \tag{46}$$

Now, if T is hermitian and λ is any positive number,

$$\|Tf\|^2 = \tfrac{1}{4}[(T(\lambda f + \lambda^{-1}Tf), \lambda f + \lambda^{-1}Tf)$$
$$- (T(\lambda f - \lambda^{-1}Tf), \lambda f - \lambda^{-1}Tf)]$$
$$\leq \tfrac{1}{4}N_T(\|\lambda f + \lambda^{-1}Tf\|^2 + \|\lambda f - \lambda^{-1}Tf\|^2)$$
$$= \tfrac{1}{2}N_T(\lambda^2\|f\|^2 + \lambda^{-2}\|Tf\|^2) \tag{47}$$

The expression $\lambda^2\|f\|^2 + \lambda^{-2}\|Tf\|^2$ is minimized by choosing

$$\lambda^2 = \|Tf\| \cdot \|f\|^{-1}$$

and, with this choice of λ and with $\|f\| = 1$, we obtain

$$\|Tf\|^2 \leq \tfrac{1}{2}N_T \cdot 2\|Tf\| = N_T\|Tf\| \tag{48}$$

and hence (even if $Tf = o$)

$$\|Tf\| \leq N_T \tag{49}$$

Since (49) holds for any unit element f, we conclude that

$$\|T\| \leq N_T \tag{50}$$

Combining (46) and (50), we conclude that $\|T\| = N_T$.

We now turn to a class of operators having an especially simple structure, the degenerate operators.

DEFINITION 6. An operator K is said to be "degenerate" if there exists a finite-dimensional subspace M such that $Kf \in M$ for every element f.

Let $\phi_1, \phi_2, \ldots, \phi_N$ and $\psi_1, \psi_2, \ldots, \psi_N$ be elements of H (not necessarily all distinct), and let

$$Kf = \sum_{i=1}^{N} (f,\psi_i)\phi_i \tag{51}$$

Clearly K is a degenerate operator, and its adjoint is also degenerate, for an elementary verification shows that

$$K^*f = \sum_{i=1}^{N} (f,\phi_i)\psi_i \tag{52}$$

Furthermore, K is completely continuous. To prove this it suffices, on account of Exercise 4-30, to show that any operator of the form $Kf = (f,\psi)\phi$ has this property. Given any bounded sequence $\{f_n\}$, the corresponding sequence of scalars $\{(f_n,\psi)\}$ is also bounded (by the Schwarz inequality), and hence possesses a convergent subsequence $\{(f_{n_k},\psi)\}$. Then the sequence $\{Kf_{n_k}\}$ is convergent, and so the proof is complete.

Now we show that this example is in no way exceptional.

Theorem 12. An operator K is degenerate if and only if it can be expressed in the form (51), from which it follows that K is completely continuous. If K is degenerate, so is K^*, which is given by (52).

Proof. Let M be a finite-dimensional subspace such that $Kf \in M$ for every element f, and let ϕ_1, ϕ_2, . . . , ϕ_N form an orthonormal basis of M. Then for every element f there exist N uniquely determined scalars $l_1(f)$, $l_2(f)$, . . . , $l_N(f)$ such that

$$Kf = \sum_{j=1}^{N} l_j(f)\phi_j \tag{53}$$

Each coefficient $l_i(f)$ may be determined by forming the inner product of both sides of (53) with ϕ_i. In this way we obtain

$$l_i(f) = (Kf,\phi_i) = (f,K^*\phi_i) \tag{54}$$

Letting $K^*\phi_i = \psi_i$, we obtain

$$l_i(f) = (f,\psi_i) \tag{55}$$

Substituting into (53), we obtain (51). The remaining assertions of the theorem have been covered by the discussion following Definition 6.

Next we prove an important approximation theorem concerning completely continuous operators.

Theorem 13. Any completely continuous operator K defined on a separable[1] Hilbert space can be expressed as the limit of degenerate operators. If K is hermitian, it can be expressed as the limit of degenerate hermitian operators.

Proof. Let ϕ_1, ϕ_2, . . . be an orthonormal basis in H. Let P_n denote the orthogonal projection on the subspace M_n spanned by ϕ_1, ϕ_2, . . . , ϕ_n,

---

[1] The theorem is true for nonseparable spaces also, but the proof given here applies only to a separable space (since a denumerable basis is employed).

and let $K_n = P_n K P_n$. Since $K_n f \in M_n$ for all elements f, it follows that K_n is degenerate. In order to show that the sequence $\{K_n\}$ converges to K, we shall prove that the contrary assumption leads to a contradiction. Suppose then that the above assertion is false; it follows that there exists a positive number ϵ such that the inequality

$$\|K - K_n\| > \epsilon \tag{56}$$

holds for infinitely many values of n. Restricting attention henceforth to such values of n, we can find elements f_n, each of unit norm, such that

$$\|Kf_n - K_n f_n\| > \epsilon \tag{57}$$

Now the sequence $\{(I - P_n)f_n\}$ is bounded, and so, by confining attention to a suitable subsequence, we may assume (because of the complete continuity of K) that the sequences $\{Kf_n\}$ and $\{K(I - P_n)f_n\}$ are both convergent, to limits g and h, respectively. We now prove three assertions: (1) $h = o$; (2) $\|P_n K(I - P_n)f_n\| \to 0$; (3) $\|(I - P_n)Kf_n\| \to 0$. We prove assertion 1 as follows:

$$(h,h) = \lim_{n \to \infty} (K(I - P_n)f_n, h) = \lim_{n \to \infty} (f_n, (I - P_n)K^*h)$$
$$\leq \lim_{n \to \infty} \|(I - P_n)K^*h\| \to 0 \tag{58}$$

(Here we have used the fact that

$$[K(I - P_n)]^* = (I - P_n)^* K^* = (I - P_n)K^*$$

and also that, for any fixed element f (in this case K^*h), $(I - P_n)f \to o$.) The proof of assertion 2 proceeds as follows:

$$\|P_n K(I - P_n)f_n\| \leq \|P_n\| \cdot \|K(I - P_n)f_n\|$$
$$= \|K(I - P_n)f_n\| \to \|h\| = 0 \tag{59}$$

Turning now to assertion 3, we write

$$\|(I - P_n)Kf_n\| = \|(I - P_n)(Kf_n - g) + (I - P_n)g\|$$
$$\leq \|(I - P_n)(Kf_n - g)\| + \|(I - P_n)g\| \leq \|I - P_n\| \cdot \|Kf_n - g\|$$
$$+ \|(I - P_n)g\| = \|Kf_n - g\| + \|(I - P_n)g\| \to 0 \tag{60}$$

Taking account of assertions 2 and 3, we obtain

$$\|(K - K_n)f_n\| = \|(I - P_n)Kf_n + P_n K(I - P_n)f_n\|$$
$$\leq \|(I - P_n)Kf_n\| + \|P_n K(I - P_n)f_n\| \to 0 \tag{61}$$

Since (61) contradicts (57), we have proved that the sequence $\{K_n\}$ converges to K. If K is hermitian, so is $P_n K P_n$ [by part (c) of Theorem 9]. The proof of the theorem is thus complete.

We now turn to operators which are both hermitian and completely continuous. First we present an important definition.

DEFINITION 7.   Given an operator K, a scalar μ is said to be a "characteristic value" of K if the equation

$$Kf = \mu f \tag{62}$$

possesses a nontrivial solution, which is termed a "characteristic vector" of K (corresponding to μ).†

Theorem 14.   If K is hermitian and completely continuous, and not the zero operator, it possesses at least one nonzero characteristic value.

Proof.   According to Theorem 11, there exists a sequence $\{f_n\}$ of unit elements such that $(Kf_n, f_n) \to \mu_1$, where $\mu_1 = \pm \|K\| \neq 0$.   Then

$$\|(K - \mu_1)f_n\|^2 = \|Kf_n\|^2 - 2\mu_1(Kf_n, f_n) + \mu_1{}^2 \tag{63}$$

Now, $\|Kf_n\|^2 \leq \|K\|^2 = \mu_1{}^2$, so that

$$\limsup_{n \to \infty} \|(K - \mu_1)f_n\|^2 = \limsup_{n \to \infty} \|Kf_n\|^2 - 2\mu_1{}^2 + \mu_1{}^2$$
$$\leq \mu_1{}^2 - 2\mu_1{}^2 + \mu_1{}^2 = 0 \tag{64}$$

Since the left side of (64) is nonnegative, it must vanish, and this is possible only if

$$Kf_n - \mu_1 f_n \to 0 \tag{65}$$

Since K is completely continuous, we can select a convergent subsequence $\{Kf_{n_k}\}$ from $\{Kf_n\}$, and it then follows from (65) that the subsequence $\{\mu_1 f_{n_k}\}$ is also convergent.   Since $\mu_1 \neq 0$, the subsequence $\{f_{n_k}\}$ must converge to an element ϕ_1 of unit norm, and by referring to (65) we find (since $f_n \to f$ implies that $Kf_n \to Kf$) that

$$K\phi_1 = \mu_1\phi_1 \tag{66}$$

With the aid of the above theorem, we shall now obtain a very precise representation of the operator K.   If g is any element orthogonal to ϕ_1, we find, taking account of (66), that

$$(Kg, \phi_1) = (g, K\phi_1) = (g, \mu_1\phi_1) = \mu_1(g, \phi_1) = 0 \tag{67}$$

Hence, K maps the subspace consisting of all elements orthogonal to ϕ_1 into itself. K may therefore be considered as an operator defined on this subspace, and it is evidently still hermitian and completely continuous.   If K reduces to the zero operator on this subspace, then Kf is given, for any element f of H, by the formula

$$Kf = \mu_1(f, \phi_1)\phi_1 \tag{68}$$

† The German terms "eigenwert" and "eigenvector" are often used, as is the hybrid term "eigenvalue."

[This may be seen by expressing f as $(f,\phi_1)\phi_1 + g$, from which we immediately conclude that $(g,\phi_1) = 0$, so that

$$Kf = K((f,\phi_1)\phi_1) = (f,\phi_1)K\phi_1 = \mu_1(f,\phi_1)\phi_1]$$

If K does not reduce to the zero operator on the subspace, we may repeat the argument used in the proof of Theorem 14 to show that there exists a unit element ϕ_2 orthogonal to ϕ_1 and a nonzero real number μ_2 such that

$$K\phi_2 = \mu_2\phi_2 \tag{69}$$

Reasoning as before, we find that either, in analogy with (68), the operator K is expressible by the formula

$$Kf = \mu_1(f,\phi_1)\phi_1 + \mu_2(f,\phi_2)\phi_2 \tag{70}$$

or we can continue the above procedure. If this procedure terminates after a finite number of steps, we obtain for K the formula

$$Kf = \sum_{k=1}^{N} \mu_k(f,\phi_k)\phi_k \tag{71}$$

If, on the other hand, the procedure does not terminate, we obtain a denumerable orthonormal system of elements ϕ_1, ϕ_2, . . . and nonzero real numbers μ_1, μ_2, . . . such that

$$K\phi_k = \mu_k\phi_k \qquad (k = 1, 2, \ldots) \tag{72}$$

Before obtaining the analogue of (71) in the latter case, we show that $\{|\mu_n|\}$ must decrease to zero. The inequalities

$$|\mu_1| \geq |\mu_2| \geq |\mu_3| \geq \cdots \tag{73}$$

are assured by the manner in which the elements $\{\phi_k\}$ were determined, so that $\{|\mu_n|\}$ must converge to a nonnegative limit δ. Now,

$$\|K\phi_j - K\phi_k\|^2 = \mu_j^2 + \mu_k^2 \geq 2\delta^2 \qquad (j \neq k) \tag{74}$$

and so, if δ were positive, the sequence $\{K\phi_n\}$ could not possess any convergent subsequence. Since this would contradict the complete continuity of K, we conclude that $\delta = 0$.

We now proceed to show that the obvious generalization of (71), namely,

$$Kf = \sum_{k=1}^{\infty} \mu_k(f,\phi_k)\phi_k \tag{75}$$

holds when the procedure described above does not terminate. If f belongs to M, the subspace spanned by the elements ϕ_1, ϕ_2, . . . , we may apply K term by term to the expansion

$$f = \sum_{k=1}^{\infty} (f, \phi_k) \phi_k \tag{76}$$

This immediately yields (75). Next we note that, according to the manner in which the elements $\{\phi_k\}$ were determined, any element f orthogonal to $\phi_1, \phi_2, \ldots, \phi_n$ must satisfy the inequality

$$|(Kf, f)| \leq |\mu_n| \cdot \|f\|^2 \tag{77}$$

Hence, if $f \in M^\perp$ (i.e., if f is orthogonal to all the ϕ's), $(Kf, f) = 0$. Since, furthermore, the reasoning following (67) shows that K may be interpreted as a hermitian operator defined on the Hilbert space $M^\perp$, we conclude from Theorem 11 that $Kf = o$. Now, given any element f of H, we may, according to Theorem 6, express it in the form $g + h$, where $g \in M$, $h \in M^\perp$. According to the foregoing remarks,

$$Kf = Kg = \sum_{k=1}^{\infty} \mu_k (g, \phi_k) \phi_k = \sum_{k=1}^{\infty} \mu_k (f, \phi_k) \phi_k \tag{78}$$

Thus the validity of (75) is established for any element f of H.

We sum up the results of the above discussion in the form of a theorem.

Theorem 15. If K is any hermitian completely continuous operator other than 0, it may be represented in the form (71) or (75), according to whether the procedure described above terminates or not. The elements $\phi_1, \phi_2, \ldots$ span H if and only if the equation $Kf = o$ possesses only the trivial solution $f = o$.

EXERCISES

17. Prove that an operator is a projection if and only if it is hermitian and idempotent.

18. Prove that every hermitian operator whose square is the identity operator I may be expressed in the form $I - 2P$, where P is a projection.

19. Prove that all eigenvalues of a hermitian operator are real, and that eigenvectors corresponding to distinct eigenvalues are orthogonal.

20. Prove the following statements: (a) The sum of two hermitian operators is hermitian. (b) A *real* scalar multiple of a hermitian operator is hermitian. (c) The inverse of a hermitian operator (if it exists) is hermitian. (d) If S and T are hermitian, then ST is hermitian if and only if S and T commute—i.e., $ST = TS$.

21. Prove that if P and Q are projections, then: (a) PQ is a projection if and only if $PQ = QP$, (b) $P + Q$ is a projection if and only if $PQ = 0$, (c) $P - Q$ is a projection if and only if $PQ = Q$.

22. Prove that a hermitian operator T possesses an inverse if there exists a positive number c such that $(Tf, f) \geq c\|f\|^2$ for every element f. *Hint:* Use Theorems 11 and 4-4.

23. Prove that the product of two operators of which at least one is degenerate is itself degenerate.

24. An operator K (other than 0) is said to be "nilpotent" if $K^n = 0$ for some integer $n > 1$. (a) Give a simple example of a nilpotent operator. (b) Prove that a hermitian operator cannot be nilpotent.

25. Although the *inequality* $\|K^2\| \leq \|K\|^2$ is always true, prove that the *equality* $\|K^2\| = \|K\|^2$ holds if K is hermitian. Then prove that, more generally, $\|K^k\| = \|K\|^k$.

26. Closely related to hermitian operators are "normal" operators, which are characterized by the relation $KK^* = K^*K$. Prove that K is normal if and only if $\|Kf\| = \|K^*f\|$ for every element f.

27. Prove directly from (75) that $\|K\|$ is equal to $|\mu_1|$.

28. A "unitary" operator U is one satisfying the conditions $UU^* = U^*U = I$. Prove that U is unitary if and only if it possesses an inverse and is "norm-preserving" —i.e., $\|Uf\| = \|f\|$ for every element f.

Prove also that in a finite-dimensional space, but *not* in an infinite-dimensional space, the norm-preserving property implies the existence of an inverse, and hence suffices to characterize unitary operators.

5. The Fredholm Alternative

The characteristic feature of the Fredholm alternative in a Hilbert space is that a close relationship is established between the equation

$$f - Kf = g \qquad (4\text{-}29)$$

and the "adjoint" homogeneous equation

$$f - K^*f = o \qquad (79)$$

It is understood throughout this section that K is completely continuous.

Theorem 16. The Fredholm Alternative in a Hilbert Space. (1) The solutions of (79) and of

$$f - Kf = o \qquad (79')$$

form subspaces of the same finite (zero or positive) dimension D. (2) If $D = 0$, the inverse $(I - K)^{-1}$ exists, so that (4-29) is solvable for every element g, and $\|f\| \leq C\|g\|$ for some sufficiently large constant C. (3) If $D > 0$, (4-29) is solvable if and only if g is orthogonal to every solution of (79), and then the solution is not unique, for to any particular solution may be added an arbitrary solution of (79').

Proof. While part 2 is covered by Theorem 4-6, it is preferable to give an independent proof of the entire theorem. First, we consider the case that the Hilbert space in question is U_n. We select the orthonormal basis $e_1, e_2, \ldots, e_n$, where e_k is the element whose kth entry is one while all the others are zero. Then, expressing the element

$$f = \{\alpha_1, \alpha_2, \ldots, \alpha_n\}$$

in the form

$$f = \sum_{i=1}^{n} \alpha_i e_i \tag{80}$$

we find that

$$Kf = \sum_{i,j=1}^{n} \alpha_i c_{ij} e_j \tag{81}$$

where the n^2 scalars c_{ij} are determined by the equalities

$$Ke_i = \sum_{j=1}^{n} c_{ij} e_j \tag{82}$$

Similarly, with the aid of n^2 scalars c_{ij}^* we can represent K^*g, for every element $g = \{\beta_1, \beta_2, \ldots, \beta_n\}$, in the form

$$K^*g = \sum_{i,j=1}^{n} \beta_i c_{ij}^* e_j \tag{83}$$

where the scalars c_{ij}^* are determined, in analogy with the c_{ij}, by the equalities

$$K^*e_i = \sum_{j=1}^{n} c_{ij}^* e_j \tag{84}$$

From (81) and (83) we obtain

$$(Kf,g) = \sum_{i,j=1}^{n} c_{ij}\alpha_i\bar{\beta}_j \qquad (f,K^*g) = \sum_{i,j=1}^{n} \overline{c_{ij}^*}\alpha_j\bar{\beta}_i = \sum_{i,j=1}^{n} \overline{c_{ji}^*}\alpha_i\bar{\beta}_j \tag{85}$$

Since the sums appearing in (85) must be equal for all choices of the $2n$ scalars $\alpha_1, \alpha_2, \ldots, \alpha_n, \beta_1, \beta_2, \ldots, \beta_n$, the n^2 equalities $\overline{c_{ji}^*} = c_{ij}$, or

$$c_{ij}^* = \bar{c}_{ji} \tag{86}$$

must hold. Equations (79'), (79), and (4-29) assume the following forms, respectively,

$$\sum_{i=1}^{n} d_{ij}\alpha_i = 0 \qquad d_{ij} = \delta_{ij} - c_{ij} \tag{87}$$

$$\sum_{i=1}^{n} \bar{d}_{ji}\alpha_i = 0 \tag{88}$$

$$\sum_{i=1}^{n} d_{ij}\alpha_i = \beta_j \tag{89}$$

and the three parts of the theorem may be stated as follows. (1) Systems (87) and (88) have the same number of linearly independent solutions.

(2) If systems (87) and (88) have only the trivial solution

$$\alpha_1 = \alpha_2 = \cdots = \alpha_n = 0$$

then system (89) has a unique solution for every choice of the scalars β_1, $\beta_2, \ldots, \beta_n$, and $\sum_{i=1}^{n} |\alpha_i|^2 \leq C^2 \sum_{i=1}^{n} |\beta_i|^2$ for some sufficiently large constant C. (3) If (87) and (88) possess nontrivial solutions, then (89) is solvable if and only if $\sum_{i=1}^{n} \alpha_i \bar{\beta}_i = 0$ for every solution $\{\alpha_1, \alpha_2, \ldots, \alpha_n\}$ of (88). Now, these three statements, which are an extended version of the first corollary of Theorem 4-6, constitute the basic theory of systems of linear (algebraic) equations, which we assume to be known.[1] (The boundedness condition of part 2 is assured by Exercise 4-29, but may also be proved from Cramer's rule by inspection.)

Next we consider the case that K is a degenerate operator defined on any Hilbert space H. Taking account of representation (51), we find that (4-29) assumes the form

$$f = g + \sum_{i=1}^{n} (f, \psi_i) \phi_i \tag{90}$$

so that any solution, if it exists, is expressible in the form

$$f = g + \sum_{j=1}^{n} \alpha_j \phi_j \tag{91}$$

Replacing f on both sides of (90) by the right side of (91), we find that (90) assumes the form

$$\sum_{j=1}^{n} \alpha_j \phi_j = \sum_{j=1}^{n} \left[(g, \psi_j) + \sum_{i=1}^{n} \alpha_i (\phi_i, \psi_j) \right] \phi_j \tag{92}$$

Since the elements $\phi_1, \phi_2, \ldots, \phi_n$ may be assumed independent,[2] we obtain from (92) the algebraic system

$$\alpha_j - \sum_{i=1}^{n} (\phi_i, \psi_j) \alpha_i = (g, \psi_j) \tag{93}$$

which reduces to (89) if we let $d_{ij} = \delta_{ij} - (\phi_i, \psi_j)$ and $\beta_j = (g, \psi_j)$. Conversely, if the scalars $\alpha_1, \alpha_2, \ldots, \alpha_n$ satisfy (93), the element f defined by (91) is readily seen to satisfy (90), and hence (4-29). In particular,

---

[1] It is of interest that in order to prove the present theorem, unlike Theorem 4-6, we must first know the simplest particular case of the theorem.

[2] Otherwise (51) could be rewritten as a sum of fewer terms.

setting $g = o$, we find that equation (79') assumes the form

$$\alpha_j - \sum_{i=1}^{n} (\phi_i, \psi_j)\alpha_i = 0 \tag{94}$$

while equation (79) assumes the form[1]

$$\alpha_j - \sum_{i=1}^{n} (\psi_i, \phi_j)\alpha_i = 0 \tag{95}$$

These equations reduce, of course, to (87) and (88), respectively, when the quantities d_{ij} are defined as above. Now, since the ϕ's are independent, any manifold of elements $\{\alpha_1, \alpha_2, \ldots, \alpha_n\}$ of U_n corresponds to a manifold of elements $\sum_{j=1}^{n} \alpha_j \phi_j$ of the same dimension in H. Thus, the solutions of (79') and (79) form manifolds of the same dimension, respectively, as the solutions of (87) and (88). Since part 1 holds in U_n, it must therefore hold in H. When $D = 0$, (93) is uniquely solvable for each set of scalars (g, ψ_j), and hence (4-29) is uniquely solvable for each element g. Furthermore, letting C have the same significance as in the discussion of U_n and using the Schwarz inequality, we obtain the n inequalities

$$|\alpha_i|^2 \le C^2 \sum_{j=1}^{n} |(g, \psi_j)|^2 \le C^2 \|g\|^2 \cdot \sum_{j=1}^{n} \|\psi_j\|^2 \tag{96}$$

and hence, from (91),

$$\|f\| \le \|g\| + C\|g\| \cdot \left(\sum_{j=1}^{n} \|\phi_j\|\right) \cdot \left(\sum_{j=1}^{n} \|\psi_j\|^2\right)^{\frac{1}{2}} \tag{97}$$

Thus, part 2, including the boundedness condition, is proved. Turning to part 3, we merely have to note that (93), and hence (4-29), is solvable if and only if $\sum_{j=1}^{n} (g, \psi_j)\bar{\alpha}_j = \left(g, \sum_{j=1}^{n} \alpha_j \psi_j\right) = 0$ for every solution α_1, $\alpha_2, \ldots, \alpha_n$ of (95). This is equivalent to asserting that $(g, f) = 0$ whenever f satisfies (79). Thus the proof is completed for the case that K is degenerate.

Finally, turning to an arbitrary K, we express it as the sum of a degenerate operator K_1 and a "small" operator K_2; more precisely, we

---

[1] Here we use (52) and the independence of the elements $\psi_1, \psi_2, \ldots, \psi_n$, which follows from that of the elements $\phi_1, \phi_2, \ldots, \phi_n$ by the following argument: If the ψ's were dependent, (52) could be rewritten as a sum of fewer terms, but then so could (51), by symmetry. (Thus, we have shown that the ranges of K and K^* have equal dimensions.)

select K_1, in accordance with Theorem 13, such that

$$\|K_2\| = \|K - K_1\| < 1$$

Then $(I - K_2)^{-1}$ exists, by Theorem 4-5, and so (4-29) may be rewritten in the form

$$f - (I - K_2)^{-1}K_1 f = (I - K_2)^{-1}g \tag{98}$$

Since $(I - K_2)^{-1}K_1$ is degenerate (cf. Exercise 23), we may assert that $I - (I - K_2)^{-1}K_1$ and its adjoint $I - K_1^*(I - K_2^*)^{-1}$ have null spaces of equal dimension. Now, on the one hand, (79′) is entirely equivalent to

$$[I - (I - K_2)^{-1}K_1]f = o$$

while, on the other hand, $I - K_1^*(I - K_2^*)^{-1}$ may be rewritten in the form $(I - K^*)(I - K_2^*)^{-1}$, from which it follows that its null space has the same dimension as that of $I - K^*$. Part 1 is thus proved. Part 2 follows, in its entirety, from applying the (bounded) operator $[I - (I - K_2)^{-1}K_1]^{-1}$ to both sides of (98). Now, when $D > 0$ we know that (4-29), being equivalent with (98), is solvable if and only if

$$(h,(I - K_2)^{-1}g) = 0$$

whenever

$$[I - K_1^*(I - K_2^*)^{-1}]h = o$$

and this is equivalent to saying that

$$((I - K_2^*)^{-1}h,g) = 0$$

whenever

$$(I - K^*)(I - K_2^*)^{-1}f = o$$

Replacing $(I - K_2^*)^{-1}h$ by f, we obtain part 3, thus completing the proof.

We now consider a simple modification of (4-29), namely,

$$f - \lambda K f = g \tag{99}$$

where λ is a (scalar) parameter. Then, by replacing K and K^* everywhere in the above proof by λK and $\bar{\lambda}K^*$, respectively, we conclude that, for each value of λ, either equation (99) is uniquely solvable for each g, and the adjoint homogeneous equation

$$f - \bar{\lambda}K^*f = o \tag{100}$$

admits only the trivial solution, or (99) fails to admit a solution for certain choices of the element g, in which case (100) and the homogeneous form of (99), namely,

$$f - \lambda K f = o \tag{101}$$

both admit nontrivial solutions, the subspace of solutions being of the same (finite) dimension for both equations. In the latter case, λ is termed a "singular value" of K. Since zero obviously cannot be a singular value, we may divide (101) by λ, and thus conclude that the singular values are simply the reciprocals of the eigenvalues of K. (Cf. Definition 7.)

We shall prove that within any disc $|\lambda| < R$ there exist only a finite number of singular values, if any. (From this it follows, in particular, that there are altogether at most a denumerable number of singular values.) If the first assertion were incorrect, there would exist a sequence $\{\lambda_n\}$ of distinct singular values within the afore-mentioned disc, and a corresponding sequence $\{\phi_n\}$ of nonzero elements such that

$$\lambda_n K \phi_n = \phi_n \tag{102}$$

These elements must be independent, for otherwise we could determine a unique index N such that $\phi_1, \phi_2, \ldots, \phi_{N-1}$ are independent while ϕ_N may be expressed in the form

$$\phi_N = \sum_{i=1}^{N-1} c_i \phi_i \tag{103}$$

Applying the operator $\lambda_N K$ to both sides of (103) and taking account of (102), we obtain

$$\phi_N = \sum_{i=1}^{N-1} c_i \left(\frac{\lambda_N}{\lambda_i}\right) \phi_i \tag{104}$$

By subtraction, we obtain

$$\sum_{i=1}^{N-1} c_i \left(1 - \frac{\lambda_N}{\lambda_i}\right) \phi_i = o \tag{105}$$

Since the first $(N - 1)$ ϕ's are independent, and since the λ's are all distinct, the coefficients $c_1, c_2, \ldots, c_{N-1}$ would have to vanish, which is impossible in view of (103). Now that the independence is established, we may apply the Gram-Schmidt process to the sequence $\{\phi_n\}$ to obtain an orthonormal sequence $\{g_n\}$. From the fact that each g_n is a linear combination of the first n ϕ's we readily find, by taking account of (102), that $\lambda_n K g_n$ is also a combination of the first n ϕ's, but that $\lambda_n K g_n - g_n$ is a combination of the first $(n - 1)$ ϕ's, which, in turn, may also be expressed as a combination of the first $(n - 1)$ g's. Therefore, for $n > m$ we find that $\lambda_n K g_n - g_n - \lambda_m K g_m$ is a combination of the first $(n - 1)$ g's, and hence

$$\lambda_n K g_n - \lambda_m K g_m = g_n + \sum_{i=1}^{n-1} \alpha_i g_i \tag{106}$$

for suitably chosen scalars α_1, α_2, . . . , α_{n-1}. Taking account of the orthonormality relation $(g_i,g_j) = \delta_{ij}$, we conclude that

$$\|\lambda_n K g_n - \lambda_m K g_m\| \geq 1$$

and hence that the sequence $\{K(\lambda_n g_n)\}$ possesses no convergent subsequence. However, this result contradicts the complete continuity of K, since the norms $\|\lambda_n g_n\|$ do not exceed R. The proof is thus complete.

We point out that the above result concerning the distribution of singular values is contained in (75) for hermitian K, and that (75) also makes possible a very simple proof of Theorem 16. By using this representation of the hermitian operator K in (99) and repeating the argument used in the portion of the proof of Theorem 16 dealing with a degenerate operator, we obtain the unique solution in the form

$$f = (I + \lambda K_\lambda)g \tag{107}$$

where K_λ, the "resolvent kernel" of K, is defined for all nonsingular values of λ by the formula[1]

$$K_\lambda g = \sum_{i=1}^{\infty} \frac{\mu_i(g,\phi_i)\phi_i}{1 - \lambda\mu_i} \tag{108}$$

If, on the other hand, λ coincides with one or more of the quantities μ_i^{-1}, the corresponding inner products (g,ϕ_i) must vanish. If this orthogonality condition is satisfied, however, (107) still furnishes a solution of (99) if those coefficients in (108) which assume the indeterminate form $0/0$ are assigned completely arbitrary values. Thus Theorem 16 is proved.

EXERCISES

29. Let the scalars c_{ij} be associated with the operator K defined on U_n as in (82). Show that K is unitary if and only if the n^2 equalities $\sum_{j=1}^{n} c_{ij}\bar{c}_{kj} = \delta_{ik}$ are satisfied. (A square matrix whose elements satisfy these conditions is therefore termed "unitary.")

30. Let two different orthonormal bases $e_1^{(1)}$, $e_2^{(1)}$, . . . , $e_n^{(1)}$ and $e_1^{(2)}$, $e_2^{(2)}$, . . . , $e_n^{(2)}$ be selected in U_n, and let the operator K be represented, in analogy with (82), in terms of each basis by the scalars $c_{ij}^{(1)}$ and $c_{ij}^{(2)}$, respectively. Show that $c_{ik}^{(2)} = \sum_{j,r=1}^{n} t_{ij}c_{jr}^{(1)}\bar{t}_{kr}$, where $t_{ij} = (e_i^{(2)},e_j^{(1)})$.

---

[1] Since the factors $\mu_i/(1 - \lambda\mu_i)$ appearing in (108) approach zero, the convergence of the series is assured by the convergence of $\sum_{i=1}^{\infty} |(g,\phi_i)|^2$. [Cf. (16) and the final paragraph of Sec. 2.]

6. Integral Equations

Our first objective in the present section is to show that there is a natural correspondence between the elements of $L_2(a,b)$ on the one hand and some (but not all) of the completely continuous operators on $L_2(a,b)$. (Henceforth we usually suppress reference to the interval $a < x < b$; also, all integrals are to be taken over this interval.) If $\phi_1, \phi_2, \ldots, \phi_n$, $\psi_1, \psi_2, \ldots, \psi_n$ are functions in L_2, then the function

$$K(x,y) = \sum_{i=1}^{n} \phi_i(x)\overline{\psi_i(y)} \tag{109}$$

belongs to $\mathbf{L}_2$. This follows from the obvious equality[1]

$$|||\phi_i(x)\overline{\psi_i(y)}||| = \|\phi_i\| \cdot \|\psi_i\|$$

and the triangle inequality. If we associate with each function f in L_2 the function

$$h(x) = \int f(y)K(x,y)\,dy \tag{110}$$

we obtain, from the Schwarz inequality, $|h(x)|^2 \leq \|f\|^2 \int |K(x,y)|^2\,dy$, and hence, by integration,

$$\|h\| \leq |||K||| \cdot \|f\| \tag{111}$$

Thus the function $K(x,y)$ determines, through (110), a linear operator on L_2; we denote this operator by K. (The use of the same letter for the function and the operator is, of course, intentional.) We note from (111) that the norm of the *operator* K is at most equal to the norm of the *function* K; it is easily shown, as in Exercise 31, that, in general, the two norms are not equal. The operator K is completely continuous, for it maps L_2 into the finite-dimensional subspace spanned by $\phi_1, \phi_2, \ldots, \phi_n$. Conversely, according to Theorem 12, every degenerate operator K may be expressed in the form (110), where the kernel $K(x,y)$ has the form (109) for suitable choices of the ϕ's and ψ's, and the adjoint operator K^* corresponds to the kernel $K^*(x,y)$, where, as is to be expected,

$$K^*(x,y) = \overline{K(y,x)} \tag{112}$$

Now, the step from (110) to (111) does not involve the assumption that $K(x,y)$ has the form[2] (109), and so it follows that every function in $\mathbf{L}_2$ determines, by (110), an operator on L_2; the formula (112) for the kernel

---

[1] We use the symbols $\|\ \ \|$ and $\|\|\ \ \|\|$ for norms in L_2 and $\mathbf{L}_2$, respectively; the former symbol will also be used for norms of operators on L_2.

[2] A function having the form (109) is, in analogy with the case of operators, termed "degenerate."

corresponding to the adjoint operator is immediately established for the general case by the identity[1]

$$(Kf,g) = \iint K(x,y)f(y)\overline{g(x)} \, dy \, dx$$
$$= \iiint f(y)\overline{K^*(y,x)g(x)} \, dx \, dy = (f,K^*g) \quad (113)$$

In order to show that the operator determined by the kernel $K(x,y)$ is completely continuous, we use a routine approximation argument. According to the definition of the Lebesgue integral, it is possible, given any $\epsilon > 0$, to approximate $K(x,y)$ by a step function $S(x,y)$ such that $|||K - S||| < \epsilon$, and it then follows from a remark following (111) that the norm of the *operator* $K - S$ cannot exceed ϵ. Since any step function is degenerate (cf. Exercise 32) and determines a degenerate operator, it follows from Theorem 4-3 that K is completely continuous.

It now follows that Theorem 16 is applicable to the integral equation

$$f(x) = g(x) + \lambda \int f(y) K(x,y) \, dy \quad (114)$$

The Fredholm alternative assumes the following form: If $K(x,y) \in \mathbf{L}_2$, then *either* (114) has a unique L_2 solution f for each L_2 function g, in which case $\|f\| \leq C\|g\|$, where C is a constant depending on λ and $K(x,y)$, *or* the homogeneous equation

$$f(x) = \lambda \int f(y) K(x,y) \, dy \quad (115)$$

possesses a finite number of independent solutions, in which case the homogeneous equation

$$h(x) = \bar{\lambda} \int h(y) \overline{K(y,x)} \, dy \quad (116)$$

possesses the same number of independent solutions, and (114) is solvable if and only if $\int g(x)\overline{h(x)} \, dx = 0$ for every solution of (116).

We stress that (114) must be interpreted in the L_2 sense. By this we mean that the integral appearing in (114) may fail to exist for a certain null set of values of x, and that there may be an additional null set for which the two sides of (114), although both well defined, have different values. If f, g, and K are replaced by equivalent functions, (114) will continue to be satisfied in this generalized sense. However, if the limits a and b are finite and if the functions $g(x)$ and $K(x,y)$ are continuous in the closed interval $a \leq x \leq b$ and square $a \leq x, y \leq b$, respectively, then (114) may be interpreted in the conventional sense of holding point-wise, without exception. This follows readily from the observation that, according to the Schwarz inequality,

$$\left| \int f(y)[K(x_1,y) - K(x_2,y)] \, dy \right|$$
$$\leq \max_{a \leq y \leq b} |K(x_1,y) - K(x_2,y)| \cdot (b - a)^{\frac{1}{2}} \cdot \|f\| \quad (117)$$

---

[1] We freely use here and elsewhere Theorem 1-8, the Fubini theorem.

Since the first factor on the right side of this inequality can be made small by making $|x_1 - x_2|$ small [this follows from the uniform continuity of $K(x,y)$], the right side of (114) must be defined and continuous throughout the closed interval.   Therefore, any solution of (114) in the L_2 sense is equivalent to a certain continuous function which satisfies the integral equation pointwise.

If the operators J and K correspond to the L_2 kernels $J(x,y)$ and $K(x,y)$, respectively, then the operator JK corresponds to the kernel $L(x,y)$, where

$$L(x,y) = \int J(x,z)K(z,y)\,dz \tag{118}$$

and $L(x,y)$ also belongs to L_2.   The first assertion follows by imitating suitably the reasoning following (110), for, just as the equality $h = Kf$ is expressed analytically in the form (110), so the equality $k = Jh = JKf$ is expressed in the form

$$k(x) = \int h(z)J(x,z)\,dz = \int J(x,z)[\int f(y)K(z,y)\,dy]\,dz \tag{119}$$

and (118) follows by inverting the order of integration.   The second assertion is proved by applying the Schwarz inequality to (118) to obtain

$$|L(x,y)|^2 \leq [\int |J(x,z)|^2\,dz][\int |K(z,y)|^2\,dz] \tag{120}$$

and then integrating (with respect to x and y) to obtain

$$|||L||| \leq |||J||| \cdot |||K||| < \infty \tag{121}$$

In particular, if $J = K$, we find that the operator K^2 is represented by a kernel constructed from $K(x,y)$ by an integration procedure.   This kernel, known as the "first iterate" of $K(x,y)$, is denoted by $K_{(2)}(x,y)$. Similarly, the operators K^n are represented for $n = 3, 4, \ldots$ by "higher iterates" $K_{(3)}(x,y)$, $K_{(4)}(x,y)$, $\ldots$.

For the purposes of the next section it is necessary to show that different kernels determine different operators.   By linearity, it suffices to show that the only kernels which determine the zero operator are the null functions.   If the kernel $K(x,y)$ "annihilates" every L_2 function, then the first equality of (113) shows that, for every pair of L_2 functions f and g, the equality

$$\int\int K(x,y)f(y)\overline{g(x)}\,dy\,dx = 0 \tag{122}$$

must hold.   By choosing f and g as the characteristic functions[1] of arbitrary (finite) subintervals of $a < x < b$, we conclude that the integral of $K(x,y)$ over any rectangle vanishes.   By a basic theorem in Lebesgue integration (cf. Exercise 34), it follows that $K(x,y)$ must vanish almost everywhere, as was to be proved.

---

[1] The characteristic function of any subset of a given set is equal to one at each point of the subset and zero elsewhere.

It is of great importance in the study of some integral equations that the kernel $L(x,y)$ defined by (118) may turn out to be an $\mathbf{L}_2$ function, even though one or both of the kernels $J(x,y)$, $K(x,y)$ may not be. In particular, it may happen that a kernel not belonging to $\mathbf{L}_2$ may have an iterate which belongs to $\mathbf{L}_2$. A simple example of such a kernel is given by

$$K(x,y) = |x - y|^{-\frac{1}{2}} \qquad (0 < x, y < 1) \qquad (123)$$

in which case the first iterate is given by the $\mathbf{L}_2$ kernel

$$K_{(2)}(x,y) = \pi$$
$$+ 2 \log \left[(x^{\frac{1}{2}} + y^{\frac{1}{2}})((1 - x)^{\frac{1}{2}} + (1 - y)^{\frac{1}{2}}) \right] - 2 \log |x - y| \quad (124)$$

(Cf. Exercise 35.) The importance of such a situation can be indicated by showing how the Fredholm theory may then be applied, by means of a simple artifice, to the integral equation (114) despite the fact that the kernel is not in $\mathbf{L}_2$. If we replace $f(y)$ on the right side of (114) by the entire right side of (114) (with suitable change of letters), we obtain

$$f(x) = g(x) + \lambda \int K(x,y)g(y) \, dy + \lambda^2 \int K_{(2)}(x,y) f(y) \, dy \qquad (125)$$

If $g(x) + \lambda \int K(x,y)g(y) \, dy$ and $K_{(2)}(x,y)$ belong to L_2 and $\mathbf{L}_2$, respectively, then the Fredholm theory applies, so that, if λ^2 is not a singular value of $K_{(2)}(x,y)$, there will exist a unique L_2 solution of (125), and (114) cannot have any L_2 solution other than this one. [The converse problem, namely, whether a solution of (125) also satisfies (114), must be studied in each case.]

EXERCISES

31. Let $a = 0$, $b = 1$, $K(x,y) = 1 + \frac{1}{2}e^{2\pi i(x-y)}$. Prove that $\|K\| = 1$, while $\||K\|| = (\frac{5}{4})^{\frac{1}{2}}$.

32. Prove that every step function of two variables is degenerate, as defined in the footnote on page 118.

33. The kernel $K(x,y)$ is said to be hermitian (cf. Sec. 7) if $K(x,y) \equiv \overline{K(y,x)}$; according to (112), a hermitian kernel determines a hermitian operator. Prove that all the iterates of a hermitian kernel are hermitian. (This is to be expected *formally*, since the powers of a hermitian *operator* are hermitian.)

34. Prove that (122) implies that $K(x,y)$ is orthogonal to every step function; from this result prove that $K(x,y)$ is orthogonal to itself, and hence is a null function.

35. Work out the details leading from (123) to (124), and prove that $K_{(2)}(x,y)$ is an $\mathbf{L}_2$ kernel but that $K(x,y)$ is not.

7. Hermitian Kernels

Let $K(x,y)$ denote any hermitian $\mathbf{L}_2$ function, and let K denote the corresponding hermitian operator on L_2. If the eigenvalues[1] $\{\mu_k\}$ and

---

[1] We assume here that there are infinitely many eigenvalues and eigenvectors, for otherwise the results of this section are trivially true.

eigenvectors $\{\phi_k\}$ of K are obtained in the manner described in Sec. 4 following Theorem 14, then the functions $\{\phi_k(x)\overline{\phi_k(y)}\}$ evidently form an orthonormal set in $\mathbf{L}_2$, and the Fourier coefficients of $K(x,y)$ with respect to this set of functions are given by

$$\iint K(x,y)\overline{\phi_k(x)\overline{\phi_k(y)}}\,dx\,dy = \iint K(x,y)\phi_k(y)\overline{\phi_k(x)}\,dy\,dx$$
$$= \mu_k \int |\phi_k(x)|^2\,dx = \mu_k \quad (126)$$

From (16) it follows that the series $\sum\limits_{k=1}^{\infty} \mu_k{}^2$ is convergent, so that the series

$$\sum_{k=1}^{\infty} \mu_k\phi_k(x)\overline{\phi_k(y)} \quad (127)$$

converges in norm to a hermitian $\mathbf{L}_2$ function $\tilde{K}(x,y)$. If we let $\tilde{K}$ denote the operator on L_2 determined by the kernel $\tilde{K}(x,y)$, it is evident from (127) that, for any element f of L_2, the expansion

$$\tilde{K}f = \sum_{k=1}^{\infty} \mu_k\phi_k(x) \int f(y)\overline{\phi_k(y)}\,dy \quad (128)$$

holds. [We emphasize that (128) holds in norm, not necessarily in the sense of pointwise convergence.] On the other hand, according to Theorem 15, the right side of (128) is also equal to Kf. Therefore, the kernels $K(x,y)$ and $\tilde{K}(x,y)$ determine the same operator on L_2, and so, according to a result stated in Sec. 6, $K(x,y)$ and $\tilde{K}(x,y)$ coincide almost everywhere. Thus we have shown that the kernel $K(x,y)$ is expressible in terms of its eigenvalues and eigenfunctions by the series expansion

$$K(x,y) = \sum_{k=1}^{\infty} \mu_k\phi_k(x)\overline{\phi_k(y)} \quad (129)$$

and that, for any element f of L_2, the element Kf admits the expansion

$$\int f(y)K(x,y)\,dy = \sum_{k=1}^{\infty} \mu_k\phi_k(x) \int f(y)\overline{\phi_k(y)}\,dy \quad (130)$$

From (129) and the Parseval equality (21) it also follows that

$$|||K|||^2 = \sum_{k=1}^{\infty} \mu_k{}^2 \quad (131)$$

Now, from (75) we obtain immediately for K^n, $n \geq 2$, the expansions

$$K^n f = \sum_{k=1}^{\infty} \mu_k{}^n(f,\phi_k)\phi_k \quad (132)$$

Since the operator K^n is determined by the iterated kernel $K_{(n)}(x,y)$ and no other kernel, it follows by a repetition of the above argument that, in analogy with (129), (130), (131),

$$K_{(n)}(x,y) = \sum_{k=1}^{\infty} \mu_k{}^n \phi_k(x) \overline{\phi_k(y)} \qquad (129')$$

$$\int f(y) K_{(n)}(x,y)\, dy = \sum_{k=1}^{\infty} \mu_k{}^n \phi_k(x) \int f(y) \overline{\phi_k(y)}\, dy \qquad (130')$$

$$|||K_{(n)}|||^2 = \sum_{k=1}^{\infty} \mu_k{}^{2n} \qquad (131')$$

We sum up these results in the form of a theorem.

Theorem 17. Expansion Theorem. Any hermitian L_2 function $K(x,y)$ and its iterates $K_{(2)}(x,y)$, $K_{(3)}(x,y)$, . . . are expressible, in terms of the eigenvalues $\{\mu_k\}$ and eigenfunctions $\{\phi_k(x)\}$ of the operator K determined on L_2 by $K(x,y)$, by the expansions (129) and (129'). The functions Kf, K^2f, K^3f, . . . are expressible, for any L_2 function $f(x)$, by the expansions (130) and (130'). The L_2 norms of $K(x,y)$ and its iterates are given by (131) and (131'). The right sides of (129), (129'), (130), (130') are convergent in norm, but not necessarily pointwise, to the functions appearing on the left.

It is to be expected that under suitable hypotheses the series appearing in (129), (129'), (130), (130') will converge pointwise to the functions appearing on the left sides. In particular, suppose that the numbers a and b are finite and that the kernel $K(x,y)$ is continuous in the closed square $a \leq x, y \leq b$. In this case, as explained in the paragraph following (116), the eigenfunctions $\{\phi_k(x)\}$ and the function Kf [for any L_2 function $f(x)$] may be considered as continuous functions. With this understanding, we can prove that the right side of (130) converges uniformly to the left side by the following argument. For any *fixed* value of x, the Fourier coefficients of $K(x,y)$ are given by

$$\int \overline{K(x,y) \phi_k(y)}\, dy = \mu_k \overline{\phi_k(x)}$$

so that, by Bessel's inequality (16),

$$\sum_{k=1}^{\infty} \mu_k{}^2 |\phi_k(x)|^2 \leq \int |\overline{K(x,y)}|^2\, dy \qquad (133)$$

From our hypotheses, the right side of (133), and hence also the left side, is bounded for all x by some constant C. By the Schwarz inequality we now have, for any positive integers n and m, $n < m$, the estimate

$$\Big| \sum_{k=n}^{m} \mu_k \phi_k(x) \int f(y)\overline{\phi_k(y)} \, dy \Big|^2 \leq C \sum_{k=n}^{m} \Big| \int f(y)\overline{\phi_k(y)} \, dy \Big|^2 \quad (134)$$

By appealing once again to Bessel's inequality we see that the right side, and hence the left side, of (134) can be made smaller than any preassigned positive number, uniformly in x, by choosing n sufficiently large. Therefore, the right side of (130) converges *uniformly*, and since each term is a continuous function of x, the sum of the series is a continuous function, which we denote by $g(x)$. In order to show that $g(x)$ coincides with the left side of (130), we note that both of these functions are the limits, in norm, of the partial sums of the series on the right side. Since a Cauchy sequence in any normed space has at most one limit element, it follows that $g(x) - \int f(y)K(x,y) \, dy$ is a null function. Since, furthermore, the last expression is the difference between two continuous functions, it must vanish identically. Thus, we have proved that (130) holds true *uniformly pointwise* as well as in norm. The same argument holds good, obviously, for (130′). Turning to (129), it is not difficult to show by examples that it does not necessarily hold in the sense of pointwise convergence under the assumptions stated at the beginning of the present paragraph. However, we can show that (129′) holds for all the iterated kernels. For any fixed value of t we replace $f(y)$ in (130) by $K(y,t)$. The left side becomes $K_{(2)}(x,t)$ and the right side assumes the form $\sum_{k=1}^{\infty} \mu_k^2 \phi_k(x)\overline{\phi_k(t)}$. Replacing t by y, we conclude that (129′) holds, with $n = 2$, uniformly in x for each fixed value of y (and, by symmetry, uniformly in y for each fixed value of x). To show that the convergence is uniform in both variables jointly, we argue as follows. Setting $x = y$, we find that the identity

$$K_{(2)}(x,x) = \sum_{k=1}^{\infty} \mu_k^2 |\phi_k(x)|^2 \quad (135)$$

holds in the sense of *pointwise* convergence. Now, according to Dini's theorem, the convergence must be uniform. Therefore, given $\epsilon > 0$, we can find $N(\epsilon)$ such that whenever $N(\epsilon) < n < m$ the inequality

$$\sum_{k=n}^{m} \mu_k^2 |\phi_k(x)|^2 < \epsilon$$

holds for all x. By the Schwarz inequality it then follows that

$$\Big| \sum_{k=n}^{m} \mu_k^2 \phi_k(x)\overline{\phi_k(y)} \Big|^2 \leq \Big[\sum_{k=n}^{m} \mu_k^2 |\phi_k(x)|^2 \Big] \Big[\sum_{k=n}^{m} \mu_k^2 |\phi_k(y)|^2 \Big] < \epsilon^2 \quad (136)$$

for all x and y, and this inequality asserts that the convergence is uniform jointly in x and y, and also that the convergence is absolute. The same reasoning obviously applies to the higher iterates. We state the results of this paragraph as a corollary of the above theorem.

CoROLLARY. If $K(x,y)$ is hermitian and continuous in the closed square $a \leq x, y \leq b$, where a and b are finite, the expansions (129'), (130), and (130'), but not necessarily (129), hold pointwise and uniformly.

The above corollary is supplemented by the following remarkable theorem.

Theorem 18. Mercer. The above corollary holds for (129) if the eigenvalues of $K(x,y)$ are all of one sign.

Proof. Without loss of generality, we assume that the eigenvalues μ_k are all positive. From the proof of Theorem 15 we then know that, for every function $f(x)$ of L_2, the inequality

$$\iint K(x,y) f(x) \overline{f(y)} \, dx \, dy \geq 0$$

must hold. If $K(x,x)$ were negative for any value of x, we could, by continuity, determine numbers α and β such that Re $K(x,y)$ would be negative throughout the square $\alpha < x, y < \beta$. By choosing $f(x)$ as the characteristic function of the interval $\alpha < x < \beta$, we would obtain the inequality

$$\iint K(x,y) f(x) \overline{f(y)} \, dx \, dy < 0$$

in contradiction with the preceding inequality. Thus, we have shown that a continuous hermitian kernel without negative eigenvalues cannot be negative along the diagonal $x = y$. Now, the kernel

$$K(x,y) - \sum_{k=1}^{n} \mu_k \phi_k(x) \overline{\phi_k(y)}$$

is also continuous, hermitian, and without negative eigenvalues, since all its eigenvalues are obviously obtained by suppressing the first n eigenvalues of $K(x,y)$. Therefore, for any positive integer n, we have the inequality

$$K(x,x) \geq \sum_{k=1}^{n} \mu_k |\phi_k(x)|^2 \tag{137}$$

and by a passage to the limit we obtain

$$K(x,x) \geq \sum_{k=1}^{\infty} \mu_k |\phi_k(x)|^2 \tag{138}$$

Since $K(x,x)$ is bounded, we may also assert that, for some constant C and all y,

$$\sum_{k=1}^{\infty} \mu_k |\phi_k(y)|^2 < C \qquad (139)$$

Given $\epsilon > 0$ and any *fixed* value of x, choose N so large that whenever $N < n < m$ the inequality

$$\sum_{k=n}^{m} \mu_k |\phi_k(x)|^2 < \epsilon \qquad (140)$$

holds. Then for all y the inequality

$$\Big| \sum_{k=n}^{m} \mu_k \phi_k(x) \overline{\phi_k(y)} \Big|^2 \leq \Big[\sum_{k=n}^{m} \mu_k |\phi_k(x)|^2 \Big] \Big[\sum_{k=1}^{\infty} \mu_k |\phi_k(y)|^2 \Big] < \epsilon C \qquad (141)$$

will hold. Therefore the series (127) converges for all x and y, *uniformly* in y for *fixed* x, so that (127) defines a function $\breve{K}(x,y)$ continuous in y for each fixed x. Furthermore, taking account of the uniform convergence (with respect to y only), we may assert that for each function of L_2 the equality

$$\int f(y) \breve{K}(x,y) \, dy = \sum_{k=1}^{\infty} \mu_k \phi_k(x) \int f(y) \overline{\phi_k(y)} \, dy \qquad (142)$$

holds. Taking account of (130), we conclude that, for each function f of L_2 and for each value of x, the equality

$$\int f(y) L(x,y) \, dy = 0 \qquad (143)$$

holds, where $L(x,y) = \breve{K}(x,y) - K(x,y)$. Since $L(x,y)$ is continuous as a function of y, we may assert that it cannot ever be different from zero; the argument is the same as that employed earlier in this proof to show that $K(x,x) \geq 0$. Therefore $K(x,y) \equiv \breve{K}(x,y)$, and we have shown that (129) holds in the sense of pointwise convergence. Finally, to show that the convergence is uniform jointly in x and y (not merely in y for each fixed value of x), we note that, since (129) is now known to hold pointwise, the equality

$$K(x,x) = \sum_{k=1}^{\infty} \mu_k |\phi_k(x)|^2 \qquad (144)$$

must hold for each value of x. We may therefore invoke Dini's theorem, exactly as in an earlier part of the present section, to show that the right side of (129) converges uniformly to $K(x,y)$. The proof is therefore complete.

At the beginning of Sec. 6 it was indicated that there exist completely continuous operators on L_2 which cannot be expressed as $\mathbf{L}_2$ kernels. The results of the present section enable us to give an example of such an operator. There exists a converse to Theorem 15 which asserts that, given in any Hilbert space H an orthonormal sequence of elements $\{\phi_n\}$ and a sequence of nonzero real numbers $\{\mu_n\}$ converging to zero, there exists a hermitian completely continuous operator K on H expressible in the form (75). (Cf. Exercise 36.) Now, let the quantities $\{\mu_n\}$ be so chosen that $\sum\limits_{n=1}^{\infty} \mu_n{}^2$ diverges. Then, taking H as L_2 and taking account of (131), we see that K cannot be represented by means of an $\mathbf{L}_2$ kernel. In fact, the convergence of the series $\sum\limits_{n=1}^{\infty} \mu_n{}^2$ is readily seen, from the developments of the present section, to be necessary and sufficient for the existence of such a kernel.

EXERCISE

36. Prove the theorem stated in the last paragraph of the present section.

8. Illustrative Example

In this section we illustrate the foregoing theory by considering in some detail a symmetric (real hermitian) kernel for which all the required computations can be carried out explicitly.

Let the kernel $K(x,y)$ be defined as follows in the square $0 \leq x, y \leq 1$:

$$K(x,y) = \begin{cases} x(1-y) & x \leq y \\ y(1-x) & x > y \end{cases} \tag{145}$$

It is obvious that $K(x,y)$ is continuous and symmetric, so that it defines a hermitian and completely continuous operator K on $L_2(0,1)$. We shall first show that all the eigenvalues of K are different from zero. If not, there would exist a function f of positive norm such that[1]

$$\int_0^1 K(x,y)f(y)\, dy \equiv 0 \tag{146}$$

Taking account of the definition of $K(x,y)$, we rewrite (146) in the more specific form

$$\int_0^x yf(y)\, dy + x\int_x^1 f(y)\, dy = x\int_0^1 yf(y)\, dy \tag{147}$$

---

[1] The left side of (146) must vanish everywhere, without the possible exception of a null set, according to the argument presented in the paragraph following (116).

We may now, by Theorem 1-9, assert that each of the two integrals on the left side of (147) defines a function which is differentiable for almost all x, and that the derivative may be found in the usual manner; thus, the equality

$$\int_x^1 f(y)\, dy = \int_0^1 yf(y)\, dy \tag{148}$$

must hold almost everywhere. Since the left side of (148) is a continuous function of x and the right side is a constant, we conclude that (148) must hold true without exception. Setting $x = 1$, we conclude that $\int_0^1 yf(y)\, dy = 0$, and hence that

$$\int_x^1 f(y)\, dy \equiv 0 \tag{149}$$

Differentiating (149), we conclude that $f(x) = 0$ for almost all x, and hence that $\|f\| = 0$, contrary to the original assumption.

We are now assured that the eigenfunctions of $K(x,y)$ span $L_2(0,1)$. Let μ be any eigenvalue of the operator K; then, since $\mu \neq 0$, the equation $Kf = \mu f$ may be written in the form

$$f(x) = \mu^{-1}\left[\int_0^x yf(y)\, dy + x\int_x^1 f(y)\, dy - x\int_0^1 yf(y)\, dy\right] \tag{150}$$

As before, we conclude that (150) holds everywhere; differentiating twice, we obtain

$$f''(x) = -\mu^{-1}f(x) \tag{151}$$

Now, from (150) we observe that $f(0) = f(1) = 0$. Hence, μ must be such that (151) admits a nontrivial solution vanishing at both ends of the interval $0 \leq x \leq 1$. By routine methods, we find that this requirement restricts μ to the values

$$\mu = \mu_n = (n^2\pi^2)^{-1} \qquad (n = 1, 2, 3, \ldots) \tag{152}$$

and that the corresponding normalized solutions are given by

$$f_n(x) = 2^{1/2}\sin n\pi x \tag{153}$$

Conversely, by direct substitution we find that $f_n(x)$ is indeed a solution of (150). Thus, we have obtained the well-known theorem from the theory of Fourier series that the functions (153) span $L_2(0,1)$.

Since the eigenvalues are all of one sign, Mercer's theorem is applicable, and we obtain the identity

$$K(x,y) = \sum_{n=1}^\infty \frac{2\sin n\pi x \sin n\pi y}{n^2\pi^2} \tag{154}$$

The equality (131) assumes in the present case the form

$$\sum_{n=1}^{\infty} \frac{1}{n^4} = \frac{\pi^4}{90} \tag{155}$$

Now let $g(x)$ be any function in $L_2(0,1)$ and let λ be any constant. Then the integral equation

$$f(x) - \lambda \int_0^1 K(x,y) f(y) \, dy = g(x) \tag{156}$$

possesses the unique solution [cf. (108)]

$$f(x) = g(x) + \sum_{n=1}^{\infty} \frac{2\lambda\mu_n \sin n\pi x \int_0^1 g(y) \sin n\pi y \, dy}{1 - \lambda\mu_n} \tag{157}$$

if λ is distinct from all the quantities μ_n^{-1}, whereas, if $\lambda = \mu_k^{-1}$, we conclude that (156) is solvable if and only if $\int_0^1 g(y) \sin k\pi y \, dy = 0$; when this condition is satisfied, (156) possesses infinitely many solutions, each of the form

$$f(x) = g(x) + \sum_{\substack{n=1 \\ n \neq k}}^{\infty} \frac{2\lambda\mu_n \sin n\pi x \int_0^1 g(y) \sin n\pi y \, dy}{1 - \lambda\mu_n} + c \sin k\pi x \tag{158}$$

where the constant c is completely arbitrary. By rewriting (156) in the form

$$f(x) - g(x) = \lambda \int_0^1 K(x,y) f(y) \, dy$$

and appealing to Theorem 17, we see that the difference $f(x) - g(x)$ possesses a uniformly convergent Fourier expansion, and hence that equations (157) and (158) hold in the sense of pointwise, and indeed uniform, convergence, not only in the sense of convergence in norm.

EXERCISES

37. Work out the details leading to (155).

38. Derive from (154) the identity

$$\sum_{m=0}^{\infty} \frac{1}{(2m+1)^2} = \frac{\pi^2}{8}$$

39. Work out $K_{(2)}(x,y)$, and obtain the analogue of (155) for this kernel.

6. ELEMENTS OF POTENTIAL THEORY

1. Introduction

The simplest of all elliptic equations, and, in many respects, the basis for the study of more general equations of elliptic type, is the so-called Laplace equation, which in n dimensions, with rectangular coordinates $x_1, x_2, \ldots, x_n$, reads

$$u_{x_1 x_1} + u_{x_2 x_2} + \cdots + u_{x_n x_n} = 0 \tag{1}$$

A function $u(x_1, x_2, \ldots, x_n)$ possessing continuous partial derivatives of second order in a domain G and satisfying (1) at all points of G will be said to be "harmonic" there. The present chapter is devoted chiefly to a study of some of the most important properties of harmonic functions, while the following chapter contains a fairly detailed study of one particular problem arising in the theory of harmonic functions—the Dirichlet problem.

Equation (1), particularly in two and three dimensions, has been one of the most carefully studied of all partial differential equations. In addition to its interest as a purely mathematical problem, it is of great importance in mathematical physics, arising in the study of such subjects as electrostatics, fluid dynamics, elasticity, and heat conduction. Indeed, much of the stimulus for the study of Laplace's equation and many of the most important concepts that have been built up around it have had their origin in physical problems. The theory of Laplace's equation and the closely related equation of Poisson (cf. Sec. 12) is usually termed "potential theory," on account of its applicability to various physical problems involving potential fields.

We remark that the left side of (1) is, in vectorial terminology, the divergence of the gradient of the scalar field $u(x_1, x_2, \ldots, x_n)$; it is frequently denoted by the symbols $\nabla \cdot \nabla u$, $\nabla^2 u$, or Δu. We shall use the last of these three notations.

EXERCISES

1. Prove that

$$\Delta u = \frac{1}{r}(ru_r)_r + \frac{1}{r^2}u_{\theta\theta} \tag{2a}$$

and

$$\Delta u = \frac{1}{r^2}(r^2 u_r)_r + \frac{1}{r^2 \sin\theta}(\sin\theta u_\theta)_\theta + \frac{1}{r^2 \sin^2\theta}u_{\phi\phi} \tag{2b}$$

in plane and spherical polar coordinates, respectively. [In $(2b)$, θ denotes the colatitude and ϕ the azimuth.]

2. Prove that (1) is unchanged under the transformation of coordinates

$$x_i' = cx_i + d_i \qquad (i = 1, 2, \ldots, n) \tag{3}$$

where the constants c ($\neq 0$) and d_i are arbitrary.

3. Prove that (1) is unchanged under any orthogonal transformation of coordinates; i.e., a linear transformation

$$x_i' = \sum_{k=1}^{n} a_{ik}x_k \qquad (i = 1, 2, \ldots, n) \tag{4}$$

for which the equality

$$\sum_{i=1}^{n}(x_i')^2 = \sum_{i=1}^{n}x_i^2 \tag{5}$$

always holds.[1]

2. Laplace's Equation and Theory of Analytic Functions

In the two-dimensional case the theory of Laplace's equation is essentially the same as that of analytic functions of a complex variable. The connection between these two theories is established by the well-known theorem stating that the real and imaginary parts of an analytic function

$$f(z) = f(x + iy) = u(x,y) + iv(x,y) \tag{6}$$

are harmonic, and that, conversely, given any function $u(x,y)$ harmonic in a domain G, there exists a (not necessarily single-valued) function $v(x,y)$, the harmonic conjugate of u, such that the function $f(z)$ defined by (6) is analytic in G. Thus, harmonic functions are the real parts of analytic functions, and the powerful techniques of analytic function theory become available for the study of harmonic functions. For one simple, but highly interesting, illustration of the use of analytic function

[1] Exercises 2 and 3, taken together, assert that a function harmonic in a domain G remains harmonic if G is subjected to any deformation which can be composed of rigid motions, reflections, and uniform stretchings.

theory in the study of harmonic functions, we refer to Exercise 29, in which an alternative derivation of (26b) is obtained. (However, the method described in this exercise is, by its very nature, confined to the plane, but the proof given in the text of Sec. 7 is, with obvious changes, applicable in any number of dimensions.) As another illustration, we refer to Exercise 30, where the Neumann problem for a plane domain is converted into a Dirichlet problem; no comparable device exists in higher dimensions.

A striking fact about Laplace's equation in the plane is that it remains invariant under analytic (conformal) mapping. More precisely, if u is a harmonic function of x and y, and if a one-to-one analytic mapping $x + iy = g(\xi + i\eta)$ is given, then $u(x(\xi,\eta), y(\xi,\eta))$ is also harmonic as a function of ξ and η. This fact, which is basic to the use of conformal mapping in determining potential fields (cf. Sec. 7-7), is easily proved as follows. We determine the harmonic conjugate v of u and form the analytic function $f(z)$, as in (6). Since an analytic function of an analytic function is itself analytic, it follows that u, as the real part of an analytic function of $\xi + i\eta$, is harmonic in ξ and η. (An alternative proof is outlined in Exercise 4.)

An important consequence of the above result is the following. If u is harmonic in a domain G, and if the function U is defined as follows:

$$U(P^*) = u(P), \ P \in G$$

where P and P^* are inverse points (cf. Sec. 7) with respect to the unit circle, then U is harmonic in the domain G^* consisting of all points inverse to those of G. This is proved by noting that the inversion may be accomplished by following the mapping $\xi + i\eta = (x + iy)^{-1}$ with a reflection in the real axis. Since each of these transformations preserves harmonicity, so does their resultant. If G contains the origin, then we find it convenient to consider G^* as including the "point at infinity." Indeed, a function U harmonic for all sufficiently large values of $r^2 = x^2 + y^2$ is defined to be "harmonic at infinity" if the function u obtained by the inversion is harmonic at the origin. (Cf. Exercise 5.)

For the sake of brevity, theorems in this and the following chapter will be formulated in two dimensions, but in order to make clear that the theorems hold (with straightforward modifications) in higher dimensions as well, the proofs presented will not use complex-variable techniques. In studying the theorems, the reader should think through the corresponding formulations and proofs in higher dimensions.

EXERCISES

4. Prove the invariance of Laplace's equation under conformal mapping by converting differentiations with respect to x and y into differentiations with respect to

ξ and η. (Of course, the Cauchy-Riemann equations $x_\xi = y_\eta$, $x_\eta = -y_\xi$ must be exploited.)

5. Prove the invariance of Laplace's equation under inversion by performing the substitution $\bar{r} = 1/r$, $\bar{\theta} = \theta$ in (2a).

3. Fundamental Solutions

An important role in the development of potential theory is played by those solutions of the Laplace equation which depend only on the distance r from some fixed point. It is readily proved (cf. Exercise 6) that in the plane the only such solutions are of the form

$$u = a \log r + b \tag{7a}$$

whereas in $n(>2)$ dimensions the solutions have the form

$$u = ar^{2-n} + b \tag{7b}$$

where a and b are arbitrary constants. These solutions are employed in establishing the most important properties of harmonic functions. This is accomplished by the use of the divergence theorem and some of its immediate corollaries. Though these are presumably familiar, we review them here briefly for convenience.

Divergence Theorem. Let G be a bounded plane domain whose boundary Γ consists of a finite number of disjoint smooth (or sectionally smooth) simple closed curves, and let the functions f and g be defined and possess continuous first derivatives in $\bar{G}$. Then

$$\iint_G (f_x + g_y)\, dx\, dy = \int_\Gamma f\, dy - g\, dx \tag{8}$$

(The integration over the boundary is understood to be in the positive sense with respect to the domain G.)

Now, if u possesses continuous first derivatives and v continuous second derivatives in $\bar{G}$, we obtain, by setting $f = uv_x$ and $g = uv_y$, Green's first identity,

$$\iint_G (u\, \Delta v + \nabla u \cdot \nabla v)\, dx\, dy = \int_\Gamma u\, \frac{\partial v}{\partial n}\, ds \tag{9}$$

where $\frac{\partial}{\partial n}$ denotes differentiation in the direction of the *outward* normal.

If u also possesses continuous second derivatives, then it is permissible to interchange u and v in (9), and by subtraction we then obtain Green's second identity,

$$\iint_G (u\, \Delta v - v\, \Delta u)\, dx\, dy = \int_\Gamma \left(u\, \frac{\partial v}{\partial n} - v\, \frac{\partial u}{\partial n} \right) ds \tag{10}$$

Several particular cases are worthy of note. If $u \equiv 1$, (9) reduces to

$$\iint_G \Delta v \, dx \, dy = \int_\Gamma \frac{\partial v}{\partial n} \, ds \tag{11a}$$

and if, furthermore, v is harmonic, (11a) simplifies to

$$\int_\Gamma \frac{\partial v}{\partial n} \, ds = 0 \tag{11b}$$

If v is harmonic and u is set equal to v, (9) yields

$$\iint_G (v_x{}^2 + v_y{}^2) \, dx \, dy = \int_\Gamma v \frac{\partial v}{\partial n} \, ds \tag{12}$$

Finally, if u and v are both harmonic in $\bar{G}$, (10) yields

$$\int_\Gamma \left(u \frac{\partial v}{\partial n} - v \frac{\partial u}{\partial n} \right) ds = 0 \tag{13}$$

Now, let Q be any point of G, let r denote the distance from Q, let $v = \log r$, and let u be any function having continuous second derivatives in $\bar{G}$. It is not permissible to apply (10), for at the point Q the function v violates the hypotheses. However, (10) is applicable to the domain G_ϵ obtained by deleting from G a disc Δ_ϵ of radius ϵ, with center at Q, which, together with its circumference Γ_ϵ, lies entirely in G. There is obtained (since v is harmonic in G_ϵ)

$$-\iint_{G_\epsilon} \log r \, \Delta u \, dx \, dy = \int_\Gamma \left(u \frac{\partial}{\partial n} \log r - \log r \frac{\partial u}{\partial n} \right) ds$$
$$- \int_{\Gamma_\epsilon} \left(u \cdot \frac{1}{\epsilon} - \log \epsilon \frac{\partial u}{\partial n} \right) ds \quad (14)$$

From the elementary inequality

$$\left| \int_{\Gamma_\epsilon} u \, ds - 2\pi\epsilon u(Q) \right| = \left| \int_{\Gamma_\epsilon} [u - u(Q)] \, ds \right| \le 2\pi\epsilon \cdot \max_{\Gamma_\epsilon} |u - u(Q)|$$

we obtain the limiting relation

$$\lim_{\epsilon \to 0} \frac{1}{2\pi\epsilon} \int_{\Gamma_\epsilon} u \, ds = u(Q) \tag{15a}$$

Similarly, from the inequality

$$\left| \log \epsilon \int_{\Gamma_\epsilon} \frac{\partial u}{\partial n} \, ds \right| = \left| \log \epsilon \iint_{\Delta_\epsilon} \Delta u \, dx \, dy \right| \le |\log \epsilon| \cdot \pi\epsilon^2 \cdot \max_{\Delta_\epsilon} |\Delta u|$$

we obtain the limiting relation

$$\lim_{\epsilon \to 0} \log \epsilon \int_{\Gamma_\epsilon} \frac{\partial u}{\partial n} \, ds = 0 \tag{15b}$$

From (14), (15a), and (15b) we immediately conclude that

$$2\pi u(Q) = \int_\Gamma \left(\frac{\partial u}{\partial n} \log \frac{1}{r} - u \frac{\partial}{\partial n} \log \frac{1}{r} \right) ds - \iint_G \log \frac{1}{r} \Delta u \, dx \, dy \quad (16)$$

This formula, expressing the value of u at any point of G in terms of the values of u and $\frac{\partial u}{\partial n}$ on Γ and of Δu in G, is of great importance, and brings out the essential role played by the function $\log (1/r)$. This function, and in higher dimensions the functions r^{2-n}, are known as fundamental solutions of Laplace's equation.

EXERCISES

6. Prove, as asserted in the text, that all harmonic functions depending only on the distance from a fixed point are given by (7a) and (7b).

7. Prove that if G possesses a sufficiently smooth boundary Γ, v is harmonic in $\bar{G}$, and v is constant on Γ, then v is constant in $\bar{G}$. (Actually, the result holds for any bounded domain, regardless of the nature of its boundary, but the proof in the general case, which is given in Sec. 5, runs quite differently.)

8. Show that the obvious generalization of (16) holds in higher dimensions, with $\log (1/r)$ replaced by r^{2-n} and the factor 2π on the left side replaced by an appropriate constant c_n. Determine the values of c_n for $n = 3, 4, \ldots$.

4. The Mean-value Theorem

If G is a disc of radius R with center at Q and if u is harmonic in $\bar{G}$, then (16) simplifies to

$$u(Q) = \frac{1}{2\pi} \log \frac{1}{R} \int_\Gamma \frac{\partial u}{\partial n} \, ds + \frac{1}{2\pi R} \int_\Gamma u \, ds \quad (17)$$

Furthermore, the first term on the right vanishes by (11b), while the second term is simply the mean value of u over Γ. Thus, we have the interesting and important

Theorem 1. Mean-value Theorem. If u is harmonic in a disc G, including the boundary Γ, then the value of u at the center is equal to the mean value of u over the circumference; i.e.,

$$u(Q) = \frac{1}{2\pi R} \int_\Gamma u \, ds \quad (18)$$

Now, suppose only that u is harmonic in the (open) disc G and continuous in $\bar{G}$. The above argument can then be applied to a slightly smaller concentric circle Γ'. Since u is uniformly continuous in $\bar{G}$, it follows that the mean value of u over Γ is equal to the limit obtained by letting Γ' expand to Γ. Since the mean value over Γ' is always equal to

$u(Q)$, it follows that the same must be true for the mean value over Γ. This result will be stated as a

COROLLARY. Theorem 1 holds under the weaker hypotheses that u is harmonic in G and continuous in $\bar{G}$.

If we multiply both sides of (18) by $2\pi R$ and integrate with respect to R, we obtain $\pi R^2 u(Q) = \iint_G u \, dx \, dy$, or

$$u(Q) = \frac{1}{\pi R^2} \iint_G u \, dx \, dy \tag{19}$$

Thus, we have obtained the following corollary, which may be termed the "areal" mean-value theorem.

COROLLARY. If u is harmonic in a disc G and continuous in $\bar{G}$, then the value of u at the center equals the mean value of u over G.

EXERCISES

9. By differentiating Laplace's equation, one is led formally to the result that all derivatives of a harmonic function are also harmonic. However, such a procedure assumes that the derivatives of higher order exist and that the necessary changes in order of differentiation are valid. Accepting these facts (cf. Sec. 7), prove the following: If u is harmonic in a disc G of radius R with center at Q and if $|u| \leq M$ throughout G, then $|u_x(Q)| \leq 4M/\pi R$. By a slight refinement of the argument, show that $[u_x{}^2(Q) + u_y{}^2(Q)]^{1/2} \leq 4M/\pi R$.

10. Let u be harmonic in a disc G of radius R with center at Q. Let

$$\iint_G u^2 \, dx \, dy = K$$

Prove that

$$|u(Q)| \leq \frac{1}{R}\left(\frac{K}{\pi}\right)^{1/2} \tag{20a}$$

and, more generally, if P is any point of G,

$$|u(P)| \leq \frac{1}{R - r}\left(\frac{K}{\pi}\right)^{1/2} \qquad r = \overline{QP} \tag{20b}$$

Hint: Consider the integral $\iint_G [u - u(Q)]^2 \, dx \, dy$.

5. The Maximum Principle

Theorem 2. First Form of Maximum Principle. If u is harmonic in a bounded domain G and continuous in $\bar{G}$, then the maximum value of u is attained at one or more points of the boundary of G.

Proof. The existence of a point where the maximum is attained is assured by the compactness of $\bar{G}$, and it is therefore required only to show that if the maximum is attained at a point Q of G, it is also attained at a boundary point. Let R be the distance from Q to the boundary, and let Γ be the circle of radius R with center at Q. By combining the first corollary of Theorem 1 with the fact that, for every point P of $\bar{G}$, $u(Q) \geq u(P)$, we obtain

$$u(Q) = \frac{1}{2\pi R} \int_\Gamma u \, ds \leq \frac{u(Q)}{2\pi R} \int_\Gamma ds = u(Q) \qquad (21)$$

from which it follows that equality must hold between the second and third terms. Since u is continuous on Γ, the equality can hold only if, at all points P of Γ, $u(P) = u(Q)$. Since, from the definition of Γ, it has at least one point in common with the boundary of G, the theorem is proved.

A second proof, quite different from the above, will also be given. Suppose that the maximum is not attained at any boundary point. Then it must be attained at a point Q of G, and, for sufficiently small positive ϵ, the function $v(x,y) = u(x,y) + \epsilon(x - x_0)^2$ is also larger at Q than at any boundary point, x_0 denoting the abscissa of Q. Therefore, the maximum of v is attained at some point Q' of G. Since

$$\Delta v = \Delta u + \epsilon \Delta (x - x_0)^2 = 2\epsilon > 0$$

it follows that at least one of the inequalities $v_{xx}(Q') > 0$, $v_{yy}(Q') > 0$ must be satisfied; but this cannot occur at a maximum.

COROLLARY. If u is harmonic but not constant in a bounded domain G and continuous in $\bar{G}$, the maximum value of u cannot be attained at any point of G.

Proof. Suppose that the maximum value of u is attained at a point Q of G. The argument employed in the first proof of the above theorem shows that u must equal $u(Q)$ at all points of any disc with center at Q which lies entirely in G. Hence, the subset of G on which the maximum of u is attained is open. On the other hand, the subset of G where the maximum of u is *not* attained is also open (by continuity), and the latter subset is nonvacuous, since u is assumed not to be constant. Thus G would be representable as the disjoint union of two nonvacuous open sets. This implies that G is not connected, and so the proof is complete.

By applying the above theorem to $-u$, we obtain a minimum principle, which need not be stated explicitly.

Theorem 3. Given a bounded domain G and a continuous function f defined on the boundary, there exists at most one function harmonic in G, continuous in $\bar{G}$, and equal to f at all boundary points.

Proof. The difference of two such functions would be harmonic in G, continuous in $\bar{G}$, and zero at all boundary points. By the maximum and minimum principles, the difference would then vanish throughout G.

It should be stressed that Theorem 3 does not assert the existence of a harmonic function with prescribed (continuous) boundary values, only its uniqueness. However, throughout much of linear analysis uniqueness goes hand in hand with existence—recall the Fredholm alternative (Chaps. 4 and 5)—and so it is plausible to expect that the given function f can be extended continuously to $\bar{G}$ in such a manner as to be harmonic in G. We are thus led to the Dirichlet problem, which will be discussed at some length in the remainder of the present chapter and the entire following one.

EXERCISE

11. Show by an explicit computation that the function

$$u = \frac{1 - x^2 - y^2}{1 - 2x + x^2 + y^2}$$

is harmonic throughout the unit disc. Since the numerator vanishes at all points of the unit circle, the maximum and minimum principles might be thought to imply that $u \equiv 0$. Find the error in this reasoning.

6. Formulation of the Dirichlet Problem

As indicated at the end of the preceding section, Theorem 3 suggests the following problem, which has been, for over a century, the subject of intensive study.

DIRICHLET PROBLEM. Given a bounded domain G and a continuous function f defined on the boundary of G, to find a function u harmonic in G and continuous in $\bar{G}$ such that, for all points Q of the boundary, $u(Q) = f(Q)$.

The condition that G be bounded can be dropped, and the condition that f be continuous can be relaxed somewhat, but we shall concern ourselves here with the more restricted formulation, as given above.

It is worthwhile to demonstrate, by means of a simple example, that, contrary to the suggestion made in the preceding section, it is possible for some domains G to choose the function f in such a manner that the Dirichlet problem is not solvable. Let G be the domain defined by the inequalities

$$0 < x^2 + y^2 < 1 \tag{22}$$

i.e., the unit disc with the origin removed (so that the origin constitutes a component of the boundary). Let f be defined on the boundary as follows:

$$f \equiv 0 \quad \text{on } x^2 + y^2 = 1 \qquad f(0,0) = 1 \tag{23}$$

By symmetry and uniqueness, it is clear that if the problem has a solution, it must possess rotational symmetry; i.e., the solution must be a function of $r = (x^2 + y^2)^{1/2}$, independent of the polar angle θ. According to Sec. 3, the solution must have the form $(7a)$,

$$u = a \log r + b$$

The first condition of (23) requires b to vanish; on the other hand, any choice of a other than 0 would cause u to be unbounded near the origin, while the choice $a = 0$ would cause u to vanish identically, so that in either case the second condition of (23) cannot be satisfied. (Cf. Exercise 12.)

A domain having the property that the Dirichlet problem can be solved for every (continuous) function f defined on the boundary is termed a "Dirichlet domain." It will be shown in the next section, by suitably exploiting the geometrical properties of the circle, that every disc is a Dirichlet domain. Then, in the next chapter, a wide variety of domains will be shown to be Dirichlet domains; in the various proofs to be presented, it will be seen that the fact that discs possess this property plays a fundamental role. Therefore, the content of the next section, and in particular the Poisson formula (27), is of basic importance in the theory of harmonic functions.

EXERCISES

12. Solve the Dirichlet problem for an annulus bounded by concentric circles of radii $R_1 < R_2$, in the case that the assigned boundary values are $f \equiv c_1$ on the inner boundary, $f \equiv c_2$ on the outer boundary. Note what happens to the solution as $R_1 \to 0$ while c_1, c_2, and R_2 are held fixed. *Hint:* Exploit the symmetry of the annulus and of the prescribed data.

13. With the aid of the maximum principle and the solution to the preceding exercise, prove: If u is harmonic inside and on the circumference Γ of a disc G, except perhaps at the center of G, if u has a finite upper bound M in the "punctured" disc G, and if $\max\limits_{\Gamma} u(P) = m$, then $m = M$.

7. Solution of the Dirichlet Problem for the Disc

In order to demonstrate many important properties of harmonic functions, and thus prepare the way for proving the solvability of the Dirichlet problem for very general domains, it is necessary to obtain, for at least one particular type of domain, an integral representation of a function which is harmonic inside and on the boundary of the domain, in terms of its values on the boundary. As might be expected, the most convenient domain for this purpose is a disc. After such a representation is obtained, the solvability of every Dirichlet problem (with continuous boundary values) for the disc will follow almost immediately.   These results will be obtained in the present section.

Formula (16) provides, if the function u is harmonic (in which case the double integral drops out), an integral representation for u at any point of G in terms of the values of u and its normal derivative on the boundary. From this representation there follows one of the most important and striking properties of harmonic functions, namely, that a harmonic function possesses derivatives of all orders, a fact certainly not directly obvious from Laplace's equation. [The proof is immediate. If u is harmonic in a domain, one covers any given point of the domain by a disc which, together with its boundary, lies in the given domain, and then represents u throughout the disc by means of (16).   Since any number of differentiations may be carried out under the integral signs, it follows that u has derivatives of all orders in the neighborhood of the selected point, and hence throughout the given domain.]   However, (16) suffers a certain defect if it is to be used as a tool for studying harmonic functions, for, according to Theorem 3, u is completely determined throughout G by its values on the boundary, and therefore the values of the normal derivative on the boundary are themselves determined by the values of u on the boundary.   This suggests that one might expect the existence of a formula expressing u at any point of G exclusively in terms of its boundary values, not involving the normal derivatives at all.   As indicated above, such a formula does exist in the case that G is a disc and, in fact, for much more general domains, as will now be shown.

The derivation of this formula is based, in the case of a disc, on a very simple geometrical fact.   Let O be the center of a circle of radius R, let Q be any point (other than O) inside the circle, and let Q^* be the inverse point of Q (i.e., $\overline{OQ} \cdot \overline{OQ^*} = R^2$ and the points Q and Q^* lie on the same half line terminating at O).   Then, as the point S describes the circumference, the ratio $\overline{SQ^*}/\overline{SQ}$ does not vary; the constant value of this ratio is therefore equal to the value which it assumes when S coincides with T, the end point of the radius through Q (cf. Fig. 6-1), in which case it evidently equals $(R^2/\rho - R)/(R - \rho)$, or R/ρ, where $\rho = \overline{OQ}$.   It then follows from $(7a)$ that the function

$$G(P,Q) = \log \frac{\overline{OQ}}{R} \cdot \frac{\overline{PQ^*}}{\overline{PQ}} = \tfrac{1}{2} \log \frac{R^2 - 2r\rho \cos(\theta - \phi) + r^2\rho^2/R^2}{r^2 - 2r\rho \cos(\theta - \phi) + \rho^2} \quad (24)$$

(where r, θ and ρ, ϕ denote polar coordinates of P and Q, respectively) possesses the following properties:

1. For any fixed point Q in the disc, $G(P,Q)$ is harmonic throughout the disc, except at Q.

2. The function $G(P,Q) + \log \overline{PQ}$ is harmonic throughout the disc, *including* Q.

3. $G(P,Q) \to 0$ as P approaches the boundary of the disc.

The function $G(P,Q)$ is known as the "Green's function" of the disc (with pole at Q). Now, by repeating the argument which led to (16), but

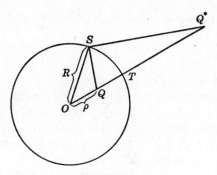

Figure 6-1

with the function $\log (1/r)$ replaced by $G(P,Q)$, we obtain

$$2\pi u(Q) = - \int_\Gamma u \frac{\partial G(P,Q)}{\partial n} \, ds - \iint_G G(P,Q) \, \Delta u \, dx \, dy \qquad (25)$$

In particular, if u is assumed harmonic in the disc (including its boundary), (25) simplifies to

$$2\pi u(Q) = - \int_\Gamma u \frac{\partial G(P,Q)}{\partial n} \, ds \qquad (26a)$$

Taking account of (24) and the fact that

$$\frac{\partial G(P,Q)}{\partial n} = \frac{\partial G(P,Q)}{\partial r}\bigg]_{r=R}$$

we find that (26a) can be written in the form

$$u(\rho e^{i\phi}) = \int_0^{2\pi} u(Re^{i\theta}) P(\rho,\phi,R,\theta) \, d\theta \qquad (26b)$$

where points are, for convenience, represented in exponential form and $P(\rho,\phi,R,\theta)$ denotes the "Poisson kernel"

$$\frac{R^2 - \rho^2}{2\pi[R^2 - 2R\rho \cos (\theta - \phi) + \rho^2]}$$

In (26b) we have a formula providing the desired representation of a function harmonic in a disc in terms of its boundary values.

We now remark, as indicated above, that (26a) is applicable for a wide class of domains. Suppose that for a domain G there exists a function $G(P,Q)$ satisfying the following obvious modifications of the above conditions 1, 2, and 3:

$1'$. For any fixed point Q of G, $G(P,Q)$ is harmonic throughout G, except at Q.

$2'$. The function $G(P,Q) + \log \overline{PQ}$ is harmonic throughout G, *including* Q, and continuous in $\bar{G}$.

$3'$. $G(P,Q) = 0$ if P lies on the boundary.

Such a function is again termed the Green's function of G. If it exists, if the boundary Γ of G is sufficiently smooth, and if $G(P,Q)$ possesses adequate differentiability properties in $\bar{G}$, then the argument leading to '(25) applies equally well to the present case, and similarly we find that (26a) holds if u is harmonic in $\bar{G}$.

It should be stressed that the derivation of (26b) does *not* in itself prove that the Dirichlet problem is solvable for the disc with arbitrary continuous boundary values; at this point (26b) only provides a method for determining the value of a harmonic function at an arbitrary point[1] of the disc from its values on the circumference. However, *if* a given continuous function $g(\theta)$ (defined on the circumference) can be extended continuously to the closed disc so as to be harmonic throughout the interior, it appears plausible that this can be accomplished by replacing $u(Re^{i\theta})$ in (26b) by $g(\theta)$. Indeed, Exercise 18 serves to show that if the Dirichlet problem can be solved for the disc, the solution must be obtainable in the manner just described. Thus we are forced, so to speak, to study the result of replacing $u(Re^{i\theta})$ by an arbitrary continuous function $g(\theta)$. This leads to the principal result of the present chapter:

Theorem 4. Let $g(\theta)$, $0 \leq \theta \leq 2\pi$, be any continuous function satisfying $g(0) = g(2\pi)$. Then the function

$$U(\rho e^{i\phi}) = \int_0^{2\pi} g(\theta) P(\rho,\phi,R,\theta) \, d\theta \tag{27}$$

is harmonic inside the circumference $\rho = R$, and for every value of θ it satisfies the condition $\lim_{\rho e^{i\phi} \to Re^{i\theta}} U(\rho e^{i\phi}) = g(\theta)$.

Proof. Since the denominator is positive for all θ and ϕ whenever $\rho < R$, the integral is meaningful. There is no difficulty in justifying differentiation under the integral sign with respect to the parameters

---

[1] Strictly speaking, (26b) was derived on the assumption that Q is not the center of the disc, but in this case (26b) gives the same result as Theorem 1, so that it is valid throughout the disc without exception.

ρ and ϕ, and so it follows that

$$\Delta U = \int_0^{2\pi} g(\theta) \left[\frac{1}{\rho} \frac{\partial}{\partial \rho} \rho \frac{\partial}{\partial \rho} + \frac{1}{\rho^2} \frac{\partial^2}{\partial \phi^2} \right] P(\rho,\phi,R,\theta) \, d\theta \tag{28}$$

(Cf. Exercise 1.) By direct computation (cf. Exercise 19) one easily shows that $P(\rho,\phi,R,\theta)$ is a harmonic function of the point $\rho e^{i\phi}$, and it follows that U is harmonic. To prove that U has the proper behavior at the circumference it clearly suffices, by symmetry considerations, to prove that

$$\lim_{\rho e^{i\phi} \to R} U(\rho e^{i\phi}) = g(0) \tag{29}$$

Now, for all ϕ and for all $\rho < R$,

$$1 = \int_0^{2\pi} P(\rho,\phi,R,\theta) \, d\theta \tag{30}$$

This can be proved, of course, by explicitly carrying out the integration, but, in fact, it is a direct consequence of (26b), with $u \equiv 1$. Multiplying both sides of (30) by $g(0)$ and subtracting from (27), we obtain

$$U(\rho e^{i\phi}) - g(0) = \int_0^{2\pi} [g(\theta) - g(0)] P(\rho,\phi,R,\theta) \, d\theta \tag{31}$$

Given any $\epsilon > 0$ we can, on account of the continuity of $g(\theta)$, choose a positive δ so small that $|g(\theta) - g(0)| < \epsilon$ for $-\delta \le \theta \le \delta$. Then, since the Poisson kernel is nonnegative, we obtain from (31) the inequality

$$|U(\rho e^{i\phi}) - g(0)| \le \epsilon \int_{-\delta}^{\delta} P(\rho,\phi,R,\theta) \, d\theta + 2M \int_{\delta}^{2\pi-\delta} P(\rho,\phi,R,\theta) \, d\theta \tag{32}$$

where $M = \max |g(\theta)|$. The first integral on the right is less than unity [by (30) and the fact that $P(\rho,\phi,R,\theta)$ is never negative]. Hence, (29) will be established if we can prove that the second integral approaches zero (for fixed δ) as $\rho e^{i\phi} \to R$. Now, for $-\delta/2 \le \phi \le \delta/2$ and $-\delta \le \theta \le 2\pi - \delta$, we have $R^2 - 2R\rho \cos(\theta - \phi) + \rho^2 \ge R^2 - 2R\rho \cos(\delta/2) + \rho^2 > [R - \rho \cos(\delta/2)]^2 > R^2[1 - \cos(\delta/2)]^2$, and hence

$$P(\rho,\phi,R,\theta) < \frac{R^2 - \rho^2}{2\pi R^2 [1 - \cos(\delta/2)]^2} \tag{33}$$

Therefore, for $-\delta/2 \le \phi \le \delta/2$,

$$\int_{\delta}^{2\pi-\delta} P(\rho,\phi,R,\theta) \, d\theta < \frac{R^2 - \rho^2}{2\pi R^2 [1 - \cos(\delta/2)]^2} \cdot (2\pi - 2\delta)$$
$$< \frac{R^2 - \rho^2}{R^2 [1 - \cos(\delta/2)]^2} \tag{34}$$

Letting $\rho \to R$, we conclude that the left side of (34) approaches zero, and the proof is complete.

Formula (27), by which the Dirichlet problem is solved for the disc, is known as the "Poisson formula."

It may be remarked that in proving (29) the continuity of $g(\theta)$ only at $\theta = 0$ was used, not the continuity of $g(\theta)$ for all θ. This suggests, as indicated in Sec. 6, the possibility of formulating in a reasonable manner a Dirichlet problem with discontinuous boundary values. We shall not discuss this possibility here, but refer to Exercise 20, which treats one particularly important case of a discontinuous boundary function.

An alternative form of (27) is obtained with the aid of the identity

$$P(\rho,\phi,R,\theta) = \frac{1}{2\pi}\left[1 + 2\sum_{n=1}^{\infty}\left(\frac{\rho}{R}\right)^n \cos n(\theta - \phi)\right] \tag{35}$$

(Cf. Exercise 22.) Since for any fixed $\rho(<R)$ the above series converges uniformly in θ and ϕ $\left[\text{being dominated by the series } 1 + 2\sum_{n=1}^{\infty}\left(\frac{\rho}{R}\right)^n\right]$,

we may multiply by $g(\theta)$ and integrate termwise, thus obtaining

$$U(\rho e^{i\phi}) = \tfrac{1}{2}a_0 + \sum_{n=1}^{\infty}\left(\frac{\rho}{R}\right)^n (a_n \cos n\phi + b_n \sin n\phi) \tag{36}$$

where $a_n = \pi^{-1}\int_0^{2\pi} g(\theta) \cos n\theta \, d\theta$, $b_n = \pi^{-1}\int_0^{2\pi} g(\theta) \sin n\theta \, d\theta$. The right side of (36) is identical with the Fourier series of $g(\phi)$, except for the appearance of the factors $(\rho/R)^n$. It should be emphasized that for $\rho = R$ the series (36) may fail to converge; that is to say, the convergence of the Fourier series of $g(\phi)$ does not enter into the derivation of (36). An important application of (36) is made in Sec. 10. Analogues of (36) exist in higher dimensions; in particular, in three dimensions one is led to expansions in spherical harmonics.

EXERCISES

14. Prove that the Green's function of a (bounded) domain is unique; i.e., there exists at most one function $G(P,Q)$ satisfying the conditions 1', 2', and 3'.

15. Prove that the Green's function must be positive throughout the domain (except at Q, where it is undefined).

16. *Symmetry of the Green's function:* Let Q_1 and Q_2 be two distinct points of a domain G. Assuming that the Green's function exists, and making appropriate assumptions about the boundary of G and the behavior of the Green's function near the boundary, prove that $G(Q_2,Q_1) = G(Q_1,Q_2)$. (The reader acquainted with the elements of electrostatics should be able to provide interesting physical interpretations of the Green's function and of the symmetry property.)

17. Work out the details leading to (26b).

18. Prove, that if u is continuous in a closed disc and harmonic in the interior (*with no differentiability assumptions on the circumference*), formula (26b) is still valid. *Hint:* Replace the disc by a slightly smaller disc and employ a continuity argument, as in the proof of the first corollary of Theorem 1.

19. Aside from a limiting argument, prove the harmonicity of the Poisson kernel without computations, by considering the manner in which it is obtained from the Green's function.

20. Let $g(\theta)$ be continuous except at $\theta = \theta_0$, and for this value let the limits from the left and right both exist. Letting $U(\rho e^{i\phi})$ be defined by (27), analyze the behavior of this function near the point $Re^{i\theta_0}$.

21. Let u be harmonic and bounded in a deleted neighborhood of a point P. Prove that u may be defined at P so as to be harmonic there. *Hint:* Use Éxercise 13, together with the fact that any disc is a Dirichlet domain.

22. Prove (35). *Hint:* Note that the right side may be rewritten in the form

$$\sum_{n=-\infty}^{\infty} \left(\frac{\rho}{R}\right)^{|n|} e^{in(\theta-\phi)}$$

23. Derive the three-dimensional analogue of $P(\rho,\phi,R,\theta)$. [This function provides the solution to the Dirichlet problem for the sphere, in complete analogy with (27).]

24. Prove that if the expansion (36) of a harmonic function is carried out twice, first in a disc of radius R and then in a concentric disc of radius R' ($\neq R$), the series obtained are identical.

25. *Poisson formula for the half plane:* Let $f(x)$ be continuous and bounded for all x. Prove that the Dirichlet problem for the half plane $y > 0$ with these boundary values is solved by the formula

$$u(x,y) = \frac{y}{\pi} \int_{-\infty}^{\infty} \frac{f(\xi)\, d\xi}{(\xi - x)^2 + y^2} \tag{37}$$

Discuss the uniqueness situation in this case. (The difference between the present case and that of the disc arises from the noncompactness of the half plane.)

26. Derive (37) from (27) by a conformal mapping.

27. Derive the three-dimensional analogue of (37).

28. Derive (26b) by exploiting the mean-value theorem in combination with a suitable conformal mapping of the disc onto itself.

29. Suppose that u is harmonic in a disc, including its circumference Γ. By extending u to an analytic function $f(z)$ and expressing $f(z)$ in terms of its values on Γ by the Cauchy integral formula, derive equation (26b). *Hint:* Set up also a second Cauchy integral with the point z replaced by its image with respect to Γ.

30. Let G be a plane domain bounded by a smooth closed curve Γ, and let a continuous function $g(s)$ be defined on Γ. The problem of determining a function u harmonic in G, continuous in $\bar{G}$, and satisfying $\dfrac{\partial u}{\partial n} \equiv g(s)$ on Γ is known as the Neumann problem. Prove that the Neumann problem is not solvable unless the condition $\displaystyle\int_{\Gamma} g(s)\, ds = 0$ is satisfied. When this condition is satisfied, show that the function u is uniquely determined up to an arbitrary additive constant. [Note that it is *not* asserted that a solution necessarily exists even when the condition $\displaystyle\int_{\Gamma} g(s)\, ds = 0$

is satisfied.] Show that the Neumann problem can be reformulated as a Dirichlet problem for the harmonic conjugate of u. (If the domain G is multiply connected, being bounded by several closed curves, the above method becomes complicated by the multiple-valuedness of the harmonic conjugate, but this difficulty can be overcome.)

8. The Converse of the Mean-value Theorem

Theorem 5. Let u be continuous in a domain D and let it possess the mean-value property; i.e., for every closed disc lying in D the mean value of u over the circumference is equal to the value of u at the center. Then u is harmonic in D.

Proof. Select any disc G which, together with its boundary Γ, lies in D. As was proved in Sec. 7, there exists a function v continuous in $\bar{G}$, harmonic in G, and coinciding with u on Γ. Then the function $w = u - v$ is continuous in $\bar{G}$, vanishes identically on Γ, and clearly possesses the mean-value property in G. The argument employed in the proof of Theorem 2 shows that w must assume its maximum and minimum values on Γ; hence w vanishes identically. Thus u coincides with the harmonic function v in G, and is therefore harmonic in G. Since every point of D can be covered by such a disc G, u is harmonic throughout D.

It may be pointed out that, as is readily seen from the above proof, it suffices to assume that for each point P of D there exists a positive number $\rho(P)$ such that the mean value of u over every circle with center at P and radius less than $\rho(P)$ is equal to $u(P)$. On the other hand, Exercise 31 shows that it does not suffice to assume merely that, for each point P, there exists *some* circle with center at P over which the mean value of u is equal to $u(P)$.

It is remarkable that no differentiability assumption is imposed on u; it turns out that the mean-value property (which is a property involving integration) suffices (together with continuity) to assure that u is harmonic, and hence indefinitely differentiable. On the other hand, the fact that the disc is known to be a Dirichlet domain plays an essential role in the above proof. It is instructive to present an alternative proof which does not make use of this fact. We begin by noting that u possesses the areal mean-value property, for the proof of the second corollary of Theorem 1 uses only the fact that u possesses the "circumferential" mean-value property (and is continuous). Given any point P of D, with coordinates x_0, y_0, let R_1 denote the distance from P to the boundary, and let G denote a disc with center at P and with radius $R < R_1$; similarly, for each point P_h with coordinates $x_0 + h$, y_0 let G_h denote the disc of radius R with center at P_h. Since both mean-value relations

$$\pi R^2 u(P) = \iint_G u\,dx\,dy \qquad \pi R^2 u(P_h) = \iint_{G_h} u\,dx\,dy \qquad (38)$$

hold, it follows that $\pi R^2[u(P_h) - u(P)]$ can be expressed as the difference between the integrals of u taken over the two lunes formed by G and G_h, as shown in Fig. 6-2. Taking account of the uniform continuity of u in any compact subset of D and the fact that each lune is of constant width

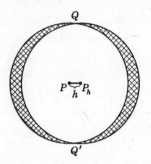

Figure 6-2

h, except for the small portions above Q and below Q', it follows readily that

$$\lim_{h \to 0} \left[\pi R^2 \frac{u(P_h) - u(P)}{h} - \int_\Gamma u \, dy \right] = 0 \tag{39}$$

where Γ denotes the circumference of G. Therefore, it follows that $u_x(P)$ exists and is given by

$$u_x(P) = \frac{1}{\pi R^2} \int_\Gamma u \, dy \tag{40}$$

Since the same representation holds for all points sufficiently close to P with a circle of the same size as Γ, it follows from the (uniform) continuity of u that u_x not only exists, but is continuous, in a neighborhood of P; since P is chosen arbitrarily, u_x (and similarly u_y) exists and is continuous throughout D. It is therefore legitimate to employ (8), with $f = u$, $g = 0$. We may therefore rewrite (40) in the form

$$u_x(P) = \frac{1}{\pi R^2} \iint_G u_x \, dx \, dy \tag{41}$$

Thus, we have shown that the first derivatives of u exist, are continuous, and possess the (areal) mean-value property. The same reasoning as that employed above now shows that the second derivatives also exist and are continuous throughout G. (In fact, they also possess the mean-value property, and by repeatedly applying the above argument we can show that u possesses derivatives of all orders. However, we shall not use this fact here.) Thus, the quantity Δu is defined and continuous throughout

D. Suppose that at some point P of D, which for convenience we take as the origin, $\Delta u \neq 0$. We express u in the neighborhood of P in the form

$$u(x,y) = u(0,0) + u_x(0,0)x + u_y(0,0)y + \tfrac{1}{2}u_{xx}(0,0)x^2 + u_{xy}(0,0)xy$$
$$+ \tfrac{1}{2}u_{yy}(0,0)y^2 + \eta(x,y)(x^2 + y^2) \quad (42)$$

where $\eta(x,y) \to 0$ as $x,y \to 0$. Integrating over the disc G defined by the inequality $x^2 + y^2 < R^2$, we obtain, by elementary computations,

$$\iint_G u \, dx \, dy - \pi R^2 u(0,0) = \frac{\pi R^4}{8} \left[u_{xx}(0,0) + u_{yy}(0,0) \right]$$
$$+ \iint_G \eta(x,y)(x^2 + y^2) \, dx \, dy \quad (43)$$

By hypothesis, the left side vanishes. Therefore,

$$\pi R^4[u_{xx}(0,0) + u_{yy}(0,0)] = -8 \iint_G \eta(x,y)(x^2 + y^2) \, dx \, dy \quad (44)$$

The absolute value of the right side of (44) is at most equal to

$$8 \cdot \max_{(x,y)\in G} |\eta| \cdot \iint_G (x^2 + y^2) \, dx \, dy = 4\pi R^4 \max_{(x,y)\in G} |\eta|$$

If we choose R so small that $4|\eta| < |u_{xx}(0,0) + u_{yy}(0,0)|$ throughout G, we obtain a contradiction. Thus, Δu must vanish everywhere, and the proof is complete.

It should be remarked that there frequently arise in the study of elliptic equations generalizations of the areal mean-value property, which are used to establish, as was done in the second proof of the above theorem, that the functions under consideration possess adequate differentiability properties.

The above theorem leads immediately to an important and striking method by which the domain in which a harmonic function is defined can be enlarged.

COROLLARY. SCHWARZ REFLECTION PRINCIPLE. Let u be harmonic in a domain D, part of whose boundary consists of a segment of a straight line L, and let D lie entirely on one side of L. If u approaches zero at all points of the segment, then the definition of u may be extended across the segment by reflection. More precisely, for each point P of D let P^* be the image of P with respect to L, and let $u(P^*) = -u(P)$. Then u is harmonic in the domain consisting of D, its reflection D^*, and the segment. (Of course, u is assigned the value zero on the segment.)

Proof. By symmetry and the fact that u possesses the property in D, it follows that u also possesses the mean-value property in D^*, while it is obvious that u also possesses the mean-value property at all points of the segment. Since u is obviously continuous throughout the enlarged domain, it follows from Theorem 5 that u is harmonic there.

EXERCISES

31. Let z_0 $(= a + ib)$ denote any root of the equation $J_0(z) = 1$, where $J_0(z)$ denotes Bessel's function of first kind and zero order, and let

$$u(x,y) = e^{bx} \cos ax \qquad (45)$$

Prove that the mean value of u over any circle of radius one is equal to the value of u at the center. *Hint:* Make use of the integral representation

$$J_0(z) = \frac{1}{2\pi} \int_0^{2\pi} \cos (z \cos \theta) \, d\theta \qquad (46)$$

[Note that for any root z_0 other than zero, $u(x,y)$ is not harmonic, thus showing, in contrast to Theorem 5, that a function possessing the "one-circle mean-value property" is not necessarily harmonic. The existence of infinitely many roots of the above equation follows easily from some of the earliest results in the theory of entire functions.]

32. In contrast to the above exercise, prove that a function which is continuous in $\bar{D}$, where D is a bounded Dirichlet domain, and possesses the "one-circle mean-value property" must be harmonic in D.

9. Convergence Theorems

A basic difficulty in many parts of analysis is that important properties may not be preserved under limiting operations. For example, the limit of a sequence of continuous functions need not be continuous; some additional hypothesis is required, the one most frequently used being, of course, that of uniform convergence. Going one step further, the convergence of the sequence may be uniform, each function of the sequence may be continuously differentiable, and yet the limit function may be nowhere differentiable. A remarkable contrast to this state of affairs is furnished by the theory of harmonic functions. It will be shown in this section that seemingly mild hypotheses turn out to assure that the limit of harmonic functions is itself harmonic, and that termwise differentiation of a sequence of harmonic functions is permissible.

Theorem 6. Harnack. If the functions w_n are harmonic in a bounded domain G, continuous in $\bar{G}$, and uniformly convergent on the boundary of G, then the sequence $\{w_n\}$ is uniformly convergent throughout $\bar{G}$, and the limit function w is harmonic in G. If δ denotes any differentiation $\left(\text{e.g., } \dfrac{\partial^3}{\partial x^2 \partial y} \right)$, then the sequence $\{\delta w_n\}$ converges throughout G to δw, uniformly in every compact subset.

Proof. For any $\epsilon > 0$, the hypothesis of uniform convergence assures the existence of $N(\epsilon)$ such that

$$|w_n(Q) - w_m(Q)| < \epsilon \qquad (n,m > N) \qquad (47)$$

at all points Q of the boundary of G. By the maximum principle (Sec. 5), inequality (47) holds at all points Q of G as well; this proves the first assertion of the theorem. Now choose any disc Δ in G, and represent each function w_n of the given sequence at any point Q of Δ by the Poisson formula (26b), the integral being extended over the circumference of Δ. Since the sequence $\{w_n\}$ is uniformly convergent and the Poisson kernel is bounded (for each fixed point Q) on the circumference of Δ, the passage to the limit may be carried out under the integral sign. The limit function w is thus represented as a Poisson integral and hence, by Sec. 7, is harmonic in Δ; since each point of G can be covered by such a disc, w is harmonic throughout G. Finally, by applying the operator δ under the integral sign in the Poisson formula, it can easily be shown that the sequence $\{\delta w_n\}$ converges, uniformly in any smaller disc Δ', to $\{\delta w\}$ (cf. Exercise 33). The Heine-Borel theorem then guarantees that the convergence is uniform in every compact subset of G.

It is perhaps instructive to give an alternative proof that w is harmonic. The uniform convergence of the sequence $\{w_n\}$ in G, together with the mean-value property of the harmonic functions w_n, shows that the function w is continuous in G and possesses the mean-value property. By Theorem 5, w is harmonic in G.

Theorem 7. Harnack's Inequalities. Let u be harmonic and non-negative in a disc Δ of radius R with center at Q. For any point P of Δ, the following inequalities hold:

$$\frac{R - r}{R + r}\, u(Q) \leq u(P) \leq \frac{R + r}{R - r}\, u(Q) \qquad (r = \overline{QP}) \qquad (48)$$

Proof. First, suppose that u is continuous in $\bar{\Delta}$, so that $u(P)$ can be expressed by the Poisson integral extended over the circumference of Δ. By integrating the obvious inequalities [which are not correct if $g(\theta) < 0$]

$$\frac{(R^2 - r^2)g(\theta)}{2\pi(R + r)^2} \leq P(r,\phi,R,\theta)g(\theta) \leq \frac{(R^2 - r^2)g(\theta)}{2\pi(R - r)^2} \qquad (49)$$

and then taking account of Theorem 1 and the Poisson formula (26b), one obtains (48).

If no assumption is made concerning the behavior of u on the boundary of Δ, one carries out the above argument in a disc of radius R' (where $r < R' < R$) and then lets R' approach R.

Theorem 8. Harnack's Theorem of Monotone Convergence.
If the functions w_n are harmonic throughout a domain G, and if at each
point P of G the sequence $\{w_n(P)\}$ is monotone-nondecreasing, then the
sequence $\{w_n\}$ either diverges (to $+\infty$) at all points of G or converges at
all points of G. In the latter case the convergence is uniform in every
compact subset of G, so that (by Theorem 6) the limit function is harmonic in G.

Proof. Suppose that the sequence $\{w_n\}$ does converge at some point
Q of the domain G. Let R denote the distance from Q to the boundary
of G, and let P be any point such that $\overline{QP} = r < R$. Given $\epsilon > 0$, we
can, on account of the convergence of the (numerical) sequence $\{w_n(Q)\}$,
choose N so large that for $n > m > N$ the following inequality holds:

$$0 \le w_n(Q) - w_m(Q) < \epsilon \frac{R - r}{R + r} \tag{50}$$

(The left-hand inequality holds on account of the monotonicity of the
sequence $\{w_n\}$.) By the right-hand inequality of (48), we obtain

$$0 \le w_n(P) - w_m(P) \le \frac{R + r}{R - r} [w_n(Q) - w_m(Q)] < \epsilon \tag{51}$$

which shows that the sequence $\{w_n\}$ is convergent at P. Thus, the
sequence $\{w_n\}$ converges in the largest disc with center at Q which lies
entirely in G, and clearly the convergence is uniform in any smaller disc.

Now, suppose that there exists a point Q' of G where the sequence $\{w_n\}$
diverges. Then the sequence must also diverge at any point P' whose
distance from Q' is less than $R'/2$, R' denoting the distance from Q' to the
boundary, for the argument employed in the preceding paragraph shows
that convergence at P' would imply convergence at Q'. (The distance
from P' to the boundary will exceed $R'/2$, and hence Q' is closer than any
boundary point to P'.) Thus, we have shown that the subsets of G on
which the sequence $\{w_n\}$ converges and diverges, respectively, are both
open. As in the proof of the corollary of Theorem 2, we conclude that the
sequence converges either nowhere or everywhere. Finally, the assertion
concerning uniform convergence follows from the Heine-Borel theorem.

EXERCISES

33. Justify in detail the termwise differentiation of the sequence $\{w_n\}$, which was
passed over in the proof of Theorem 6.

34. Obtain an analogue of the inequalities (48) in three dimensions.

35. Prove that a function harmonic in the whole plane and bounded either above
or below reduces to a constant. (This theorem is essentially the same as Liouville's
theorem in the theory of analytic functions.)

10. Strengthened Form of the Maximum Principle

The following theorem provides an interesting and important extension of Theorem 2 and its corollary.

Theorem 9. Let u be harmonic in a domain G, and let u assume a local maximum at some point Q of G; i.e., for some $\delta > 0$, $u(P) \leq u(Q)$ whenever $\overline{QP} < \delta$. Then u is constant throughout G.

Proof. Without loss of generality, it may be assumed that $u(Q) = 0$. Let Δ_1 be the disc of radius δ with center at Q. By the afore-mentioned corollary, $u \equiv 0$ in Δ_1. Let R be the distance from Q to the boundary of G, and let Δ_2 be the disc of radius R with center at Q. Then the Fourier series representation (36) of u obtained for the disc Δ_1 must be valid in the larger disc Δ_2 as well (cf. Exercise 24). However, all the coefficients obtained with the smaller disc obviously vanish; hence $u \equiv 0$ in Δ_2. Now let Q' be any point of G other than Q. Connect Q to Q' by a simple arc C. Since C is compact, its distance from the boundary of G is a positive quantity, which we denote by ϵ. Now, we select a finite number of points $P_1, P_2, \ldots, P_n$ on C such that each of the distances $\overline{QP_1}$, $\overline{P_1P_2}, \ldots, \overline{P_{n-1}P_n}, \overline{P_nQ'}$ is less than ϵ. Then, by the construction, u vanishes identically in some neighborhood of P_1, and therefore, by the above argument, at least in the disc of radius ϵ with center at P_1. Since P_2 lies in this disc, we may carry out the same argument again. Thus, in a finite number of steps, we conclude that $u(Q') = 0$, completing the proof.

COROLLARY. UNIQUE CONTINUATION PROPERTY OF HARMONIC FUNCTIONS. If u and v are harmonic in a domain G and if $u \equiv v$ throughout a subdomain G', then $u \equiv v$ throughout G.

Proof. Apply Theorem 9 to the function $u - v$, taking as the point Q any point of the subdomain G'.

11. Single and Double Layers

Equation (16), which is of great importance in potential theory, admits a striking physical interpretation which we shall now discuss briefly.

Letting r denote, as usual, the distance from a fixed point Q in the plane, the function $u = m \log (1/r)$ represents the potential created by placing a mass[1] m at Q; i.e., the force exerted on a unit (positive) mass at any point P (distinct from Q) is given by the negative gradient[2] of u. [In

---

[1] Frequently one speaks of (electric) *charge* rather than mass, since both positive and negative values of m are to be allowed. However, we prefer to speak of mass.

[2] The gradient is understood to be the "field gradient"; i.e., the vector whose components are the derivatives with respect to the coordinates of P. The "source gradient," whose components are the derivatives with respect to the coordinates of Q, is evidently the negative of the field gradient.

the three-dimensional case, which is of more immediate physical significance, log $(1/r)$ is replaced by $1/r$.] If the mass, instead of being concentrated at a single point, is continuously distributed over a domain D with density f, the potential is given by

$$u(x,y) = \iint_D f(\xi,\eta) \log \frac{1}{r} \, d\xi \, d\eta \qquad r = [(\xi - x)^2 + (y - \eta)^2]^{\frac{1}{2}} \quad (52)$$

Similarly, if the mass is distributed along a (sufficiently smooth) curve Γ with linear density (mass per unit length) σ, the potential is given by

$$u(x,y) = \int_\Gamma \sigma(s) \log \frac{1}{r} \, ds \tag{53}$$

The function $u(x,y)$ defined by (53) is known as a "single-layer" potential. We now proceed to define a "double-layer" potential. Let masses $-m$ and $+m$ be placed at points Q and Q', respectively. Now let Q' approach Q along a fixed direction $\mathbf{n}$, and let m increase according to the rule

$$m \cdot \overline{QQ'} = \text{constant} = M \tag{54}$$

The potential at a point P is given, in the limit, by

$$u(P) = u(x,y) = M \frac{\partial \log (1/r)}{\partial n_Q} \tag{55}$$

where $\dfrac{\partial}{\partial n_Q}$ denotes differentiation with respect to Q in the direction $\mathbf{n}$; this is, of course, the component of the source gradient in the direction $\mathbf{n}$. The configuration of masses just described is termed a "dipole" of strength M. Now let dipoles be distributed continuously with density τ along the curve Γ, the direction of the dipole at each point being normal to Γ. In this manner we obtain a double layer, which gives rise to a potential

$$u(P) = u(x,y) = \int_\Gamma \tau(s) \frac{\partial \log (1/r)}{\partial n_Q} \, ds_Q \tag{56}$$

By taking account of (52), (53), and (56), we obtain for (16) the following physical interpretation: Every function u defined and sufficiently differentiable in a domain having a sufficiently smooth boundary can be expressed as the sum of three potentials; one produced by a mass of density $-(2\pi)^{-1} \Delta u$ distributed over D, one produced by a single layer of density $(2\pi)^{-1} \dfrac{\partial u}{\partial n}$ distributed on the boundary of D, and one produced by a double layer of density $u/2\pi$, also distributed on the boundary of D. If, in particular, u is harmonic in D, it can be expressed as the sum of a single-layer and a double-layer potential.

Conversely, if the function f is selected more or less arbitrarily, equation (52) defines, as will be shown in the following section, a function in the entire plane which is harmonic in the complement of $\bar{D}$ and which satisfies at any point of D the Poisson equation

$$\Delta u = u_{xx} + u_{yy} = -2\pi f \tag{57}$$

Similarly, if σ and τ are chosen arbitrarily,[1] then (53) and (56) are easily seen to define functions harmonic everywhere in the plane except on Γ; this is easily proved by differentiating each of these equations under the integral sign. Of particular importance is the question, to which we now turn, of the behavior of these potentials on and near Γ.

A simple example, in which all the calculations can easily be carried out in closed form, will help to suggest the general situation. Let a single layer of constant density (which may be taken, for convenience, equal to unity) be deposited on the line segment $-1 \leq x \leq 1$, $y = 0$. The potential arising from this mass distribution is given by

$$u(x,y) = -\frac{1}{2} \int_{-1}^{1} \log \left[(\xi - x)^2 + y^2 \right] d\xi \tag{58}$$

For any point (x,y) not lying on the line segment this integral works out to

$$u(x,y) = \frac{x-1}{2} \log \left[(x-1)^2 + y^2 \right] - \frac{x+1}{2} \log \left[(x+1)^2 + y^2 \right]$$
$$+ 2 - y \left(\arctan \frac{x+1}{y} - \arctan \frac{x-1}{y} \right) \tag{59}$$

(The principal branch of the inverse tangent is understood, and for points on the x axis, but not on the segment containing the mass, the last term is understood to vanish.) For any point lying on the segment the integral (58) becomes improper, but convergent, and works out to

$$u(x,0) = (x-1) \log (1-x) - (x+1) \log (x+1) + 2$$
$$[u(\pm 1,0) = 2 - 2 \log 2] \tag{60}$$

This result could have been obtained *formally* by setting $y = 0$ in (59), but doing so is tantamount to assuming that the potential remains continuous at the points occupied by the mass distribution. It must be emphasized that (60) can actually be obtained by *first* setting $y = 0$ in the integrand appearing in (58), and *then* evaluating the resulting improper integral. In this manner, the potential $u(x,y)$ is shown to exist and to be continuous everywhere in the plane, even at points occupied by mass.

[1] Subject, of course, to mild conditions which suffice to assure that the integrals in question are meaningful.

Next let us consider the gradient of the potential u.   For all points not lying on the segment we obtain from (59)

$$u_x = \tfrac{1}{2} \log \frac{(x-1)^2 + y^2}{(x+1)^2 + y^2} \tag{61a}$$

$$u_y = \tfrac{1}{2} \arctan \frac{x-1}{y} - \tfrac{1}{2} \arctan \frac{x+1}{y} \qquad (u_y = 0 \text{ for } y = 0,\ |x| > 1) \tag{61b}$$

From (60) we obtain

$$u_x(x,0) = \log \frac{1-x}{1+x} \qquad (-1 < x < 1) \tag{62}$$

which coincides with (61a) when y is replaced by zero.   Therefore, u_x is continuous over the entire plane, except at the end points of the segment on which the mass is deposited.   The behavior of u_y is quite different; if we choose any sequence of points converging to a point $(x,0)$ of the segment (the end points excepted) *from above*, we obtain

$$\lim u_y = -\pi \qquad (y \to +0,\ -1 < x < 1) \tag{63a}$$

but, in contrast to this, if the sequence of points converges to $(x,0)$ *from below*, we obtain

$$\lim u_y = +\pi \qquad (y \to -0,\ -1 < x < 1) \tag{63b}$$

Now we consider the potential corresponding to a *double* layer of constant unit density on the same segment.   This potential may be expressed by the integral

$$v(x,y) = -\frac{1}{2} \int_{-1}^{1} \frac{\partial}{\partial \eta_{\eta=0}} \log\left[(\xi - x)^2 + (y - \eta)^2\right] d\xi \tag{64}$$

The integral can, of course, be computed explicitly, but it is worth noting that, since $\left[\dfrac{\partial g(y - \eta)}{\partial \eta}\right]_{\eta=0} = -g'(y)$, (64) can be written in the alternative form (valid for points not on the segment)

$$v(x,y) = \frac{1}{2} \frac{\partial}{\partial y} \int_{-1}^{1} \log\left[(\xi - x)^2 + y^2\right] d\xi \tag{65}$$

Therefore, $v(x,y)$ must coincide, except for sign, with the right side of (61b).   Thus,

$$v(x,y) = \tfrac{1}{2} \arctan \frac{x+1}{y} - \tfrac{1}{2} \arctan \frac{x-1}{y} \tag{66}$$

Comparison with (63) immediately shows that v, like u_y, approaches limits from above and below which differ by 2π.   For any point on the segment, the integral (64) exists and has the value zero.

The components of the gradient are given by

$$v_x = \frac{1}{2}\left[\frac{y}{(x+1)^2 + y^2} - \frac{y}{(x-1)^2 + y^2}\right] \qquad (67a)$$

$$v_y = \frac{1}{2}\left[\frac{x-1}{(x-1)^2 + y^2} - \frac{x+1}{(x+1)^2 + y^2}\right] \qquad (67b)$$

Again it should be emphasized that these formulas have been derived for points not lying on the segment. If we set $y = 0$ in (67a), we obtain (formally) $v_x = 0$, in agreement with the previously mentioned fact that v vanishes at all inner points of the segment. More important, however, is the fact that, as the point (x,y), $y \neq 0$, approaches any inner point of the segment, v_y approaches a limit, namely, $1/(x^2 - 1)$, *independent* of the manner of approach (from above, from below, or "mixed").

We sum up the results as follows: For the single layer, the potential exists and is continuous everywhere, the tangential component of the gradient exists on the segment (except at the end points) and is continuous, but the normal component changes by 2π as the segment is crossed. For the double layer, both components of the gradient vary continuously as the segment is crossed (at an inner point), but the potential changes by 2π, and on the segment it has a value equal to the mean of the limits from the two sides.

These relationships were found by explicit computation. It could again be shown by explicit computation that the numerical values 2π of the two kinds of discontinuities that were encountered are independent of the length of the segment, but it is more instructive to see this, without any computation, as follows. Take any segment s_1, and select a portion s_2 of it. The potential arising from a layer, either single or double, on s_1 can be divided into the sum of the potentials arising from s_2 and $s_1 - s_2$, respectively. As the segment is crossed at any (inner) point of s_2, the potential arising from $s_1 - s_2$ behaves regularly, so that any departure from regularity of the entire potential is reflected in the same behavior in the potential arising from s_2; i.e., any "singular" behavior of the potential can be ascribed to that part of the potential arising from any segment, however small, containing the point where the original segment is crossed. Thus the value 2π which was obtained is independent of the length of the segment. Finally, we remark that if the density of the layer is an arbitrary constant σ, the discontinuity obtained is simply $2\pi\sigma$.

The argument used in the last paragraph is obviously restricted neither to straight-line segments nor to layers of uniform density. Any potential arising from a single- or double-layer distribution on a curve must be harmonic, and hence differentiable infinitely often, at all points off the curve, and any discontinuities in the potential or its derivatives occurring when the curve is crossed can be ascribed entirely to the portion of the curve *restricted to an arbitrarily small neighborhood* of the crossing point. Furthermore, it appears plausible that the behavior of the potential in the immediate vicinity of a point on the layer is essentially unaltered by approximating the layer by a straight-line segment with uniform density. This last idea is rendered more precise in the following theorem, whose proof is left to the reader.

Theorem 10. Let a single or double layer of continuous density σ be distributed along a smooth arc Γ [i.e., one which can be parametrized in terms of arc length s by functions $x(s)$, $y(s)$ having continuous first derivatives]. Then, in the vicinity of any inner point P of Γ, the potential exhibits the following behavior:

Single Layer. Potential and tangential components of gradient are continuous; normal component changes by $2\pi\sigma(P)$ across Γ.

Double Layer. Potential changes by $2\pi\sigma(P)$ across Γ; value of potential at P is equal to mean of limits from both sides; tangential and normal components of gradient are both continuous.

EXERCISES

36. Prove that the potential produced by a double layer of *uniform* density σ distributed on any (sufficiently smooth) simple closed curve Γ is equal, at any point P, to 0, $\pi\sigma$, or $2\pi\sigma$, according as P lies outside, on, or inside Γ. (The density is considered positive if the positive masses face inward.)

37. Prove, as a simple corollary of the above exercise, that the potential arising from a double layer of uniform density σ distributed along an open curve Γ is equal to $\sigma\theta$, where θ is the angle subtended at the "field point" by Γ. (Discuss carefully the ambiguity in measuring θ, and also the value of the potential when the field point lies on Γ.)

12. Poisson's Equation

In this section we shall study some of the properties of potentials of the form (52). It is understood that the domain D is bounded, and the density $f(\xi,\eta)$ is both continuous and bounded; these restrictions can be somewhat relaxed without much difficulty, but we shall not concern ourselves with this matter. The integral (52) obviously converges for all points (x,y) outside $\bar{D}$, and $u(x,y)$ may be differentiated any number of times under the integral sign; in this way it is established, in particular, that u is harmonic. When the point P lies in $\bar{D}$, the integral becomes improper, and a more detailed analysis is required. The absolute convergence of the integral follows from the inequalities

$$\iint_D \left| f(\xi,\eta) \log \frac{1}{r} \right| d\xi \, d\eta \le M \iint_\Delta \left| \log \frac{1}{r} \right| d\xi \, d\eta < \infty \qquad (68)$$

[Here, and also later, $M = \sup |f(\xi,\eta)|$, and Δ denotes a disc with center at P which contains D.] The continuity of the potential can also be easily proved directly, but will instead be established later as a consequence of the differentiability properties of $u(x,y)$.

Formally, the first derivative u_x is obtained, as in the case that (x,y) is outside $\bar{D}$, by differentiating the right side of (52) under the integral sign. In this manner, we are led to the formula

$$u_x = \iint_D \frac{f(\xi,\eta)(\xi - x)}{(\xi - x)^2 + (\eta - y)^2} \, d\xi \, d\eta \tag{69}$$

The first step in justifying (69) is to establish the convergence of the integral; this follows easily from the inequalities

$$\iint_D \left| \frac{f(\xi,\eta)(\xi - x)}{(\xi - x)^2 + (\eta - y)^2} \right| d\xi \, d\eta$$

$$\leq M \iint_\Delta [(\xi - x)^2 + (\eta - y)^2]^{-\frac{1}{2}} \, d\xi \, d\eta < \infty \tag{70}$$

Now we select an arbitrary point, which without loss of generality we take as the origin. From (52) we obtain, for any $h \neq 0$, the difference quotient

$$h^{-1}[u(h,0) - u(0,0)] = -\frac{1}{2} \iint_D f(\xi,\eta) h^{-1} \log \frac{(\xi - h)^2 + \eta^2}{\xi^2 + \eta^2} \, d\xi \, d\eta \tag{71}$$

The correctness of (69) is therefore equivalent to the limiting relationship

$$\lim_{h \to 0} \iint_D f(\xi,\eta) \left[\frac{2\xi}{\xi^2 + \eta^2} + h^{-1} \log \frac{(\xi - h)^2 + \eta^2}{\xi^2 + \eta^2} \right] d\xi \, d\eta = 0 \tag{72}$$

The above integral is dominated by

$$M \iint_\Delta \left| \frac{2\xi}{\xi^2 + \eta^2} + h^{-1} \log \frac{(\xi - h)^2 + \eta^2}{\xi^2 + \eta^2} \right| d\xi \, d\eta \tag{73}$$

Disregarding the factor M and introducing polar coordinates ($\xi = hr \cos \theta$, $\eta = hr \sin \theta$), we are led to the integral

$$|h| \int_0^{2\pi} \int_0^{R/|h|} \left| 2 \cos \theta + r \log \frac{r^2 - 2r \cos \theta + 1}{r^2} \right| dr \, d\theta \tag{74}$$

(R = radius of Δ.) Since h is to approach zero, we may split the above expression into two parts:

$$|h| \int_0^{2\pi} \int_0^3 + |h| \int_0^{2\pi} \int_3^{R/|h|} \tag{75}$$

The first term vanishes as $h \to 0$, since the integrand does not involve h. To estimate the second term, we apply the extended mean-value theorem to the function $\log (1 - \alpha)$, obtaining (for any real value of $\alpha < 1$)

$$\log (1 - \alpha) = -\alpha - \frac{\alpha^2}{2(1 - \gamma\alpha)^2} \tag{76}$$

where γ denotes some number between zero and unity. Setting

$$\alpha = r^{-2}(2r \cos \theta - 1)$$

and noting that $|\alpha| \leq \frac{7}{9}$ for $r \geq 3$ [so that (76) is applicable], we obtain

$$\left| 2 \cos \theta + r \log \frac{r^2 - 2r \cos \theta + 1}{r^2} \right| = \left| \frac{1}{r} - \frac{r\alpha^2}{(1 - \gamma\alpha)^2} \right| \leq \frac{1}{r} + \frac{r\alpha^2}{(1 - \frac{7}{9})^2} \tag{77}$$

Note that we have employed the afore-mentioned inequality for $|\alpha|$ only in the denominator; in the numerator we now employ the stronger inequality $|\alpha| \leq 7/3r$, and thus obtain

$$\left| 2 \cos \theta + r \log \frac{r^2 - 2r \cos \theta + 1}{r^2} \right| \leq \frac{C}{r} \tag{78}$$

where C is a certain constant whose value is of no importance. Thus, the second term of (75) is dominated by

$$|h| \int_0^{2\pi} \int_3^{R/|h|} \frac{C}{r} \, dr \, d\theta = 2\pi C |h| \log \frac{R}{3|h|} \tag{79}$$

Since the last expression approaches zero with h, we have established both the existence of u_x and the validity of formula (69).

The continuity of the first derivatives can also be easily established. Since, as explained above, the potential is harmonic outside $\bar{D}$, it suffices to establish the continuity at an arbitrary point P of $\bar{D}$; without loss of generality, we may assume that P is located at the origin. Then, from the equation

$$u_x(h,k) - u_x(0,0) = \iint_D f(\xi,\eta) \left[\frac{\xi - h}{(\xi - h)^2 + (\eta - k)^2} - \frac{\xi}{\xi^2 + \eta^2} \right] d\xi \, d\eta \tag{80}$$

and the analogous equation for $u_y(h,k) - u_y(0,0)$, we obtain

$$|[u_x(h,k) - u_x(0,0)] - i[u_y(h,k) - u_y(0,0)]| \leq M \iint_\Delta \left| \frac{1}{\zeta - \omega} - \frac{1}{\zeta} \right| d\xi \, d\eta$$

$$= M|\omega| \iint_\Delta \frac{1}{|\zeta| \cdot |\zeta - \omega|} \, d\xi \, d\eta \tag{81}$$

(Here $\zeta = \xi + i\eta$, $\omega = h + ik$.) Now, let Δ be divided into four parts: the discs Δ_1 and Δ_2, each of radius $|\omega|/2$, with centers at $(0,0)$ and (h,k), respectively; the region Δ_3 consisting of the disc $|\zeta| < 3|\omega|/2$ minus the afore-mentioned discs; and, finally, Δ_4, the remainder of Δ. Then we readily obtain the inequalities[1]

$$\iint_{\Delta_1} = \iint_{\Delta_2} \leq \int_0^{2\pi} \int_0^{|\omega|/2} \frac{1}{r|\omega/2|} \, r \, dr \, d\theta = 2\pi \tag{82a}$$

$$\iint_{\Delta_3} \leq \iint_{\Delta_3} \frac{1}{|\omega/2|^2} \, d\xi \, d\eta < \frac{\pi|3\omega/2|^2}{|\omega/2|^2} = 9\pi \tag{82b}$$

$$\iint_{\Delta_4} \leq \int_0^{2\pi} \int_{3|\omega|/2}^{R} \frac{1}{r(r - |\omega|/2)} \, r \, dr \, d\theta = 2\pi \log \frac{2R - |\omega|}{2|\omega|} \tag{82c}$$

It now follows that the left side of (81) approaches zero with $|\omega|$, and, hence, that u_x and u_y are continuous. It follows, of course, that, as asserted earlier, the potential itself is continuous everywhere. It is of interest to note, however, that this fact follows more directly from (69), for from this equation we easily obtain the inequality (cf. Exercise 38)

$$|u_x(P)| \leq M \cdot 2\pi d \tag{83}$$

where d denotes the maximum distance between P and any point of D. Hence u_x (and similarly u_y) is bounded in any bounded subset of the plane, and from this it follows[2] that u is continuous everywhere.

We now consider what can be said about Δu at a point P of D. First we remark that the value of Δu, and even its existence, is determined wholly by the behavior of the density $f(\xi, \eta)$ in the immediate vicinity of P. To see this, let D be divided into a disc D_1 containing P and D_2, the remainder of D, and let u be considered as the sum $u_1 + u_2$ of the potentials arising from the mass distributed in the two parts of D. Then, since P is an exterior point of D_2, it follows that Δu_2 vanishes at P, and hence Δu exists if and only if Δu_1 does, in which case their values are equal. Since the disc D_1 may be chosen arbitrarily small, the statement made at the beginning of this paragraph is proved.

Now, let us consider the particular case of a uniform mass distribution of unit density over a disc of radius R, whose center may be taken at the origin. In this case, (52) assumes the form

$$u(x,y) = -\frac{1}{2} \iint_{\xi^2 + \eta^2 < R^2} \log [(\xi - x)^2 + (\eta - y)^2] \, d\xi \, d\eta \tag{84}$$

[1] $\iint$ denotes $\iint \frac{1}{|\zeta| \cdot |\zeta - \omega|} \, d\xi \, d\eta$ over the indicated region. R [in (82c)] denotes, as above, the radius of Δ. The equality of $\iint_{\Delta_1}$ and $\iint_{\Delta_2}$ follows by symmetry.

[2] By the theorem of mean value, $u(x + h, \, y + k) - u(x,y) = hu_x(x + \theta_1 h, \, y) + ku_y(x + h, \, y + \theta_2 k)$, where $0 < \theta_1, \theta_2 < 1$. Hence $|u(x + h, \, y + k) - u(x,y)|$ is at most equal to $(|h| + |k|)$ times an upper bound on both $|u_x|$ and $|u_y|$ in the rectangle determined by the points (x,y) and $(x + h, \, y + k)$.

Although the integration can be carried out explicitly by tedious but elementary methods, we shall evaluate (84) very easily by using some of the ideas that have been introduced in this chapter. Since u is harmonic outside the disc and obviously rotationally symmetric, it follows that u must, for $r[= (x^2 + y^2)^{1/2}] > R$, be of the form $(7a)$, namely, $a \log r + b$. To determine the values of a and b, we note that

$$(r - R)^2 < (\xi - x)^2 + (\eta - y)^2 < (r + R)^2 \tag{85}$$

for all points (ξ, η) of the disc and all points (x, y) outside, and hence, taking logarithms and integrating over the disc, we obtain

$$-\pi R^2 \log (r + R) < u = a \log r + b < -\pi R^2 \log (r - R) \tag{86}$$

Dividing by $\log r$ (we make the additional restriction $r > 1$ if necessary) and letting $r \to \infty$, we find that the end terms both approach $-\pi R^2$, which must therefore be the value of a. To determine b, we subtract $a \log r$ $(= -\pi R^2 \log r)$ from each term of (86) and obtain

$$-\pi R^2 \log \left(1 + \frac{R}{r}\right) < b < -\pi R^2 \log \left(1 - \frac{R}{r}\right) \tag{87}$$

Letting $r \to \infty$, we conclude that $b = 0$. This proves, incidentally, the remarkable fact that the potential at any point outside the disc is the same as would be obtained by concentrating all the mass at the origin. Next we consider the potential produced at any point inside the inner boundary of an annulus by a mass uniformly distributed with density one over the annulus. As before, we conclude that the potential must be of the form $a \log r + b$, but in this case a must vanish, since the potential must be finite at the origin. Therefore, the potential is equal, everywhere inside the inner circle, to its value at the center, in which case the potential is given by

$$u = - \int_0^{2\pi} \int_{R_1}^{R_2} r \log r \, dr \, d\theta = -\pi[r^2 \log r - \tfrac{1}{2}r^2]_{R_1}^{R_2} \tag{88}$$

where R_1 and R_2 denote the radii of the annulus. Returning now to the potential at an inner point P of the disc, we consider the latter divided into a smaller disc, containing P on its boundary, and an annulus. Applying the formulas developed above for the disc and for the annulus, we obtain

$$u(P) = -\pi r^2 \log r - \pi[r^2 \log r - \tfrac{1}{2}r^2]_r^R$$

$$= - \frac{\pi}{2} r^2 + \frac{\pi}{2} R^2 - \pi R^2 \log R \tag{89}$$

We conclude that Δu exists (as expected) at each interior point of the disc and has the value -2π; if the density has the constant value ρ_0, the value of Δu is equal to $-2\pi\rho_0$. Thus, the Poisson equation (57) has been shown to hold at a point P if the density is constant in some neighborhood of P, however small. (Of course, the fact that u is harmonic outside $\bar{D}$ corresponds simply to the case $\rho_0 = 0$.)

Turning now to the general case where $f(\xi, \eta)$ is not constant in a neighborhood of P, one might expect that the hypothesis that $f(\xi, \eta)$ is continuous would suffice to permit a proof that (57) holds. In fact, this is not true, as will be shown later by an example. A more stringent version of continuity, known as Hölder continuity, does suffice, however. A function g defined in a neighborhood of P is said to be "Hölder-continuous" at P if there exist positive numbers c and α such that, for all points Q in the neighborhood, $|g(Q) - g(P)| \le c\overline{PQ}^\alpha$. Assuming that the density is Hölder-continuous at a point P of D, which we again may place at the origin, we may, in view of the foregoing discussion, confine attention to the case that D is a disc with center at P and radius R and f vanishes at P; for if f does not vanish at P, we consider u as the sum of two potentials, one produced by the constant density $f(P)$ and the other by the density $f - f(P)$. Thus, we must prove that u_{xx} and u_{yy} both exist at P, and that their sum vanishes. Now, according to (69),

$$h^{-1}[u_x(h,0) - u_x(0,0)]$$

$$= h^{-1} \iint_D f(\xi, \eta) \left[\frac{\xi - h}{(\xi - h)^2 + \eta^2} - \frac{\xi}{\xi^2 + \eta^2} \right] d\xi\, d\eta$$

$$= \iint_D f(\xi, \eta) \frac{\xi^2 - \eta^2 - h\xi}{(\xi^2 + \eta^2)((\xi - h)^2 + \eta^2)}\, d\xi\, d\eta \qquad (h \ne 0) \quad (90)$$

Formally, therefore, we are led to the formula

$$u_{xx}(0,0) = \iint_D f(\xi, \eta) \frac{\xi^2 - \eta^2}{(\xi^2 + \eta^2)^2}\, d\xi\, d\eta \tag{91a}$$

First, we must show that the integral appearing in (91a) is meaningful. Because of our assumption that the density vanishes and is Hölder-continuous at $(0,0)$, the integral is dominated by

$$c \int_0^{2\pi} \int_0^R r^\alpha \cdot \frac{1}{r^2} \cdot r\, dr\, d\theta$$

which converges, since $\alpha > 0$. Subtracting the right sides of (90) and (91a), we find that the validity of (91a) is equivalent to the limiting relationship

$$\lim_{h \to 0} h \iint_D f(\xi, \eta) \operatorname{Re} \left[\frac{1}{\zeta^2(\zeta - h)} \right] d\xi\, d\eta = 0 \tag{92}$$

Taking account of the Hölder continuity of f at the origin, we conclude that it will suffice to prove the limiting relation

$$\lim_{h \to 0} h \iint_D \frac{|\zeta|^{\alpha-2}}{|\zeta - h|}\, d\xi\, d\eta = 0 \tag{93}$$

Dividing D into four parts and using simple upper bounds on the integrand in each of these parts, as was done previously in establishing the continuity of u_x and u_y, we readily show that (93) does indeed hold. The details are left as Exercise 40.

Thus, (91a) has been established; similarly,

$$u_{yy}(0,0) = \iint_D f(\xi,\eta) \frac{\eta^2 - \xi^2}{(\xi^2 + \eta^2)^2} \, d\xi \, d\eta \qquad (91b)$$

Adding (91a) and (91b), we conclude that Δu vanishes at the origin, and this is equivalent to proving, as explained above, that the potential (52) satisfies (57) at any point where the density is Hölder-continuous.

We now prove, as indicated previously, that it would not be possible to establish (57) under the hypothesis that $f(\xi,\eta)$ is merely continuous. Consider the density

$$f(\xi,\eta) = \frac{\xi^2 - \eta^2}{2(\xi^2 + \eta^2) \log (\xi^2 + \eta^2)} \qquad (94)$$

in the disc $\xi^2 + \eta^2 < R^2$, where $R < 1$. Clearly $f(\xi,\eta)$ is continuous everywhere in the disc [with the understanding that $f(0,0) = 0$], but it is not Hölder-continuous at the origin. [This may be seen by setting $\eta = 0$ and letting ξ approach zero. For any positive constant α, we then have $f(\xi,\eta)/|\xi|^\alpha = (|\xi|^\alpha \log \xi^2)^{-1} \to \infty$.] The right side of (90) assumes in this case the form (after transforming to polar coordinates)

$$\int_0^{2\pi} \int_0^R \frac{\cos 2\theta}{\log r} \cdot \frac{r \cos 2\theta - h \cos \theta}{r^2 - 2hr \cos \theta + h^2} \, dr \, d\theta \qquad (95)$$

Now (cf. Exercise 41)

$$\int_0^{2\pi} \frac{\cos 2\theta(r \cos 2\theta - h \cos \theta)}{r^2 - 2hr \cos \theta + h^2} \, d\theta = \begin{cases} -\pi r^3 h^{-4} & 0 \le r < |h| \\ \pi r^{-1} & r > |h| \end{cases} \qquad (96)$$

Hence (95) is equal to

$$-\pi h^{-4} \int_0^{|h|} \frac{r^3}{\log r} \, dr + \pi \int_{|h|}^R \frac{dr}{r \log r}$$

The first integral is easily shown to approach zero with h, but the second integral is equal to $[\log \log (1/R) - \log \log (1/|h|)]$ and, hence, becomes infinite. Thus, u_{xx} does not exist at the origin in this example.

We sum up the results of this section up to this point as follows.

Theorem 11. The potential (52) is defined and continuous everywhere and possesses continuous first partial derivatives, given by (69) and its analogue for u_y. The potential is harmonic at all points outside $\bar{D}$, and satisfies the Poisson equation (57) at every point of D where the density is Hölder-continuous.

We now consider a kind of converse to the problem discussed above. Given a function $f(x,y)$ defined in a bounded domain D, we seek a function $v(x,y)$ satisfying equation (57) and vanishing everywhere on the boundary. The question of uniqueness is immediately settled by the observation that the difference of two solutions would be harmonic in D and identically zero on the boundary, hence zero everywhere in D. The question of existence is more delicate. If we assume that f is bounded and Hölder-continuous[1] in D, then (52) furnishes a solution u of (57) which is continuous in $\bar{D}$; if we assume further that D is a Dirichlet domain, we may determine a function w harmonic in D and agreeing with u on the boundary and then obtain the unique solution to the problem in the form $v = u - w$. Suppose, however, that f fails to be Hölder-continuous at a single point P of D; as has been shown by an example in this section, the function $u(x,y)$ may fail to be sufficiently differentiable to satisfy (57) at P. The question arises whether a function $\bar{u}$ satisfying (57) everywhere in D, including P, might nevertheless exist. That the answer is negative is easily shown as follows: The difference $\bar{u} - u$ would be continuous throughout D and harmonic everywhere in D except at P. However, this is an impossibility, according to Exercise 21. We therefore specifically assume that f is bounded and Hölder-continuous in D, and that D is a Dirichlet domain. Then the problem under consideration is solvable.

Now, instead of setting up the integral (52) and then solving a suitable Dirichlet problem, one can, at least formally, solve the problem under consideration by setting up the integral[2]

$$v(P) = \iint_D f(Q)G(P,Q)\, d\xi\, d\eta \tag{97}$$

Since $G(P,Q)$ is equal to $\log 1/\overline{PQ}$ plus a function $H(P,Q)$ harmonic in P, we might argue (by differentiation under the integral sign) that the integral

$$\iint_D f(Q)H(P,Q)\, d\xi\, d\eta$$

defines a harmonic function, and hence that

$$\Delta v = \Delta \iint_D f(Q) \log 1/\overline{PQ}\, d\xi\, d\eta = -2\pi f$$

Secondly, we might argue that, as P approaches the boundary,

$$v(P) \to \iint_D f(Q) \lim G(P,Q)\, d\xi\, d\eta = 0$$

---

[1] The index α would not have to be constant.

[2] For convenience, we denote the points with rectangular coordinates x, y and ξ, η by P and Q, respectively.

However, we must justify carrying out the above operations under the integral sign. Rather than undertake this task, we shall justify the formula (97) only in the specific case that D is a disc—without loss of generality, the unit disc $x^2 + y^2 < 1$. In this case, (97) assumes the form

$$v(P) = \iint_D f(Q) \log \frac{(\overline{QP^*})(\overline{OP})}{\overline{QP}} \, d\xi \, d\eta \qquad (98)$$

[Here we have employed the first form of the Green's function given in (24) and exploited the symmetry in P and Q; P^* denotes, of course, the image of P in the unit circle.] Since the Green's function is never negative (cf. Exercise 16), we obtain from (95) the inequality

$$|v(P)| \leq M \left(\iint_D \log \overline{OP} \, d\xi \, d\eta + \iint_D \log \frac{1}{\overline{QP}} \, d\xi \, d\eta \right.$$
$$\left. - \iint_D \log \frac{1}{\overline{QP^*}} \, d\xi \, d\eta \right) \qquad (99)$$

The first integral is simply $\pi \log \overline{OP}$; the second integral represents the potential of a uniform mass distribution at P; while the last integral represents the potential of the same distribution at P^*. Since the potential is continuous over the entire plane, it follows that the last two integrals cancel each other out as $\overline{OP}$ (and hence $\overline{OP^*}$) approaches unity. Finally, $\log \overline{OP}$ approaches zero, and hence $|v(P)|$ approaches zero, as P approaches the circumference. It remains to show that $\Delta v = -2\pi f$ throughout. Breaking up the integral (98) into three parts, as in (99), we observe that it is necessary to prove merely that $\iint_D f(Q) \log \overline{OP} \, d\xi \, d\eta$ $+ \iint_D f(Q) \log \overline{QP^*} \, d\xi \, d\eta$ is harmonic. The first integral is simply $\log \overline{OP} \iint_D f(Q) \, d\xi \, d\eta$, and hence is harmonic.[1] The second integral represents a potential *outside* the unit disc, and hence is a harmonic function of the point P^*. As shown in Sec. 2, the property of harmonicity is preserved under inversion, so that the second integral defines a harmonic function of P. Thus, we have shown that formula (97) furnishes the solution to the problem under consideration, at least for a disc. Actually, a more delicate analysis of the behavior of the Green's function near the boundary shows that (97) may be employed for domains whose boundary satisfies very mild hypotheses.

---

[1] A separate consideration is necessary when $P = O$, for then both integrals become meaningless, but this is a minor matter.

EXERCISES

38. Prove (83).

39. Prove that u_x and u_y are bounded in the entire plane, not merely in any bounded set. *Hint:* Obtain an inequality similar to (83) valid for points outside $\bar{D}$.

40. Prove (93) in the manner indicated in the text.

41. Prove (96). *Hint:* Exploit the resemblance to the integral (27).

42. In contrast to the procedure used in the text, derive the potential of a disc of uniform density by exploiting the Poisson equation.

43. Determine the potential of a (three-dimensional) sphere of uniform density in two ways, analogous to those used in the text and in Exercise 42, respectively.

7. THE DIRICHLET PROBLEM

In this chapter we present a fairly extensive treatment of the Dirichlet problem, which was formulated in Sec. 6-6. In addition to the great significance of this problem, whether considered from either the mathematical or the physical viewpoint, the study of a large number of different approaches to this problem is justified by the fact that it may serve to acquaint the reader with some of the most important techniques of analysis, which can be used in a wide variety of problems.

Most of the methods to be discussed in this chapter are applicable in any number of dimensions. Although we shall formulate all the methods only in two dimensions, the reader should consider in each case how the generalization to higher dimensions, if possible, is carried out.

1. Subharmonic Functions

Among the methods for studying the Dirichlet problem which will be presented in this chapter, there are two which use subharmonic functions. These methods will be presented in the next two sections, and the present section is devoted to a brief development of those properties of subharmonic functions which will be needed for the presentation of the afore-mentioned two methods. A definition of subharmonicity will be used which is somewhat simpler than is needed in more comprehensive treatments of this topic, but it will be adequate for the present purposes.

DEFINITION 1. A function u is said to be "subharmonic" in a domain D if it is continuous in D and if, for each point P of D, there exists a positive number $\rho(P)$ such that for all r, $0 < r < \rho(P)$, the inequality

$$u(P) \le \frac{1}{2\pi} \int_0^{2\pi} u(P + re^{i\theta}) \, d\theta \tag{1}$$

is satisfied.

167

The reason for the use of the term subharmonic will be readily apparent from Exercise 7, or from a comparison of (1) with Theorem 6-1. Clearly any function harmonic in a domain is subharmonic there, and a function subharmonic in a domain is also subharmonic in any subdomain. The properties of subharmonic functions that will be needed later are developed in the following sequence of theorems.

Theorem 1. If $u_1, u_2, \ldots, u_n$ are each subharmonic in a domain D, and $c_1, c_2, \ldots, c_n$ are nonnegative constants, then the functions $c_1u_1 + c_2u_2 + \cdots + c_nu_n$ and $u = \max(u_1, u_2, \ldots, u_n)$ are also subharmonic in D.

Proof. The first part is obvious. As for the second part, we first consider the case $n = 2$. The continuity of u follows from the explicit representation

$$u = \max(u_1, u_2) = \tfrac{1}{2}(u_1 + u_2 + |u_1 - u_2|) \tag{2}$$

At any point P of D we have either $u(P) = u_1(P)$ or $u(P) = u_2(P)$ (or both). In the former case we have, for sufficiently small r,

$$u(P) = u_1(P) \le \frac{1}{2\pi} \int_0^{2\pi} u_1(P + re^{i\theta})\, d\theta \le \frac{1}{2\pi} \int_0^{2\pi} u(P + re^{i\theta})\, d\theta \tag{3}$$

and similarly if $u(P) = u_2(P)$. For $n > 2$, the desired result is obtained by an induction in which the fact that

$$\max(u_1, u_2, \ldots, u_n) = \max(u_n, \max(u_1, u_2, \ldots, u_{n-1}))$$

is used.

Theorem 2. First Form of Maximum Principle. If u is subharmonic in D and assumes a local maximum at some point P of D, then u is constant in some neighborhood of P. (Cf. Exercise 6.)

Proof. Since u assumes a local maximum at P, there exists $\delta > 0$ such that

$$u(Q) \le u(P) \qquad \overline{PQ} < \delta \tag{4}$$

Then for all positive $r < \min[\delta, \rho(P)]$, we obtain, by combining (1) and (4),

$$u(P) \le \frac{1}{2\pi} \int_0^{2\pi} u(P + re^{i\theta})\, d\theta \le \frac{1}{2\pi} \int_0^{2\pi} u(P)\, d\theta = u(P) \tag{5}$$

The equalities must therefore hold in (5), and then, as in the proof of Theorem 6-2, we conclude that $u(Q)$ must equal $u(P)$ for $\overline{PQ}$ sufficiently small; this completes the proof.

Theorem 3. Second Form of Maximum Principle. If u is subharmonic in a bounded domain D and continuous in $\bar{D}$, then u attains its maximum on the boundary; if u also attains its maximum at a point of D, then u is constant throughout $\bar{D}$.

Proof. It clearly suffices to prove the second assertion. Let A be the set of points of D at which u assumes its maximum value. Since any point of A is, a fortiori, a local maximum (but not conversely, as shown by Exercise 6), it follows from Theorem 2 that the set A is open. The set $D - A$, on which u is less than its maximum value, is also open (by continuity). As in the proof of the corollary of Theorem 6-2, we conclude that either A or $D - A$ is void. This statement is equivalent to the second assertion.

Theorem 4. Let u be subharmonic in a domain D, and let Δ be any disc such that $\bar{\Delta} \subset D$. Define the function v in D as follows: v continuous in D, $v \equiv u$ in $D - \Delta$, v harmonic in Δ. (By Theorem 6-4, v is uniquely determined by these conditions.) Then v is also subharmonic in D, and $u \leq v$ throughout D.

Proof. For all points of $D - \bar{\Delta}$ the subharmonicity of v follows immediately from the fact that $v \equiv u$ there. For any point P of Δ the inequality (1), with u replaced by v, becomes an equality for all sufficiently small values of r (by the mean-value theorem). It remains only to consider the circumference Γ of Δ. Now, $u - v = u + (-v)$ is subharmonic in Δ, by Theorem 1, and vanishes identically on Γ. By Theorem 3, $u - v \leq 0$ at all points of Δ, and hence at all points of D. Therefore, for any point P of Γ and any sufficiently small r,

$$v(P) = u(P) \leq \frac{1}{2\pi} \int_0^{2\pi} u(P + re^{i\theta})\, d\theta \leq \frac{1}{2\pi} \int_0^{2\pi} v(P + re^{i\theta})\, d\theta \quad (6)$$

This completes the proof.

The procedure described in Theorem 4 of altering a subharmonic function u by redefining it inside a disc Δ as the solution of the Dirichlet problem with the boundary values derived from u (i.e., by replacing u in the disc Δ by the Poisson integral formed with the values of u on the circumference) will be used systematically in the following two sections. We shall refer to this procedure as "harmonization," and denote the function thus obtained as u^Δ. Thus, if u is subharmonic in D, so is u^Δ, and $u \leq u^\Delta$ throughout D.

EXERCISES

1. Prove that a function u is harmonic in a domain D if and only if both u and $-u$ are subharmonic in D.

2. Prove, by explicitly evaluating the right side of (1) for all P and all positive r, that the function $u = x^2 + y^2$ is subharmonic in the entire plane. (Cf. Exercise 11.)

3. Prove that the square of a nonnegative subharmonic function is also subharmonic. *Hint:* Apply the Schwarz inequality to (1).

4. Prove, by a simple example, that the assertion of Exercise 3 may be false if the word "nonnegative" is omitted.

5. Prove that the square of a harmonic function is subharmonic. (Note that, in contrast to Exercise 3, it is not necessary to assume in this case that the given function is nonnegative.)

6. Show, by a simple example, that (in contrast to harmonic functions) a subharmonic function can attain a local maximum (and hence, by Theorem 2, be locally constant) without being constant throughout the domain. (Note the contrast with Theorem 3, where the function is assumed to attain not merely a local maximum, but the maximum for the entire domain, at a point of the domain.)

7. Let u and v be subharmonic and harmonic, respectively, in a bounded domain D and continuous in $\bar{D}$, and suppose that $u \leq v$ on the boundary. Prove that $u \leq v$ throughout D, and that the equality holds either everywhere or nowhere in D.

8. A seemingly more restrictive definition is obtained by modifying Definition 1 as follows: " . . . for each point P of D and each positive r less than the distance of P from the boundary of D, the inequality . . . ." Show that in reality the two definitions are equivalent.

9. Prove, by a simple example, that if the continuity requirement is dropped, Definition 1 is no longer equivalent to the modified definition given in the preceding exercise.

10. Formulate a definition of subharmonicity based on Exercise 7, and prove its equivalence with Definition 1.

11. Prove that a function of class C^2 (i.e., possessing continuous second partial derivatives) in a given domain D is subharmonic if and only if the inequality

$$u_{xx} + u_{yy} \geq 0 \tag{7}$$

holds throughout D, by applying Green's second identity over an annulus to the functions u and $\log r$, where r denotes the distance from the center of the annulus.

12. Prove the assertion of Exercise 11 by using Taylor's formula with remainder. *Hint:* Consider the second proof of Theorem 6-5.

13. Prove the assertion of Exercise 3 for functions of class C^2 by applying condition (7).

14. Formulate an appropriate definition of subharmonicity for a function of one variable and the analogues of the theorems of this section.

15. *Extension of Theorem* 3: Let v be subharmonic in a bounded domain D, and let a function g be defined on the boundary as follows:

$$g(Q) = \lim_{P \to Q} \sup v(P) \qquad (P \in D) \tag{8}$$

(The values $\pm \infty$ must be allowed.) Show that, for all $P \in D$,

$$v(P) \leq \sup g(Q) \tag{9}$$

[It may be assumed that $\sup g(Q)$ is finite, for the case $\sup g(Q) = +\infty$ is trivial and the case $\sup g(Q) = -\infty$ cannot occur. The latter assertion should, of course, be proved.] As in Theorem 3, show that equality holds in (9) either everywhere or nowhere in D.

2. The Method of Balayage

As explained in Sec. 6-11, a (sufficiently differentiable) function can be represented in a domain (with sufficiently smooth boundary) as the potential resulting from a suitably chosen distribution of mass over the domain and its boundary; in particular, if the function is harmonic the mass is distributed entirely on the boundary. In the method of *balayage* (sweeping), due to Poincaré, one begins with a function having the prescribed boundary values, but not harmonic, so that its associated mass distribution is deposited, at least in part, within the domain. This function undergoes a succession of modifications such that at each stage the modified function is harmonic in some disc; the mass is thus removed from this disc. In subsequent modifications of the original function, however, new mass distributions are deposited in the disc. In the limit a harmonic function is obtained, so that the entire process may be looked upon as a scheme for *sweeping* the original mass distribution onto the boundary.

Let a bounded domain D be given, and let a continuous function f be defined on the boundary. We assume that there exists a function continuous in $\bar{D}$, subharmonic in D, and coinciding with f on the boundary. We shall presently show that the process of balayage leads to a harmonic function which, under certain conditions, provides the solution to the Dirichlet problem.

While the assumption made in the preceding paragraph, namely, that f can be extended into D as a subharmonic function, appears extremely restrictive, it is noteworthy that if the Dirichlet problem is solvable for every such boundary function, it is solvable for every continuous boundary function. This is easily seen as follows. First, according to Exercise 20, every Dirichlet problem whose boundary values can be extended into D as a polynomial will be solvable. Then, given any continuous function f defined on the boundary, we can, according to the Weierstrass approximation theorem, select a sequence of polynomials $\{P_n(x,y)\}$ which converge uniformly to f. For each of these polynomials, we determine the solution of the corresponding Dirichlet problem. According to Theorem 6-6, these solutions will converge to a function which solves the given Dirichlet problem.

We select a denumerable sequence of discs $\Delta_1, \Delta_2, \ldots$ whose union is precisely the domain D. This may be accomplished, for example, by enumerating the "rational" points of D (those whose coordinates x, y are both rational) and constructing about each such point the disc with center at this point and radius equal to half the distance from the point to the boundary. A second enumeration of these discs is also introduced, in which each disc appears infinitely often. This new enumeration can be described as follows:

$$\tilde{\Delta}_1 = \Delta_1, \quad \tilde{\Delta}_2 = \Delta_2, \quad \tilde{\Delta}_3 = \Delta_1, \quad \tilde{\Delta}_4 = \Delta_2, \quad \tilde{\Delta}_5 = \Delta_3, \quad \tilde{\Delta}_6 = \Delta_1, \quad \ldots \quad (10)$$

It is seen that the first two discs are enumerated, then the first three, the first four, and so on.

A sequence of functions (the "modifications" referred to above) is now defined as follows:

$$f_1 = f \qquad f_n = f_{n-1}^{\tilde{\Delta}_n} \qquad (n > 1) \tag{11}$$

By induction it is readily seen that the functions f_n are well defined, subharmonic in D (by Theorem 4), and continuous in $\bar{D}$, and that they all coincide on the boundary with the original function f. Furthermore (again by Theorem 4), these functions constitute a nondecreasing sequence, and (by Theorem 3) each f_n is bounded (above) everywhere in $\bar{D}$ by the maximum of f. Therefore, the sequence $\{f_n\}$ converges throughout D to a function u whose value nowhere exceeds the maximum of f.

Next it will be shown that the limit function u is harmonic throughout D. We choose any disc Δ_k of the set that was used in the balayage, and consider the subsequence of $\{f_n\}$ consisting of those functions which were obtained by "harmonizing" with respect to Δ_k (e.g., if $\Delta_k = \Delta_3$, we choose the subsequence $f_5, f_8, f_{12}, \ldots$). Since the functions of this subsequence constitute a monotone sequence of functions harmonic in Δ_k, Harnack's theorem (6-8) applies. The function u is therefore harmonic in Δ_k; since this argument can be applied to any one of the discs, and since these discs exhaust D, it follows that u is harmonic throughout D. (We are, of course, exploiting the fact that every subsequence of a convergent sequence is itself convergent to the same limit.)

Although it is not essential for the rest of the discussion, it is instructive to settle a question that may be raised in connection with the process of balayage which has been described. The function u which has been constructed seems to depend not only on the function f originally defined on the boundary, but also on the particular (subharmonic) extension of f that is selected and on the choice of the set of discs $\Delta_1, \Delta_2, \ldots$. It is noteworthy that the dependence of u on these two choices is only apparent; a different extension of f, or a different choice of the discs, or both, has no effect on u. We can prove this statement as follows. Let two subharmonic extensions of f, say $f^{(1)}$ and $f^{(2)}$, and two sets of discs, $\Delta_1^{(1)}, \Delta_2^{(1)}, \ldots$ and $\Delta_1^{(2)}, \Delta_2^{(2)}, \ldots$, be given, and let balayage be performed twice, first with the extension $f^{(1)}$ and the discs $\Delta_n^{(1)}$ and then with the extension $f^{(2)}$ and the discs $\Delta_n^{(2)}$. We thus obtain harmonic functions $u^{(1)}$ and $u^{(2)}$, respectively. For $n = 1, 2, 3, \ldots$, $\nu = 1, 2$ we have, for any boundary point Q,

$$\lim_{P \to Q} f_n^{(\nu)}(P) = f(Q) \tag{12}$$

Since $f_n^{(\nu)}(P) \leq u^{(\nu)}(P)$, we also have

$$\liminf_{P \to Q} u^{(\nu)}(P) \geq f(Q) \tag{13}$$

Setting $\nu = 1$ in (13) and $\nu = 2$ in (12) and subtracting, we obtain

$$\limsup_{P \to Q} [f_n^{(1)}(P) - u^{(2)}(P)] \leq 0 \tag{14}$$

Applying the result of Exercise 15 (with v replaced by $f_n^{(1)} - u^{(2)}$), we obtain

$$f_n^{(1)}(P) - u^{(2)}(P) \leq 0 \tag{15}$$

Letting n become infinite, we obtain

$$u^{(1)}(P) - u^{(2)}(P) \leq 0 \tag{16}$$

By symmetry, we can also obtain

$$u^{(2)}(P) - u^{(1)}(P) \leq 0 \tag{17}$$

From (16) and (17) we conclude that

$$u^{(1)} \equiv u^{(2)} \tag{18}$$

Thus, the assertion that u depends only on the boundary function f has been proved.

It now remains to decide whether the function u assumes the prescribed values on the boundary, or, more precisely, whether

$$\lim_{P \to Q} u(P) = f(Q) \tag{19}$$

In order to settle this question, it is necessary to introduce a new and important concept, that of a barrier.

DEFINITION 2. Let D be a bounded domain and Q a boundary point of D. A function ω defined in $\bar{D}$ is said to be a "barrier for D at Q" if it satisfies the following conditions:

$$\omega \text{ continuous in } \bar{D} \tag{20a}$$
$$\omega \text{ harmonic in } D \tag{20b}$$
$$\omega(Q) = 0 \tag{20c}$$
$$\omega(P) > 0 \qquad P \in \bar{D} - Q \tag{20d}$$

Before explaining the role played by the barrier in the study of the Dirichlet problem, we give one simple but highly important example of a barrier. Suppose that there exists a disc Δ such that Δ and $\bar{D}$ are disjoint, but $\bar{\Delta}$ and $\bar{D}$ have a single point Q in common. Then the function

$$\omega(P) = \log \frac{\overline{OP}}{\overline{OQ}} \tag{21}$$

where O denotes the center of Δ, is readily seen to be a barrier at Q. Boundary points which admit a barrier of this type were termed "regular"

by Poincaré.   (However, we shall assign a different meaning to this term
in Definition 3.)   More delicate conditions guaranteeing the existence of a
barrier are known, but need not concern us here.   On the other hand,
there do exist domains some of whose boundary points do not admit a
barrier.   For example, an isolated boundary point cannot admit a
barrier.   (Cf. Exercises 17 and 18.)

DEFINITION 3.   A boundary point Q of a domain D is said to be regular
if it admits a barrier.

We are now in position to state the following important result.

Theorem 5.   The function u defined by balayage satisfies condition
(19) at every regular boundary point of D.
Proof.   We shall prove the inequalities

$$\liminf_{P \to Q} u(P) \geq f(Q) \qquad \limsup_{P \to Q} u(P) \leq f(Q) \qquad (22)$$

which, taken together, are equivalent to (19).   To prove the first
inequality, we recall that

$$u(P) \geq f(P)\dagger \qquad (23)$$

for all points P of D, and hence

$$\liminf_{P \to Q} u(P) \geq \liminf_{P \to Q} f(P) \qquad (24)$$

However, since f is continuous in $\bar{D}$, the right side of (24) is equal to $f(Q)$,
and so the first inequality of (22) is established.   (Note that this half of
the proof does not depend on the existence of a barrier.)
Now let ω be a barrier at Q.   Given $\epsilon > 0$, choose a disc Δ with center
at Q such that $|f(Q') - f(Q)| < \epsilon$ for all boundary points Q' of D lying
within Δ; also, we require that Δ should not include all of D.   Let
$\omega_0 = \min_{P \notin \Delta} \omega(P)$; by the definition of a barrier, ω_0 must be positive.   Also,
let $M = \max |f|$.   Now we define the harmonic function W as follows:

$$W(P) = f(Q) + \epsilon + \frac{\omega(P)}{\omega_0} [M - f(Q)] \qquad (25)$$

Everywhere in $\bar{D}$, $W \geq f(Q) + \epsilon$, and, in particular, for all boundary
points Q' lying in Δ it follows, from the definition of Δ, that $W(Q') >
f(Q')$.   On the other hand, by the definition of ω_0, we see that for all

† We denote the (subharmonic) extension of the boundary function f by the same
letter.

boundary points Q'' *not* lying in Δ,

$$W(Q'') \geq f(Q) + \epsilon + [M - f(Q)] = M + \epsilon > f(Q'')$$

Thus $f - W < 0$ everywhere on the boundary, and since the function is subharmonic in D it must, by Theorem 3, be negative throughout D. By Theorem 4, this inequality is preserved under any sequence of harmonizations of f. Therefore, for any point P of D and for $n = 1, 2, 3, \ldots$, the inequality

$$f_n(P) < W(P) \tag{26}$$

must hold. Hence,

$$u(P) = \lim_{n \to \infty} f_n(P) \leq W(P) \tag{27}$$

Letting $P \to Q$ in (27) and taking account of (25), we obtain $\limsup\limits_{P \to Q} u(P)$ $\leq f(Q) + \epsilon$. Since ϵ is arbitrary, we conclude that the second inequality of (22) holds, and the proof is complete.

The following theorem is an easy corollary of Theorem 5. It states, without making reference to balayage, the results that have been established by the use of that method.

Theorem 6. A necessary and sufficient condition that a domain be a Dirichlet domain is that its boundary consist entirely of regular points.

Proof. The sufficiency follows from Theorem 5 together with the approximation argument presented earlier in the present section. As for the necessity, let Q_0 be any boundary point of a Dirichlet domain. Then, in particular, the Dirichlet problem with boundary values $f(Q) = \overline{Q_0Q}$ possesses a solution which clearly constitutes a barrier at Q_0; i.e., Q_0 is a regular boundary point.

EXERCISES

16. Explain why it would not suffice, in the selection of the sequence of discs $\Delta_1, \Delta_2, \ldots$, to take the "rational" points (or some other denumerable dense set) and construct about each of these points *some* disc.

17. Let D be the punctured unit disc, $0 < x^2 + y^2 < 1$, and let f be defined as follows: $f \equiv 1$ on the circumference, $f = 0$ at the origin. Determine the function which balayage assigns to this Dirichlet problem. (Cf. Sec. 6-6.)

18. Prove the assertion made in the text that an isolated boundary point of any domain cannot be regular. *Hint:* Consider a Dirichlet problem similar to that posed in the preceding exercise.

19. Prove that a bounded domain cannot possess a boundary consisting entirely of irregular points; in fact, the boundary must contain a nondenumerable set of regular points.

20. Prove that every polynomial can be expressed in a given bounded domain as the difference of two subharmonic polynomials. *Hint:* Add and subtract a suitable multiple of r^2 and use the result of Exercise 11.

3. The Perron-Remak Method

The treatment of the Dirichlet problem to be presented in this section is closely related to the method which was presented in the preceding section. Both methods use subharmonic functions and assign to every Dirichlet problem a harmonic function which coincides with the solution whenever the latter exists. It is of interest to note that both methods assign the same harmonic function to any Dirichlet problem; this fact will be proved at the end of this section. The outstanding differences between the two methods are that balayage is constructive while the method of the present section is not, and that the latter method, unlike the former, does not require any extension of the boundary function to the domain itself. The latter method is thus independent of the Lebesgue extension theorem and the Weierstrass approximation theorem, and may therefore be applied without additional complications to a domain situated on a Riemann surface. It may also be added that the latter method can be readily adapted to the treatment of Dirichlet problems with discontinuous boundary values. However, we shall not pursue these two remarks further.

Given a continuous function f defined on the boundary of a bounded domain D, we define S_f to be the family of all functions w continuous in $\bar{D}$, subharmonic in D, and satisfying at all boundary points the inequality

$$w \leq f \tag{28}$$

The class S_f is nonvoid, for it certainly contains the constant function $w \equiv \min f$. For any point P of D and any function w of the family S_f, the inequality

$$w(P) \leq \max f \tag{29}$$

holds, by virtue of Theorem 3 and the inequality (28). Therefore, the following definition is meaningful:

$$u(P) = \sup_{w \in S_f} w(P) \qquad (P \in D) \tag{30}$$

and the function u also satisfies the inequality (29). It is by no means obvious that u is harmonic, or even continuous, in D, but this will now be shown.

Let Δ be any disc such that $\bar{\Delta} \subset D$, and let P be any point of Δ. From the definition (30) of u it follows that there exists a sequence of functions $u_1, u_2, \ldots$ of the family S_f such that

$$u_n(P) \to u(P) \tag{31}$$

From Theorem 1, it follows that the functions

$$U_n = \max (u_1, u_2, \ldots, u_n) \tag{32}$$

are also members of S_f, and, by their very definition, the U_n satisfy the inequalities

$$U_{n+1} \geq U_n \tag{33}$$

throughout D. Therefore the sequence $\{U_n\}$ converges monotonely throughout D, and, since $U_n(P) \geq u_n(P)$, (31) holds with u_n replaced by U_n. It follows from Theorem 4 and the minimum principle for harmonic functions that the sequence of functions U_1^Δ, U_2^Δ, $\ldots$ obtained by harmonizing the U_n with respect to Δ also constitutes a monotone sequence of members of S_f. By a repetition of the previous argument, it follows that

$$U_n^\Delta(P) \to u(P) \tag{34}$$

By Harnack's theorem of monotone convergence, the limit of the sequence $\{U_n^\Delta\}$, which we denote as U^Δ, is harmonic throughout Δ. If we can show that $U^\Delta \equiv u$ throughout Δ, the harmonicity of u will be established in Δ, and hence at all points of D. Since each of the functions U_n^Δ is a member of S_f, it follows that $U^\Delta \leq u$ everywhere in D (even though u might not itself be a member of S_f). Thus it will suffice to prove that

$$U^\Delta \geq u \qquad \text{throughout } \Delta \tag{35}$$

Suppose, on the contrary, that at some point Q of Δ the following inequality holds:

$$U^\Delta(Q) < u(Q) \tag{36}$$

Now a sequence $\{v_n\}$ of members of S_f is selected which plays the same role at Q as the sequence $\{u_n\}$ did at P; i.e.,

$$v_n(Q) \to u(Q) \tag{31'}$$

However, instead of imitating the definition of the functions U_n, we employ a different definition, namely,

$$V_n = \max (u_1, v_1, u_2, v_2, \ldots, u_n, v_n) \tag{37}$$

It is evident from this definition that the sequence $\{V_n\}$ converges monotonely, and that *both* of the following relations hold:

$$V_n(P) \to u(P) \qquad V_n(Q) \to u(Q) \tag{38}$$

Next, we replace the sequence $\{V_n\}$ by $\{V_n^\Delta\}$. As before, we see that the functions V_n^Δ converge monotonely to a function V^Δ which is harmonic throughout Δ.   Furthermore, the following three relations hold:

$$V^{\Delta} \geq U^{\Delta} \qquad \text{throughout } D \tag{39a}$$
$$V^{\Delta}(P) = U^{\Delta}(P) = u(P) \tag{39b}$$
$$V^{\Delta}(Q) = u(Q) > U^{\Delta}(Q) \tag{39c}$$

From (39a) and (39b) we conclude that the function $V^{\Delta} - U^{\Delta}$ is harmonic and nonnegative in Δ, but vanishes at the point Q of Δ.   By the minimum principle for harmonic functions, we conclude that

$$V^{\Delta} - U^{\Delta} \equiv 0 \qquad \text{throughout } \Delta \tag{40}$$

which contradicts (39c).   We therefore have shown that

$$U^{\Delta} \equiv u \qquad \text{throughout } \Delta \tag{41}$$

and hence, as was to be proved, u is harmonic throughout D.

It remains to analyze the behavior of u near the boundary.   The analysis is similar to that given in the preceding section, as might be expected, and the result is the same, namely:

Theorem 7.   If the boundary point Q is regular, then

$$\lim_{P \to Q} u(P) = f(Q) \tag{42}$$

Proof.   The proof of the second part of (22) applies without change up to and including the words: "Thus $f - W < 0$ everywhere on the boundary."   The rest of the paragraph does not apply, because f has not been extended to D, but we can now argue as follows.   For any function w of the family S_f, the function $w - W$ is continuous in $\bar{D}$, subharmonic in D, and negative on the boundary.   By Theorem 3, $w - W$ is negative throughout D, and hence, by the definition of u, we have, for all $P \in D$, the inequality

$$u(P) \leq W(P) \tag{43}$$

Therefore,

$$\limsup_{P \to Q} u(P) \leq \limsup_{P \to Q} W(P) = \lim_{P \to Q} W(P) = f(Q) + \epsilon \tag{44}$$

On account of the arbitrariness of ϵ, we obtain

$$\limsup_{P \to Q} u(P) \leq f(Q) \tag{45}$$

Turning to the first part of (22), we consider the harmonic function

$$V(P) = f(Q) - \epsilon - \frac{\omega(P)}{\omega_0} [M + f(Q)] \tag{46}$$

Proceeding as in the previous section, we find that $V < f$ everywhere on the boundary. Therefore V is a member of S_f, and hence, for all $P \in D$,

$$V(P) \leq u(P) \tag{47}$$

From this we immediately obtain

$$\liminf_{P \to Q} u(P) \geq \liminf_{P \to Q} V(P) = \lim_{P \to Q} V(P) = f(Q) - \epsilon \tag{48}$$

Again exploiting the arbitrariness of ϵ, we obtain

$$\liminf_{P \to Q} u(P) \geq f(Q) \tag{49}$$

Combining (45) and (49) we obtain (42), and the proof is complete.

We now prove that the Perron-Remak method leads to the same function as that obtained by balayage. (Of course, this statement is trivially true if D is a Dirichlet domain, for in this case both methods furnish the unique solution of the Dirichlet problem.) Let f be defined and continuous on the boundary, and let it, as in the previous section, possess a subharmonic extension to D, which we also denote by f. Let u_1 denote the function obtained by balayage and u_2 the function obtained by the Perron-Remak method. Then, since the function f (defined in $\bar{D}$) belongs to S_f, it follows that all the functions of the sequence $\{f_n\}$ also belong to S_f. Thus, $f_n \leq u_2$ throughout D, and hence $u_1 \leq u_2$. Now suppose that $u_1(P) < u_2(P)$ at some point P of D. According to the definition of u_2, there would exist a function w belonging to S_f such that $u_1(P) < w(P)$. By Theorem 1, the function max (f, w) represents a subharmonic extension of the function f originally defined on the boundary. Then the function u_3 which balayage would furnish with this new extension of f would exceed u_1 at P. However, it was shown in Sec. 2 that the function furnished by balayage is independent of the particular subharmonic extension of f which is employed. Hence, we conclude that $u_1 \equiv u_2$, as asserted.

EXERCISE

21. Solve Exercise 17 by studying the behavior of the sequence of functions $(x^2 + y^2)^{1/n}$, $n = 1, 2, 3, \ldots$.

4. The Method of Integral Equations

In this section we employ the results of Sec. 6-11 to show that, for a domain with sufficiently smooth boundary, the Dirichlet problem can be formulated as an integral equation for an unknown function defined on the boundary. Let the domain D be the interior of a curve Γ possessing continuous curvature; i.e., Γ can be parametrized in terms of arc length by functions $x(s)$, $y(s)$ possessing continuous second derivatives. Let the continuous function f be defined on Γ. We make the assumption, to be justified below, that the Dirichlet problem with boundary values f possesses a solution u which is expressible as the potential of a double layer with continuous density τ distributed on Γ.

To justify the above assumption we shall prove that: (1) the function τ must satisfy a certain integral equation; (2) this integral equation possesses a unique continuous solution; (3) the potential produced by the double layer whose density satisfies the afore-mentioned integral equation does, in fact, constitute a solution to the given Dirichlet problem.

Starting with the representation (6-56) of a double-layer potential, namely,

$$u(P) = \int_{\Gamma} \tau(Q)\, \frac{\partial \log (1/r)}{\partial n_Q}\, ds \qquad (r = \overline{PQ}) \tag{50}$$

we consider any point P on Γ. According to Theorem 6-10,

$$u(P) - \pi\tau(P) = \lim_{P' \to P} u(P') \qquad (P' \in D) \tag{51}$$

From (50), (51), and the fact that the right side of (51) is supposed to equal $f(P)$, we find that the density τ of the double layer must satisfy the integral equation

$$\tau(P) = \frac{-f(P)}{\pi} + \int_{\Gamma} \tau(Q) K(P,Q)\, ds_Q \tag{52}$$

where $K(P,Q) = -\pi^{-1} \dfrac{\partial \log r}{\partial n_Q}$. This completes the proof of part 1.

Next, we note that $K(P,Q)$ is jointly continuous in P and Q; this follows from a straightforward computation, which is left as Exercise 22. Therefore, in order to establish the unique solvability of (52) for every continuous f, it suffices, according to the Fredholm alternative, to show that the corresponding homogeneous equation

$$\bar{\tau}(P) = \int_{\Gamma} \bar{\tau}(Q) K(P,Q)\, ds_Q \tag{53}$$

admits no nontrivial continuous solution. Let $\bar{\tau}$ be any continuous solution of (53), and let v be defined as the potential arising from the double layer of density $\bar{\tau}$:

$$v(P) = \int_{\Gamma} \bar{\tau}(Q)\, \frac{\partial \log (1/r)}{\partial n_Q}\, ds_Q \tag{54}$$

If the point P' approaches any boundary point P from within D, the relation

$$v(P') \to v(P) - \pi\bar{\tau}(P) \tag{55}$$

must hold (by Theorem 6-10), while by (53) the right side of (55) must vanish. Hence, the function v is harmonic in D and approaches zero at the boundary, so that, by the maximum and minimum principles, $v \equiv 0$ in D. The interior normal derivative of v therefore vanishes everywhere on Γ, and by appealing once again to Theorem 6-10 we see that the *exterior* normal derivative also vanishes everywhere on Γ. Now let Δ be a disc containing D, and let Γ_1 be its boundary. By (6-12),

$$\iint_{\Delta-D} (v_x{}^2 + v_y{}^2)\, dx\, dy = \int_\Gamma v\, \frac{\partial v}{\partial n}\, ds + \int_{\Gamma_1} v\, \frac{\partial v}{\partial n}\, ds \tag{56}$$

The first integral on the right vanishes on account of the vanishing of $\dfrac{\partial v}{\partial n}$

on Γ. Now, as the radius R of Γ_1 increases, the center remaining fixed, v and its gradient approach zero at least as rapidly as c/R and c/R^2, respectively, where c is a sufficiently large constant. This may be seen as follows. Introducing rectangular coordinates x, y and ξ, η for P and Q, respectively, we write (54) in the form

$$v(x,y) = \int_\Gamma \bar{\tau}(Q) \cdot \frac{1}{r^2} [(\eta - y)\, d\xi - (\xi - x)\, d\eta] \tag{54'}$$

where $r^2 = (x - \xi)^2 + (y - \eta)^2$. By the Schwarz inequality,

$$[(\eta - y)\, d\xi - (\xi - x)\, d\eta]^2 \le [(\xi - x)^2 + (\eta - y)^2] \cdot (d\xi^2 + d\eta^2) \tag{57}$$

and we now conclude from (54') that

$$|v(x,y)| \le \int_\Gamma |\bar{\tau}(Q)| \cdot \frac{1}{r}\, ds \tag{58}$$

This, in turn, yields the further inequality

$$|v(x,y)| \le (\min_{Q \in \Gamma} \overline{PQ})^{-1} \int_\Gamma |\bar{\tau}(Q)|\, ds \tag{59}$$

which implies the assertion following (56) concerning the behavior of v. By differentiating the right side of (54) under the integral sign with respect to x (or y), we obtain, very much as in the above argument, an upper bound on $|v_x|$ (or $|v_y|$). In this manner it is shown that the gradient of v has the behavior asserted after (56); the details of the computation are left as Exercise 24. It now follows that $\left| v\, \dfrac{\partial v}{\partial n} \right|$ approaches zero at least as rapidly as c^2/R^3, and hence that the right side of (56) approaches zero as R becomes infinite. Thus, we have shown that $\iint (v_x{}^2 + v_y{}^2)\, dx\, dy$ vanishes when the integration is extended over the entire exterior of Γ. Since $v_x{}^2 + v_y{}^2$ is nonnegative and continuous, v_x and v_y must vanish identically outside Γ. This shows that v must be constant outside Γ, and from its behavior "at infinity" it follows that v vanishes identically outside Γ. Thus, v is continuous across Γ, and by Theorem 6-10 we conclude that $\bar{\tau} \equiv 0$, as was to be shown. Thus part 2 is proved. Turning now to part 3, we note that the function u defined by (50) is harmonic throughout D and that, by Theorem 6-10, it satisfies the limiting condition

$$\lim_{P' \to P} u(P') = u(P) - \pi\tau(P) \qquad (P' \in D, P \in \Gamma) \tag{60}$$

By comparing (52) and (60), we conclude that

$$\lim_{P' \to P} u(P') = f(P) \tag{61}$$

The proof is thus completed.

We illustrate the method of integral equations by showing that when applied to the unit disc, it leads to the Poisson formula. Letting θ and ϕ denote the polar angles of P and Q, respectively, we obtain

$$K(P,Q) = \frac{1}{\pi} \frac{\partial \log (1/r)}{\partial \rho} \bigg]_{\rho=1} \qquad r^2 = 1 - 2\rho \cos (\theta - \phi) + \rho^2 \tag{62}$$

This works out to

$$K(P,Q) \equiv -\frac{1}{2\pi} \tag{63}$$

Substituting into (52), we obtain for the density the integral equation

$$\tau(\theta) = -\frac{f(\theta)}{\pi} - \frac{1}{2\pi} \int_0^{2\pi} \tau(\phi) \, d\phi \tag{64}$$

Hence, $\tau(\theta) = c - f(\theta)/\pi$, and the constant c is determined by substitution into (64):

$$c - \frac{f(\theta)}{\pi} = -\frac{f(\theta)}{\pi} - \frac{1}{2\pi} \int_0^{2\pi} \left[c - \frac{f(\phi)}{\pi} \right] d\phi \tag{65}$$

From (65) we obtain

$$c = \frac{1}{4\pi^2} \int_0^{2\pi} f(\phi) \, d\phi \tag{66}$$

and thus

$$\tau(\theta) = \frac{1}{4\pi^2} \int_0^{2\pi} f(\phi) \, d\phi - \frac{f(\theta)}{\pi} \tag{67}$$

Substituting from (67) into (50) and performing a few elementary manipulations, we obtain for any point P of the disc the formula[1]

$$u(P) = \frac{1}{\pi} \int_0^{2\pi} f(\phi) \left[\frac{1 - r \cos (\theta - \phi)}{1 - 2r \cos (\theta - \phi) + r^2} \right.$$
$$\left. - \frac{1}{4\pi} \int_0^{2\pi} \frac{1 - r \cos (\alpha - \theta)}{1 - 2r \cos (\alpha - \theta) + r^2} \, d\alpha \right] d\phi \tag{68}$$

[1] In (68) the symbol r denotes the radius vector of P, not the length $\overline{PQ}$.

By elementary computation it can be shown that

$$\int_0^{2\pi} \frac{1 - r \cos (\alpha - \theta)}{1 - 2r \cos (\alpha - \theta) + r^2} \, d\alpha = 2\pi \tag{69}$$

and (68) therefore simplifies to

$$u(P) = \frac{1}{2\pi} \int_0^{2\pi} f(\phi) \frac{1 - r^2}{1 - 2r \cos (\theta - \phi) + r^2} \, d\phi \tag{70}$$

in agreement with equation (6-27).

A few remarks may be made concerning the scope of the method of integral equations. In addition to providing an existence proof, this method provides, in (50) and (52), a pair of equations which are, for some domains, amenable to accurate computational techniques. A second advantage is that the basic ideas which were used in this case can be generalized to provide existence proofs for boundary-value problems for certain classes of elliptic equations. On the other hand, a serious disadvantage of this method is that it is tied very strongly to the smoothness properties of the boundary. If the boundary consists of a single closed curve formed by two smooth arcs which form an angle at one or both of their common end points, the kernel of equation (52) becomes discontinuous. In cases such as this the method of integral equations can be modified, but requires, for example, a proof that a finite number of iterations leads to a continuous kernel, and that the solution of the integral equation obtained by iterating (52) this number of times is also a solution of (52). However, if the boundary is not "almost" smooth, in some sense, the method of integral equations becomes completely inapplicable.

EXERCISES

22. Prove that the kernel $K(P,Q)$ is continuous (jointly) for all P and Q on Γ. (Recall that Γ is assumed to possess continuous curvature.)

23. Prove that the Neumann problem (Exercise 6-30) can be formulated as an integral equation similar to (52), with a single layer instead of a double layer.

24. Prove, with the aid of (54'), the assertion made in the text concerning the behavior of the gradient of v "at infinity."

25. Work out the details leading to (68), (69), and (70).

5. The Dirichlet Principle

The Dirichlet problem is readily converted, by a simple argument, into a problem of the calculus of variations. Though the approach thus introduced into the study of the Dirichlet problem appeared at first to offer an elementary method of establishing the existence of a solution, it was subsequently realized that the argument employed was defective. However, after a lapse of several decades it was shown that the variational approach could, in fact, be used to study the Dirichlet problem and, more generally, boundary-value problems involving a broad class of linear elliptic equations.

The argument leading from the Dirichlet problem to the "Dirichlet principle," as Riemann named the variational principle to be presented here, proceeds as follows.[1] Given a domain G and a function f defined on the boundary, associate with each sufficiently differentiable function u defined in $\bar{G}$ and assuming the prescribed boundary values the number

$$D(u) = \iint_G (u_x{}^2 + u_y{}^2) \, dx \, dy \tag{71}$$

This quantity is termed the Dirichlet integral of the function u. Since $D(u)$ is, as the integral of the sum of squares, nonnegative, there is a finite (in fact, nonnegative) lower bound on $D(u)$. Let v denote a function for which this lower bound is attained. Then, for any function w which vanishes on the boundary and for any constant ϵ, the inequality

$$D(v + \epsilon w) \geq D(v) \tag{72}$$

must hold, since $v + \epsilon w$ satisfies the prescribed boundary condition. Writing out (72) in the form

$$D(v) + 2\epsilon D(v,w) + \epsilon^2 D(w) \geq D(v) \tag{73}$$

where

$$D(v,w) = \iint_G (v_x w_x + v_y w_y) \, dx \, dy \tag{74}$$

we easily show that $D(v,w)$ must vanish. Suppose, for example, that $D(v,w)$ were positive. Then we could choose ϵ negative and so close to zero that $2D(v,w) + \epsilon D(w)$ would also be positive. Then $2\epsilon D(v,w) + \epsilon^2 D(w)$ would be negative, contradicting (73). Similarly, the possibility that $D(v,w)$ might be negative is ruled out. Hence, taking account of (6-9), with u replaced by w, we obtain

$$D(v,w) = - \iint_G w \, \Delta v \, dx \, dy = 0 \tag{75}$$

since the vanishing of w causes the boundary integral to vanish. Now, suppose that Δv is different from zero at any point P of G. We select a disc which contains P and is so small that Δv maintains the same sign throughout the disc, and then construct a function w which is positive in the (open) disc and zero outside. (Such a function is easily constructed; cf. Exercise 26.) Then (75) is obviously violated. We conclude that v is harmonic throughout G, and hence that the function minimizing (71) subject to the given boundary condition also provides the solution to the Dirichlet problem with the same boundary condition. The uniqueness

---

[1] The reader acquainted with the calculus of variations will recognize that we are simply deriving here the Euler-Lagrange equation associated with the integral (71).

of v, incidentally, would follow either from Theorem 6-3 or from the following argument. Suppose that there were two functions which minimize (71), say v and $\tilde{v}$. Letting $w = \tilde{v} - v$ and taking account of the fact that $D(v,w)$ vanishes, we obtain

$$D(\tilde{v}) = D(v) + D(w)$$

and hence

$$D(w) = 0$$

Therefore w reduces to a constant, which must be zero, since v and $\tilde{v}$ coincide on the boundary.

There are a number of defects in the above arguments. First, we note that the use of (6-9) tacitly assumes that v possesses second derivatives, that the boundary of G is sufficiently smooth, and that the behavior of v near the boundary is "reasonable." Also, the argument involving the disc tacitly assumes the continuity of Δv. Aside from these objections, however, there is one which is absolutely basic, namely, that the fact that the integral (71) possesses a lower bound does not guarantee the existence of a function for which this lower bound is actually attained. Related to this basic objection is another one, namely, that it is possible to prescribe continuous boundary values such that, for every function continuously differentiable in G, continuous in $\bar{G}$, and possessing the prescribed boundary values, the integral (71) is divergent. (Cf. Exercise 27.)

We devote the remainder of this section to showing how it is possible, by suitably modifying the formulation of the Dirichlet problem, to approach this problem along the lines suggested by the foregoing incorrect arguments.

Until otherwise indicated, we make no assumption concerning the domain G except that it is bounded. We associate with G two normed linear spaces, as follows.

DEFINITION 4. H denotes the space of real-valued functions continuous and quadratically integrable in G; $\|u\|_H$ is defined as $\left(\iint_G u^2 \, dx \, dy\right)^{\frac{1}{2}}$

DEFINITION 5. D denotes the subset of H consisting of functions whose first derivatives also belong to H; $\|u\|_D$ is defined as[1]

$$\left[\iint_G (u_x{}^2 + u_y{}^2) \, dx \, dy\right]^{\frac{1}{2}}$$

We remark that H and D each become (real) inner-product spaces with the following definitions of inner products:

$$H(u,v) = \iint_G uv \, dx \, dy \qquad D(u,v) = \iint_G (u_x v_x + u_y v_y) \, dx \, dy \quad (76)$$

---

[1] A slight complication arises from the fact that all constant functions, not only the zero function, have zero norm in D. This causes no difficulty in the following developments.

The proof that either inner product is meaningful when the functions u and v belong to the appropriate space is entirely elementary and may be omitted.

We next define two linear manifolds of D.

DEFINITION 6.   By $\dot{D}$ we denote the subset of D consisting of those functions which vanish in some boundary strip;[1] that is, for each such function u there exists a positive number ϵ such that u vanishes at all points of G whose distance from the boundary is less than ϵ.   By $\mathring{D}$ we denote the subset of D consisting of those functions which are expressible as the limit in *both* norms of a sequence of functions of $\dot{D}$; that is, $u \in \mathring{D}$ if there exists a sequence $\{u_n\}$ of functions of $\dot{D}$ such that $\|u - u_n\|_H \to 0$, $\|u - u_n\|_D \to 0$.

Theorem 8. $\mathring{D}$ is relatively closed in D; that is, if there exists a sequence $\{u_n\}$ of functions of $\mathring{D}$ such that $\|u - u_n\|_H \to 0$, $\|u - u_n\|_D \to 0$, then $u \in \mathring{D}$.

Proof.   For each u_n we can, according to Definition 6, find an element v_n of $\dot{D}$ such that

$$\|u_n - v_n\|_H < \frac{1}{n} \qquad \|u_n - v_n\|_D < \frac{1}{n}$$

By the triangle inequality,

$$\|u - v_n\|_H \leq \|u - u_n\|_H + \|u_n - v_n\|_H < \|u - u_n\|_H + \frac{1}{n}$$

Hence $\|u - v_n\|_H \to 0$, and, similarly, $\|u - v_n\|_D \to 0$.   By Definition 6, $u \in \mathring{D}$.

We now proceed to reformulate the Dirichlet problem.   Actually, we find it convenient to define two problems, which will subsequently be shown to have the *same* unique solution.   Then it will be shown that, under certain circumstances, this solution also solves the Dirichlet problem as originally formulated in Sec. 6-6.

GENERALIZED DIRICHLET PROBLEM.   Given a function $g \in D$, to find a harmonic function $u \in D$ such that $u - g \in \mathring{D}$.

VARIATIONAL DIRICHLET PROBLEM.   Given $g \in D$, to find $u \in D$ such that $u - g \in \mathring{D}$ and $\|u\|_D$ is as small as possible.

It should be stressed that the condition $u - g \in \mathring{D}$ is indeed a boundary condition, in the sense that only the behavior of u and g near the boundary of the domain G is involved; for, if u_1 and u_2 coincide in some boundary

---

[1] Such functions are said to be "of compact support," the support of u being the closure with respect to G of the set on which u does not vanish.

strip, and similarly for g_1 and g_2, then, from the equality

$$u_2 - g_2 = (u_1 - g_1) + (u_2 - u_1) + (g_1 - g_2)$$

it follows that $u_2 - g_2$ belongs to $\mathring{D}$ if and only if $u_1 - g_1$ does. (Of course, the functions u_1, u_2, g_1, g_2 are required to belong to D.) It may also be noted that it is not assumed that g can be extended continuously from G to $\bar{G}$; such an assumption will have to be made only in order to show that the common solution of the two problems formulated above possesses the boundary behavior required in the original formulation of the Dirichlet problem.

Before proceeding with the details, we indicate briefly the general outline. From the class of "competing" functions (i.e., the functions of D which satisfy the "boundary condition") we select a sequence $\{\phi_n\}$ such that $\|\phi_n\|_D$ approaches the minimum value. Then it is shown that $\{\phi_n\}$ is a Cauchy sequence with respect to both the H and the D norms. Although this does not suffice to guarantee convergence (either pointwise or in norm) of the sequence, it will be shown that a rather simple averaging process may be applied to the functions $\{\phi_n\}$ to define a function u, which is then shown to be harmonic throughout G. Then it is shown that u is, in fact, the limit in the D norm of the sequence $\{\phi_n\}$, from which it follows that $u - g \in \mathring{D}$ and that $\|u\|_D$ is actually equal to the minimum possible value. Thus, u provides the solution to both of the problems formulated above. Finally, as indicated above, it will be shown that, if the domain G and the "boundary function" g satisfy certain conditions, u also solves the Dirichlet problem as originally formulated.

Several simple theorems will be proved before the function u is constructed.

Theorem 9. If $\phi \in D$ and is harmonic throughout G, and if $\psi \in \mathring{D}$, then

$$D(\phi,\psi) = 0 \tag{77}$$

Proof. If $\psi \in \mathring{D}$, then $D(\phi,\psi)$ is equal to $\iint_{G'} (\phi_x\psi_x + \phi_y\psi_y)\, dx\, dy$, where G' is a smoothly bounded domain obtained from G by discarding a suitable portion of the boundary strip in which ψ vanishes. Since (6-9) is then applicable (with u and v replaced by ψ and ϕ, respectively), and since ψ vanishes on the boundary of G', we immediately obtain (77). If $\psi \notin \mathring{D}$, we select a sequence $\{\psi_n\}$ of functions belonging to $\mathring{D}$ and converging to ψ in the D norm. Then, using the Schwarz inequality, we obtain

$$|D(\phi,\psi)| = |D(\phi,\psi_n) + D(\phi,\psi - \psi_n)| = |D(\phi,\psi - \psi_n)|$$
$$\leq \|\phi\|_D \cdot \|\psi - \psi_n\|_D \to 0$$

so that (77) holds in this case also.

Theorem 10.   For every function ϕ of $\dot{D}$, the inequality[1]

$$H(\phi) \leq \gamma D(\phi) \tag{78}$$

holds, γ denoting a constant determined by G.

Proof.   Let G be contained in the square $|x| < \alpha$, $|y| < \alpha$, and let $\gamma = 4\alpha^2$.   Then if $\phi \in \dot{D}$ and is defined to be $\equiv 0$ outside G, we obtain (since ϕ is now continuously differentiable in the entire square), at any point (x,y) of the square, the equality $\phi(x,y) = \int_{-\alpha}^{x} \phi_\xi(\xi,y)\, d\xi$, and hence, by the Schwarz inequality,

$$\phi^2(x,y) \leq \left(\int_{-\alpha}^{x} 1^2\, d\xi \right) \left(\int_{-\alpha}^{x} \phi_\xi^2(\xi,y)\, d\xi \right) \leq 2\alpha \int_{-\alpha}^{\alpha} \phi_\xi^2(\xi,y)\, d\xi \tag{79}$$

Integrating with respect to y, we obtain

$$\int_{-\alpha}^{\alpha} \phi^2(x,y)\, dy \leq 2\alpha \int_{-\alpha}^{\alpha} \int_{-\alpha}^{\alpha} \phi_\xi^2(\xi,y)\, d\xi\, dy \leq 2\alpha D(\phi) \tag{80}$$

Finally, integrating with respect to x, we obtain the desired result, for

$$H(\phi) = \int_{-\alpha}^{\alpha} \int_{-\alpha}^{\alpha} \phi^2(x,y)\, dx\, dy \leq 4\alpha^2 D(\phi) = \gamma D(\phi) \tag{81}$$

If $\phi \notin \dot{D}$, we select a sequence $\{\phi_\nu\}$ of functions belonging to $\dot{D}$ and converging to ϕ in both norms.   Then, from what has already been shown, the inequality $H(\phi_n) \leq \gamma D(\phi_n)$ holds for $n = 1, 2, \ldots$. Passing to the limit and recalling that the norms of a convergent sequence converge to the norm of the limit element, we obtain (78).

Theorem 11.   The "generalized Dirichlet problem" has at most one solution.

Proof.   Suppose there were two solutions, u and v.   Then both ϕ and ψ in (77) could be replaced by $u - v$, and we would conclude that $D(u - v) = 0$.   Then (78) would imply that $H(u - v) = 0$, or $u \equiv v$.

Theorem 12.   The "variational Dirichlet problem" has at most one solution.

Proof.   Suppose that both u and v provide solutions; then, for all values of the parameter ϵ,

$$D(u + \epsilon(v - u)) = D(u) + 2\epsilon D(u, v - u) + \epsilon^2 D(u - v) \geq D(u) \tag{82}$$

since $v - u \in \dot{D}$.   Thus, the quadratic function of ϵ appearing in (82) attains its minimum value for two distinct values of ϵ, namely, 0 and 1, but this is possible only if $D(u - v) = 0$.   As in the preceding proof, we conclude that $u \equiv v$.

---

[1] $H(\phi)$ and $D(\phi)$ are abbreviations for $H(\phi,\phi)$ and $D(\phi,\phi)$, respectively.

Theorem 13. The solution of the generalized Dirichlet problem, if it exists, also solves the variational Dirichlet problem.

Proof. Let u solve the former problem and let w be any function satisfying the boundary condition; i.e., $w - g \in \mathring{D}$. Then, by Theorem 9, $D(u, w - u) = 0$, and therefore

$$D(w) = D(u) + D(w - u) \geq D(u) \tag{83}$$

showing that u solves the latter problem as well.

Since $D(\phi)$ cannot be negative, it is meaningful to define the quantity d as follows:

$$d = \inf D(\phi) \qquad (\phi \in D, \ \phi - g \in \mathring{D}) \tag{84}$$

Although it is not obvious that there exists a function for which the lower bound is actually attained (this is, of course, the essential difficulty of the entire analysis), there must in any case exist a "minimizing sequence" of competing functions $\{\phi_n\}$ such that

$$D(\phi_n) \to d \tag{85}$$

Henceforth $\{\phi_n\}$ denotes a fixed minimizing sequence.

Theorem 14. Let $\{\zeta_n\}$ be any sequence of elements of $\mathring{D}$ with bounded D norms, say $\|\zeta_n\|_D \leq M$. Then $D(\phi_n, \zeta_n) \to 0$.

Proof. For any value of the parameter ϵ, $\phi_n + \epsilon\zeta_n$ satisfies the boundary condition, and therefore

$$D(\phi_n + \epsilon\zeta_n) - d = [(D(\phi_n) - d)] + 2\epsilon D(\phi_n, \zeta_n) + \epsilon^2 D(\zeta_n) \geq 0 \tag{86}$$

The discriminant of the quadratic function of ϵ must then be nonpositive, and so

$$D^2(\phi_n, \zeta_n) \leq [(D(\phi_n) - d)]D(\zeta_n) \leq M^2[D(\phi_n) - d] \to 0 \tag{87}$$

Theorem 15. $\{\phi_n\}$ is a Cauchy sequence with respect to both norms.

Proof. From (85) it follows that the quantities $\|\phi_n\|_D$ are bounded, say by $M/2$. Hence

$$\|\phi_n - \phi_m\|_D \leq \|\phi_n\|_D + \|\phi_m\|_D \leq M$$

and, since $\phi_n - \phi_m \in \mathring{D}$, we can, by the preceding theorem, given $\delta > 0$, choose N so large that

$$|D(\phi_{n'}, \phi_n - \phi_m)| < \delta$$

whenever $n' > N$. In particular, if n and m also exceed N, we obtain

$$|D(\phi_n, \phi_n - \phi_m)| < \delta \qquad |D(\phi_m, \phi_n - \phi_m)| < \delta$$

and hence

$$D(\phi_n - \phi_m) = D(\phi_n, \phi_n - \phi_m) - D(\phi_m, \phi_n - \phi_m) < 2\delta$$

Thus the Cauchy convergence is established for the D norm, and Theorem 10 now enables us to draw the same conclusion for the H norm.

This theorem suggests an interesting proof of Theorem 12. If there were two solutions of the variational problem, say u and v, then u, v, u, v, . . . would be a minimizing sequence and hence, by Theorem 15, a Cauchy sequence. Then $D(u - v) = 0$, and therefore, as in the proof of Theorem 11, $u \equiv v$.

We now proceed to construct the function which solves both of the afore-mentioned problems. Let G' denote any closed subset of G, and let ρ denote the distance between G' and the boundary of G. Then, for any positive number $R < \rho$, the following definition is meaningful at all points P_0 of G':

$$U_{n,R}(P_0) = \iint_{\Delta_R} \phi_n \, dx \, dy \tag{88}$$

where Δ_R denotes the disc of radius R with center at P_0. By the Schwarz inequality we obtain

$$[U_{n,R}(P_0) - U_{m,R}(P_0)]^2 \leq \left(\iint_{\Delta_R} 1^2 \, dx \, dy \right) \left(\iint_{\Delta_R} (\phi_n - \phi_m)^2 \, dx \, dy \right)$$
$$\leq \pi R^2 H(\phi_n - \phi_m) \to 0 \tag{89}$$

The uniform convergence of the sequence $\{U_{n,R}\}$ throughout G' is thus assured; since each function $U_{n,R}$ is continuous (this follows from the continuity of ϕ_n), it follows that the limit function, which we denote as U_R, is also continuous. Now, suppose that the same construction is carried out with R replaced by a smaller radius, say R'. (The corresponding function $U_{R'}$ could then be defined in a set larger than G', but this need not concern us.) With any point P_0 of G' we associate the function

$$\zeta(P, P_0) = \begin{cases} 2 \log \dfrac{R'}{R} + r^2 \left(\dfrac{1}{R'^2} - \dfrac{1}{R^2} \right) & r < R' \\[2ex] 2 \log \dfrac{r}{R} + \left(1 - \dfrac{r^2}{R^2} \right) & R' \leq r < R \qquad (r = \overline{P_0 P}) \\[2ex] 0 & r \geq R \end{cases} \tag{90}$$

From the continuity of ζ and of its first derivatives (cf. Exercise 28) and the fact that it vanishes for $r \geq R$, it follows that $\zeta \in \dot{D}$ (in fact, $\zeta \in \dot{D}$), and from Theorem 14, taking each ζ_n as ζ, we conclude that $D(\phi_n, \zeta) \to 0$. Now, by (6-9),†

$$D(\phi_n, \zeta) = - \iint_G \phi_n \, \Delta \zeta \, dx \, dy \tag{91}$$

---

† Although the second derivatives of ζ fail to exist on the circles $r = R'$ and $r = R$, the applicability of (6-9) follows readily by dividing G into the regions $r < R'$, $R' \leq r < R$, and the remainder of G, and then taking account of the continuity of the *first* derivatives of ζ on the afore-mentioned circles.

Since $\Delta\zeta$ equals $4(1/R'^2 - 1/R^2)$ for $r < R'$ and $-4/R^2$ for $R' < r < R$, (91) may be rewritten as follows:

$$D(\phi_n, \zeta) = \frac{4}{R^2} \iint_{\Delta_R} \phi_n \, dx \, dy - \frac{4}{R'^2} \iint_{\Delta_{R'}} \phi_n \, dx \, dy \qquad (92)$$

Thus, the assertion that $D(\phi_n, \zeta) \to 0$ is equivalent to the following one:

$$(\pi R^2)^{-1} U_{n,R}(P_0) - (\pi R'^2)^{-1} U_{n,R'}(P_0) \to 0 \qquad (P_0 \in G') \qquad (93)$$

This, in turn, is equivalent to the assertion that the function $(\pi R^2)^{-1} U_R$ is independent of R, so that we may denote it simply as u. Since R may be chosen arbitrarily small, we may consider u to be defined throughout G. Thus, u is defined at each point P of G by averaging ϕ_n over an arbitrary disc (independent of n) with center at P and then letting $n \to \infty$.

Theorem 16. The function u defined above is harmonic throughout G.

Proof. Select any point P of G and a disc Δ of radius R with center at P such that $\bar{\Delta} \subset G$. Given $\epsilon > 0$, it is possible to choose $\delta > 0$ such that $|u(Q_2) - u(Q_1)| < \epsilon$ for every pair of points Q_1, Q_2 in Δ satisfying the inequality $\overline{Q_1 Q_2} < \delta$, and also such that it is possible to express Δ as the union of a finite number of nonoverlapping discs whose radii are each less than δ together with a remaining set S of area not exceeding ϵ. The first part of the above statement follows from the fact that u is uniformly continuous on any closed subset of G, and in particular $\bar{\Delta}$, while the second fol'ows from a simple geometrical argument.[1] Let the small discs be denoted by $\Delta_1, \Delta_2, \ldots, \Delta_N$; their radii by $r_1, r_2, \ldots, r_N$; and their centers by $P_1, P_2, \ldots, P_N$. Since $\left| \iint_S u \, dx \, dy \right| \leq \epsilon \cdot \max_{Q \in \Delta} |u(Q)|$ and $\left(\iint_S \phi_n \, dx \, dy \right)^2 \leq \left(\iint_S 1^2 \, dx \, dy \right) \left(\iint_S \phi_n^2 \, dx \, dy \right) \leq \epsilon H(\phi_n)$, we obtain the inequality[2]

$$\left| \iint_S (u - \phi_n) \, dx \, dy \right| \leq \epsilon \cdot \max_{Q \in \Delta} |u(Q)| + \epsilon^{1/2} \cdot \max \|\phi_n\|_H \qquad (94)$$

---

[1] Since an arbitrarily large fraction of the area of Δ can be exhausted by constructing a sufficiently fine, equally spaced mesh and retaining those squares of the mesh lying entirely within Δ, it suffices to prove that an arbitrarily large fraction of the area of a given square can be exhausted by nonoverlapping discs. If the inscribed disc is removed from the given square, less than one-fourth of the original area A remains. The remainder is then exhausted to a sufficiently great extent by a set of squares, as above, and then the inscribed disc of each of these new squares is removed. Now the remaining area amounts to less than $A/4^2$. If this procedure is repeated a sufficient number of times, the original square is exhausted to the desired extent.

[2] The existence of max $\|\phi_n\|_H$ follows from Theorem 15.

Next, expressing $u - \phi_n$ in Δ_i as $[u - u(P_i)] + [u(P_i) - \phi_n]$, and taking account of the manner in which the discs Δ_i were chosen and of the definitions of U_{n,r_i} and U_{r_i}, we obtain

$$\left| \iint_{\Delta_i} (u - \phi_n) \, dx \, dy \right| \leq \epsilon \cdot \pi r_i^2 + |U_{r_i}(P_i) - U_{n,r_i}(P_i)| \qquad (95)$$

Summing (95) over the discs Δ_i, we obtain

$$\left| \iint_{\Delta-S} (u - \phi_n) \, dx \, dy \right| \leq \epsilon \cdot \pi R^2 + \sum_{i=1}^{N} |U_{r_i}(P_i) - U_{n,r_i}(P_i)| \qquad (96)$$

Combining (94) with (96), and taking account of the arbitrariness of ϵ, we conclude that $\iint_\Delta (u - \phi_n) \, dx \, dy \to 0$, or

$$\pi R^2 u(P) = \iint_\Delta u \, dx \, dy \qquad (97)$$

Since the above argument can be carried out for all points P of G and all values of R less than the distance between P and the boundary of G, we may invoke Theorem 6-5 to conclude that u is harmonic throughout G, as was to be proved. (Note that we are not exploiting here the fact that every disc is a Dirichlet domain, for the second proof of Theorem 6-5 does not use this fact.) We remark that this proof may be looked upon as a rigorous version of the following heuristic argument: Since u was constructed as a generalized limit of the sequence $\{\phi_n\}$ (recall that the generalized limiting operation consists of averaging the functions ϕ_n before letting $n \to \infty$), it appears plausible that the result of applying the averaging operation to u is to yield u over again. Thus u should possess the areal mean-value property, and it may therefore be expected to be harmonic.

Theorem 17. Let f be continuous in G and let G' be any subdomain (or, more generally, any open subset) of G such that $\overline{G'} \subset G$. Then $\iint_{G'} (u - \phi_n) f \, dx \, dy \to 0$.

Proof. Given $\epsilon > 0$, we can exhaust the area of G' to within at most ϵ with a finite number of nonoverlapping discs which are so small that $|f(Q) - f(Q')| < \epsilon$ for any two points Q, Q' of the same disc. (The argument given in the preceding proof for exhausting a disc obviously applies equally well to an arbitrary open set.) Then, employing the same notation as in the preceding proof and the equality

$$\iint_{G'} (u - \phi_n) f \, dx \, dy = \sum_{i=1}^{N} f(P_i) \iint_{\Delta_i} (u - \phi_n) \, dx \, dy$$

$$+ \sum_{i=1}^{N} \iint_{\Delta_i} (u - \phi_n)[f - f(P_i)] \, dx \, dy + \iint_S (u - \phi_n) f \, dx \, dy \qquad (98)$$

we proceed to analyze each term on the right. First,

$$\iint_{\Delta_i} (u - \phi_n) \, dx \, dy \to 0$$

This follows from (97) and the definition of u. The first sum therefore approaches zero. We estimate the second sum by noting that it is dominated by

$$\epsilon \iint_{G'} |u - \phi_n| \, dx \, dy$$

and, a fortiori, by

$$\epsilon \left(\iint_{G'} |u| \, dx \, dy + \iint_{G} |\phi_n| \, dx \, dy \right)$$

which cannot exceed, by the Schwarz inequality,

$$\epsilon \left[\iint_{G'} |u| \, dx \, dy + \|\phi_n\|_H \cdot (\text{area of } G)^{1/2} \right]$$

Since the sequence $\{\|\phi_n\|_H\}$ is bounded (cf. preceding footnote), we have shown that the second term is dominated by ϵC, where C is independent of ϵ and n. The last term on the right side of (98) is dominated, similarly, by

$$\max_{P \in S} |f(P)| \cdot \left[\iint_S |u| \, dx \, dy + \|\phi_n\|_H \cdot (\text{area of } S)^{1/2} \right]$$

and, a fortiori, by

$$\max_{P \in G'} |f(P)| \cdot [\epsilon \max_{P \in G'} |u(P)| + \epsilon^{1/2} \max \|\phi_n\|_H]$$

Combining the above results, we conclude that

$$\limsup_{n \to \infty} \left| \iint_{G'} (u - \phi_n) f \, dx \, dy \right| \leq C_1 \epsilon + C_2 \epsilon^{1/2} \tag{99}$$

where C_1 and C_2 are independent of ϵ. Since ϵ is arbitrarily small, the left side of (99) must vanish, and the theorem is proved. It should be stressed that the condition $\bar{G}' \subset G$ is essential in the proof, for otherwise u and f might be unbounded in G'.

Theorem 18. $\|u - \phi_n\|_H \to 0$; that is, the sequence $\{\phi_n\}$ converges to u in the H norm.

Proof. First we must prove that $u \in H$. Let G' be chosen as in Theorem 17, and let N be chosen such that $\|\phi_N - \phi_n\|_H < \epsilon$ for all $n > N$. Then we write

$$\iint_{G'} (u - \phi_n)^2 \, dx \, dy = \iint_{G'} (u - \phi_n)(u - \phi_N) \, dx \, dy$$
$$+ \iint_{G'} (u - \phi_n)(\phi_N - \phi_n) \, dx \, dy \tag{100}$$

and note that the first term on the right vanishes as $n \to \infty$, by the preceding theorem. We apply the Schwarz inequality to the second term and conclude that it is dominated by

$$\left(\iint_{G'} (u - \phi_n)^2 \, dx \, dy \right)^{1/2} \cdot \| \phi_N - \phi_n \|_H$$

and hence by

$$\epsilon \left(\iint_{G'} (u - \phi_n)^2 \, dx \, dy \right)^{1/2}$$

The latter expression is dominated, according to the triangle inequality, by

$$\epsilon \cdot \left[\left(\iint_{G'} u^2 \, dx \, dy \right)^{1/2} + \| \phi_n \|_H \right]$$

Taking account of the boundedness of $\| \phi_n \|_H$ and the arbitrariness of ϵ, we conclude that

$$\iint_{G'} (u - \phi_n)^2 \, dx \, dy \to 0$$

Since the norm is known to be a continuous function in a normed space (we momentarily associate with G' a space analogous to H), it follows that

$$\iint_{G'} u^2 \, dx \, dy = \lim_{n \to \infty} \iint_{G'} \phi_n{}^2 \, dx \, dy \leq \lim_{n \to \infty} \iint_{G} \phi_n{}^2 \, dx \, dy \qquad (101)$$

Since G' may exhaust G to within an arbitrarily small remainder, we conclude that $\iint_{G} u^2 \, dx \, dy$ exists (so that $u \in H$) and, in fact, is not greater than $\lim_{n \to \infty} H(\phi_n)$.

Now, given $\epsilon > 0$, we choose N as before, and then choose the open subset G' of G such that (in addition to the earlier condition $\overline{G'} \subset G$) both of the following conditions are satisfied:

$$\iint_{G-G'} u^2 \, dx \, dy < \epsilon^2 \qquad \iint_{G-G'} \phi_N{}^2 \, dx \, dy < \epsilon^2 \qquad (102)$$

For convenience we introduce the notation $\| w \|_{H'}$ and $\| w \|_{H''}$ for $\left(\iint_{G'} w^2 \, dx \, dy \right)^{1/2}$ and $\left(\iint_{G-G'} w^2 \, dx \, dy \right)^{1/2}$, respectively. Since

$$H(u - \phi_n) = \| u - \phi_n \|_{H'}^2 + \| u - \phi_N + (\phi_N - \phi_n) \|_{H''}^2 \leq \| u - \phi_n \|_{H'}^2$$
$$+ (\| u \|_{H''} + \| \phi_N \|_{H''} + \| \phi_N - \phi_n \|_{H''})^2 < \| u - \phi_n \|_{H'}^2 + (3\epsilon)^2$$

(for $n > N$), it follows from Theorem 17 and the arbitrariness of ϵ that $\| u - \phi_n \|_H \to 0$, as was to be proved.

Theorem 19. Let Δ be any disc such that $\bar{\Delta} \subset G$. Then

$$\iint_{\Delta} (u - \phi_n)_x \, dx \, dy \to 0$$

Proof. Let f be any function which is continuously differentiable in $\bar{\Delta}$ and vanishes on the boundary of Δ. Then since $\iint_{\Delta} [(u - \phi_n)f]_x \, dx \, dy$ vanishes,[1] we may write

$$\iint_{\Delta} (u - \phi_n)_x f \, dx \, dy = - \iint_{\Delta} (u - \phi_n) f_x \, dx \, dy \qquad (103)$$

From Theorem 17 (with f replaced by f_x), we conclude that the right side, and hence the left side, of (103) approaches zero as $n \to \infty$. Given $\epsilon > 0$, choose f so as to satisfy, in addition to the afore-mentioned conditions, the following:

(1) $f \equiv 1 \qquad 0 \le r < R - \epsilon$
(2) $0 \le f \le 1 \qquad R - \epsilon \le r \le R$

(Here R denotes the radius of Δ and r the distance from the center of Δ.) It is obvious that such a function exists; for example, we may define f in the annulus $R - \epsilon \le r \le R$ as

$$\sin^2 \frac{\pi}{2\epsilon} (r - R)$$

Then we write

$$\iint_{\Delta} (u - \phi_n)_x \, dx \, dy - \iint_{\Delta} (u - \phi_n)_x f \, dx \, dy$$
$$= \iint_{\tilde{\Delta}} (u - \phi_n)_x (1 - f) \, dx \, dy \qquad (104)$$

where $\tilde{\Delta}$ denotes the annulus $R - \epsilon < r < R$. We now apply the Schwarz inequality to the right side and thus conclude that it is dominated by $\|(u - \phi_n)_x\|_{\tilde{\Delta}} \cdot (\text{area of } \tilde{\Delta})^{1/2}$, which is dominated in turn by $(2\pi R \epsilon)^{1/2} \cdot (\|u_x\|_{\tilde{\Delta}} + \max \|\phi_n\|_D)$. Therefore the left side, and hence the first integral, must approach zero as $n \to \infty$, and the proof is complete.

Now, since u is harmonic in G, so is u_x; furthermore, the fact that $\{\phi_n\}$ is a Cauchy sequence in the D norm implies that $\{\phi_{n,x}\}$ is a Cauchy sequence in the H norm. Once Theorem 19 has been established, it is clear that the line of reasoning which culminated in Theorem 18 may be repeated with u and ϕ_n replaced everywhere by u_x and $\phi_{n,x}$. We thus conclude that u_x, and similarly u_y, belongs to H, and that

$$\|u_x - \phi_{n,x}\|_H \to 0 \qquad \|u_y - \phi_{n,y}\|_H \to 0 \qquad (105)$$

Now (105) immediately implies that

$$\|u - \phi_n\|_D \to 0 \qquad (106)$$

Thus we have shown that u is the limit of the sequence $\{\phi_n\}$ in both

---

[1] Replace f and g in (6-8) by $(u - \phi_n)f$ and 0, respectively, and recall that f vanishes on the boundary of Δ.

norms. Therefore, $\|u\|_D = \lim\limits_{n \to \infty} \|\phi_n\|_D = d$, and, furthermore, since $\{\phi_n - g\}$ converges to $\{u - g\}$ in both norms, it follows from Theorem 8 that $u - g \in \dot{D}$. Thus we have proved the following theorem, which is the principal result of the present section.

Theorem 20. The "generalized" and "variational" Dirichlet problems are both solvable, and the unique solution of each is furnished by the function u which was constructed from the minimizing sequence $\{\phi_n\}$ in the manner described following Theorem 15.

Finally, we return to the original formulation of the Dirichlet problem, and establish the following theorem.

Theorem 21. Suppose that G has the property that there exists a positive constant R such that, for every positive constant R' less than R and for every boundary point Q, there exists a boundary point Q' whose distance from Q is exactly R'. Then, if the function g (belonging to D) can be extended continuously from G to $\bar{G}$, the function $u - g$ approaches zero at the boundary, so that u provides the solution to the Dirichlet problem as originally formulated.

Proof. Let P be any point of G whose distance, $2h$, from the boundary is less than $2R/3$, let Δ be the disc of radius h with center at P, let Q be a boundary point such that $\overline{PQ} = 2h$ (at least one such point must exist), and let S be the set consisting of all points of G whose distance from Q is less than $3h$. Then S is open, contains Δ, and each circle C_k of radius k less than $3h$ with center at Q lies partially, but not entirely, in S. We denote the portion of C_k lying in S as Γ_k; Γ_k consists of one or more (perhaps infinitely many) circular arcs whose end points[1] are boundary points of G.

We shall show later that, for any function ϕ belonging to $\dot{D}$, the following inequality holds:

$$\left(\iint_\Delta \phi \, dx \, dy \right)^2 \le 36\pi^3 h^4 \iint_S (\phi_x{}^2 + \phi_y{}^2) \, dx \, dy \qquad (107)$$

Accepting (107) momentarily, noting that $\iint_S (\phi_x{}^2 + \phi_y{}^2) \, dx \, dy$ is increased if S is replaced by the set of all points of G whose distance from the boundary is less than $3h$, and that the new integral thus obtained must approach zero with h (this is merely a restatement that $\phi \in D$), we conclude that $(\pi h^2)^{-1} \iint_\Delta \phi \, dx \, dy$ approaches zero as P approaches the boundary. Replacing ϕ by $u - g$ (recalling that $u - g \in \dot{D}$), we con-

---

[1] Γ_k may consist of all of C_k except for a single point; consider, for example, the domain obtained by removing from a disc one of its radii.

clude that, as P approaches the boundary,

$$\frac{1}{\pi h^2} \iint_\Delta u \, dx \, dy - g(P) - \frac{1}{\pi h^2} \iint_\Delta [g - g(P)] \, dx \, dy \to 0 \qquad (108)$$

By the (areal) mean-value theorem, the first term on the left equals $u(P)$. The last term is dominated by $\max_{P' \in \Delta} |g(P') - g(P)|$, which is dominated in turn by

$$\max |g(P') - g(P)| \qquad (P \in G, P' \in G, \overline{PP'} < h)$$

(It is at this point that we use the fact that g can be extended continuously to $\bar{G}$, for this guarantees the *uniform* continuity of g in G, and hence the existence of the above maximum.) Since this maximum must approach zero with h (again by uniform continuity), we conclude that $u - g$ approaches zero at the boundary, as was to be proved.

It remains to establish (107). It suffices to confine attention to the case that $\phi \in \dot{D}$, for the more general case that $\phi \in \dot{D}$ may then be treated by approximating ϕ suitably by functions in $\dot{D}$, as in the proof of Theorem 10. Now, by the Schwarz inequality, the left side of (107) is dominated by $\left(\iint_\Delta 1^2 \, dx \, dy \right) \left(\iint_\Delta \phi^2 \, dx \, dy \right)$, and hence by $\pi h^2 \iint_S \phi^2$ $dx \, dy$. It will therefore suffice to prove that

$$\iint_S \phi^2 \, dx \, dy \leq 36\pi^2 h^2 \iint_S (\phi_x{}^2 + \phi_y{}^2) \, dx \, dy \qquad (109)$$

Let polar coordinates r, θ be introduced with origin at Q; then ϕ is expressible at any point T of S in the form

$$\phi(T) = \int \phi_\theta \, d\theta \qquad (110)$$

where the integration is performed along an arc of Γ_r whose two end points are T and a boundary point of G. (Here $r = \overline{QT}$, and we are using the fact that $\phi \in \dot{D}$ and also the fact that Γ_r meets the boundary.) By applying the Schwarz inequality and lengthening the path of integration to all of Γ_r, we obtain

$$\phi^2(T) \leq \left(\int_{\Gamma_r} 1^2 \, d\theta \right) \left(\int_{\Gamma_r} \phi_\theta{}^2 \, d\theta \right) \leq 2\pi \int_{\Gamma_r} \phi_\theta{}^2 \, d\theta \qquad (111)$$

Since (111) holds for all points T on Γ_r, we obtain, by integrating both sides,

$$\int_{\Gamma_r} \phi^2(T) \, d\theta \leq (2\pi)^2 \int_{\Gamma_r} \phi_\theta{}^2 \, d\theta \qquad (112)$$

Multiplying both sides of (112) by $r \, dr$ and integrating from 0 to $3h$, we obtain

$$\iint_S \phi^2 \, dx \, dy \leq (2\pi)^2 \iint_S \phi_\theta{}^2 \, dx \, dy \qquad (113)$$

The right side of (113) is increased when the integrand is multiplied by $(3h/r)^2$, since $r < 3h$ throughout S. Thus,

$$\iint_S \phi^2 \, dx \, dy \leq 36\pi^2 h^2 \iint_S (r^{-1}\phi_\theta)^2 \, dx \, dy \tag{114}$$

Finally, noting that $(r^{-1}\phi_\theta)^2 \leq (r^{-1}\phi_\theta)^2 + \phi_r^2 = \phi_x^2 + \phi_y^2$, we find that (114) implies (109), and the proof is completed.

We conclude this section by pointing out two simple but highly important consequences of the above theorem. First, since every polynomial $p(x,y)$ belongs to D (for any bounded domain G), and since, as pointed out in Sec. 2, a bounded domain is a Dirichlet domain if the Dirichlet problem can be solved whenever the boundary values are expressible as a polynomial, it follows that G is a Dirichlet domain if it satisfies the condition imposed in Theorem 21. Secondly, any domain of finite connectivity whose boundary contains no isolated points is readily seen to satisfy the afore-mentioned condition, and so is a Dirichlet domain. (This is consistent with the example given in Sec. 6-6 of a domain which is not a Dirichlet domain.) In particular, we obtain the remarkable result that every bounded simply connected domain is a Dirichlet domain.

EXERCISES

26. Let Δ denote any disc. Define a function which possesses continuous derivatives of all orders throughout the plane, vanishes everywhere outside Δ, and is positive throughout Δ.

27. *Hadamard's example:* Let u be defined for $x^2 + y^2 \leq 1$ by the series

$$u(re^{i\theta}) = \sum_{n=1}^{\infty} \frac{r^{n!} \cos n!\theta}{n^2}$$

Prove that for this function, which, according to (6-36), solves the Dirichlet problem with the continuous boundary values

$$g(\theta) = \sum_{n=1}^{\infty} \frac{\cos n!\theta}{n^2}$$

the Dirichlet integral (71) is divergent.

28. Prove that the function $\zeta(P,P_0)$ defined in (90) is completely determined, except for a constant factor, by the following requirements: (a) ζ should depend only on $\overline{P_0P}$; (b) ζ should possess continuous first derivatives everywhere; (c) ζ should vanish for $\overline{P_0P} > R$; (d) $\Delta\zeta$ should be constant in the disc $\overline{P_0P} < R'$ and in the annulus $R' < \overline{P_0P} < R$.

29. Let u and v be harmonic in G and continuous in $\bar{G}$, and suppose that the relation $u = |v|$ is satisfied everywhere on the boundary. Prove that $D(u) \leq D(v)$. [This implies, in particular, that the finiteness of $D(v)$ guarantees that of $D(u)$.]

Assume that the boundary of G and the behavior of u and v near the boundary are such as to permit free use of the "transformation theorems" of Sec. 6-3.

30. Prove that $D(u)$ is invariant under (one-to-one) conformal mapping. (It is not assumed that u is harmonic.)

6. The Method of Finite Differences[1]

As in the case of ordinary differential equations, partial differential equations are often approximated by replacing the derivatives with corresponding difference quotients. The idea underlying such a procedure is, of course, the intuitively plausible one that, if the increments of the independent variables are chosen sufficiently small, an algebraic system of equations will be obtained whose solutions will be close, in some sense, to solutions of the given differential equation. An extensive literature has grown up about this idea. Not only has it been possible to analyze, for many classes of differential equations, the relationship between the solutions of the differential equations and those of the approximating systems of algebraic equations, but in many cases finite-difference methods have been successfully employed in establishing the existence of solutions. A comparatively simple, but highly instructive and important, example of the use of such methods will be presented in this section.

For any positive number h we denote by L_h the set of all points in the plane whose rectangular coordinates are integral multiples of h. Two points of L_h are termed "neighbors" if their abscissas coincide and their ordinates differ by h, or vice versa. Thus, each point $P = (nh, mh)$ of L_h has four neighbors, which we denote P^N, P^E, P^S, P^W, as follows:[2]

$$P^N = (nh, (m+1)h) \qquad P^E = ((n+1)h, mh)$$
$$P^S = (nh, (m-1)h) \qquad P^W = ((n-1)h, mh) \tag{115}$$

A point P is termed an "inner" point of a subset S of L_h if P and its four neighbors all belong to S; a point of S which is not an inner point is termed a "boundary" point.

Let a (real-valued) function u be defined on a subset S of L_h. We define the difference quotient u_x at each point P of S as follows:

$$u_x(P) = \frac{1}{h}[u(P^E) - u(P)] \qquad \text{if } P^E \in S \tag{116a}$$
$$u_x(P) = 0 \qquad \text{if } P^E \notin S$$

[1] This section consists of a somewhat amplified presentation of a portion of an important paper by R. Courant, K. O. Friedrichs, and H. Lewy appearing in *Mathematische Annalen*, vol. 100, 1928.

[2] We employ this notation in order to suggest the four directions of the compass.

Three more (first-order) difference quotients are defined analogously:

$$u_{\bar{x}}(P) = \frac{1}{h}[u(P) - u(P^W)] \qquad \text{if } P^W \in S$$
$$u_{\bar{x}}(P) = 0 \qquad \text{if } P^W \notin S \tag{116b}$$

$$u_y(P) = \frac{1}{h}[u(P^N) - u(P)] \qquad \text{if } P^N \in S$$
$$u_y(P) = 0 \qquad \text{if } P^N \notin S \tag{116c}$$

$$u_{\bar{y}}(P) = \frac{1}{h}[u(P) - u(P^S)] \qquad \text{if } P^S \in S$$
$$u_{\bar{y}}(P) = 0 \qquad \text{if } P^S \notin S \tag{116d}$$

We can then define difference quotients of higher order, in particular $u_{x\bar{x}}$, $u_{\bar{x}x}$, $u_{y\bar{y}}$, $u_{\bar{y}y}$, defined respectively as $(u_x)_{\bar{x}}$, $(u_{\bar{x}})_x$, $(u_y)_{\bar{y}}$, $(u_{\bar{y}})_y$. We readily find that $u_{x\bar{x}}$ and $u_{\bar{x}x}$ are equal at any inner point P of S, and that their common value is equal to $h^{-2}[u(P^E) + u(P^W) - 2u(P)]$. Combining this expression with the analogous one for $u_{y\bar{y}}$ ($= u_{\bar{y}y}$), we obtain for the "discrete Laplacian" the formula

$$\Delta u = u_{x\bar{x}} + u_{y\bar{y}} = \frac{1}{h^2}[u(P^N) + u(P^E) + u(P^S) + u(P^W) - 4u(P)] \tag{117}$$

Thus, a "harmonic" function defined on S is one whose value at any inner point is equal to the arithmetic mean of its values at the four neighboring points. (Cf. Theorem 6-1.) In analogy with Theorem 6-2, we obtain the following.

Theorem 22.   Principle of Maximum and Minimum.   A function u harmonic on a finite subset S of L_h attains its maximum and minimum values at boundary points of S.

Proof.   As in the proof of Theorem 6-2, it suffices to confine attention to the maximum, and to show that if the maximum value is attained at an inner point, it is also attained at a boundary point. If the maximum value, M, is attained at an inner point P, it follows immediately from (117) that u has this same value at all four neighbors of P. By repeating this argument a finite number of times, we ultimately find a boundary point at which u has the value M.

We now show that the discrete analogue of the Dirichlet problem is always uniquely solvable (in contrast with Theorem 6-3, which settles only the uniqueness question), and that, furthermore, the problem may be formulated as the discrete analogue of the variational Dirichlet problem which was formulated in the preceding section.

Theorem 23.   Let a function f be defined on the boundary of a finite subset S of L_h.   Then there exists a unique function u harmonic at all inner points of S and coinciding with f at all boundary points. If $\bar{u}$ is any other function satisfying the "boundary condition," then

$$\sum_{S} (\bar{u}_x^2 + \bar{u}_y^2) - \sum_{S} (u_x^2 + u_y^2) > 0 \tag{118}$$

Proof. The uniqueness is established by applying the preceding theorem to the difference of two presumed solutions. As for existence, we note that by equating the right side of (117) to zero for each inner point P we obtain a system of linear equations equal in number to the number of unknowns. By "Cramer's rule" we conclude from the uniqueness that a solution must exist. (Cf. Sec. 4-5.)

To prove (118), we note that, for any two functions v and w defined on S, the following analogue of (6-9) holds:[1]

$$\sum_{S} (v_x w_x + v_y w_y) = - \sum_{S}' w \Delta v + \cdots \tag{119}$$

where $\sum_{S}'$ denotes summation over the inner points of S and $\cdots$ denotes an expression in which the values of w at the boundary points appear *linearly*, and hence vanishes if w vanishes at all boundary points. Identifying v and w with u and $\bar{u} - u$, respectively, and taking account of the equalities $\Delta u = 0$, $\bar{u} - u = 0$ at inner and boundary points, respectively, we find that the left side of (118) is equal to $\sum_{S} (\bar{u} - u)_x^2 + (\bar{u} - u)_y^2$, which is positive unless $\bar{u} - u$ is constant throughout S;[†] taking account of the assumed equality of u and $\bar{u}$ on the boundary, we find that equality holds in (118) only if $u \equiv \bar{u}$.

We now associate with a given Dirichlet problem a sequence of algebraic problems of the type described above. Given a bounded domain

---

[1] Letting S_E and S_W denote the subsets of S consisting of those points whose "eastern" and "western" neighbors, respectively, also belong to S, we note that $\sum_{S} v_x w_x = \sum_{S_E} v_x w_x$ [cf. (116a)], and hence

$$\sum_{S} v_x w_x = h^{-1} \sum_{S_E} v_x(P)[w(P^E) - w(P)] = h^{-1} \Big[\sum_{S_W} v_x(P^W)w(P) - \sum_{S_E} v_x(P)w(P) \Big]$$

$$= - \sum_{S_E \cap S_W} v_{x\bar{x}}(P)w(P) + \text{boundary term}$$

Adding the analogous formula for $\sum_{S} v_y w_y$, we obtain (119).

[†] More precisely, $\bar{u} - u$ would be constant on each "component" of S, this term denoting, in analogy with its customary use, a maximal subset of S such that any two of its points P, Q can be joined by a sequence $P, P_1, P_2, \ldots, P_n, Q$ of points which belong to the subset and where every successive pair of points in the sequence are neighbors.

G, let a continuous function f be defined on the boundary. By the Lebesgue extension theorem (cf. Sec. 1-2), we can extend (in an unlimited number of ways) f continuously to $\bar{G}$; any function thus obtained is denoted as f. For each positive integer n, let G_n be the portion of $L_{2^{-n}}$ lying in G, so that $G_n \subset G_{n+1}$. Any point P of G whose coordinates are both dyadic rationals (quotients of integers in which the denominator is a power of 2) lies in all but a finite number (depending on P) of the sets G_n. Note that the set of all such points, which we denote as B, is dense in G. For each n we solve the "algebraic Dirichlet problem" with boundary values given by the values of f at the boundary points of G_n; the functions thus obtained will be denoted as u_n. It will now be shown that, as might be expected, the sequence $\{u_n\}$ converges, under suitable hypotheses, at all points of B, and that the limit function can be extended to all of G as a harmonic function. The proof of these assertions will be presented in the form of a series of lemmas.

LEMMA 1. The sums $h^2 \sum_{G_n} u^2(P)$ are bounded; here $h = 2^{-n}$, and we write u for u_n.

Proof. Let G be covered by a closed square R of edge a, and let M denote the maximum of $|f|$ in $\bar{G}$. Then, since the boundary values of u lie between $-M$ and M, the same assertion is true at inner points of G_n, by Theorem 22. Since the number of points in G_n cannot exceed the number of mesh points in R, which is at most $(1 + a/h)^2$, it follows that the sum $h^2 \sum_{G_n} u^2(P)$ is dominated by $M^2(a + h)^2$, which approaches $M^2 a^2$, and hence remains bounded as n increases, as was to be proved.

LEMMA 2. Let G^* be any (open) square such that $\overline{G^*} \subset G$. Then the sums $\sum_n = h^2 \sum_{G_n^*} (u_x{}^2 + u_{\bar{x}}{}^2 + u_y{}^2 + u_{\bar{y}}{}^2)$ remain bounded as n increases.

Proof. We assume, for convenience, that the vertices of G^* possess coordinates which are all dyadic rationals. (This is permissible because G^* can be replaced by a larger square possessing this property, and the sums $\sum_n$ will be dominated by the corresponding sums for the larger square.) We then select an (open) square G^{**} such that $\overline{G^*} \subset G^{**}$, $\overline{G^{**}} \subset G$ and such that the vertices of G^{**}, like those of G^*, all have dyadic rational coordinates. Then, for all sufficiently large n, G_n^{**} can be expressed as the union of G_n^* and a sequence of "shells," $S_0, S_1, \ldots, S_N$, each of which lies inside the next; S_0 lies on the boundary of G^*, and S_N lies on the boundary of a square slightly smaller than G^{**}. [The value of N is easily seen to be equal to $2^{n-1}(a - b) - 1$, where a and b denote the edges of G^{**} and G^*, respectively.] From the elementary equalities

$$h^2[u_x{}^2(P) + u_{\bar{x}}{}^2(P) + u_y{}^2(P) + u_{\bar{y}}{}^2(P)] = u^2(P^N) + u^2(P^S)$$
$$+ u^2(P^E) + u^2(P^W) - 2u(P)[u(P^N) + u(P^S) + u(P^E) + u(P^W)]$$
$$= u^2(P^N) + u^2(P^S) + u^2(P^E) + u^2(P^W) - 4u^2(P)\dagger$$

we obtain, by summing over all points P of G_n and S_0, the identity

$$\sum_{G_n * \cup S_0} (u_x{}^2 + u_{\bar{x}}{}^2 + u_y{}^2 + u_{\bar{y}}{}^2) = \sum_{S_1} u^2 - \sum_{S_0} u^2 - \sum_{0,1}' u^2 \tag{120}$$

where $\sum_{0,1}'$ denotes summation over the vertices of both S_0 and S_1. Similarly, we can write analogues of (120) with the summation on the left extended over G_n^* and the shells $S_0, S_1, \ldots, S_k$, where k may take any of the values $1, 2, \ldots, N - 1$. Since the left side of (120) and each of the analogous equations dominate the quantity $\sum_n$, we obtain the N inequalities

$$\sum_n \le \sum_{S_j} u^2 - \sum_{S_{j-1}} u^2 - \sum_{j-1,j}' u^2 \tag{121}$$

and, a fortiori,

$$\sum_n \le \sum_{S_j} u^2 - \sum_{S_{j-1}} u^2 \qquad (j = 1, 2, \ldots, N) \tag{122}$$

Summing the first m inequalities of (122), we obtain

$$m \sum_n \le \sum_{S_m} u^2 - \sum_{S_0} u^2 \le \sum_{S_m} u^2 \qquad (m = 1, 2, \ldots, N) \tag{123}$$

We now sum the inequalities (123) over all values of m and obtain

$$\tfrac{1}{2}N(N + 1) \sum_n \le \sum_{m=1}^{N} \sum_{S_m} u^2 \tag{124}$$

Observing that the right side of (124) is increased if u^2 is summed over all of G_n and taking account of the formula given earlier for N, we obtain the further inequality

$$\sum_n \le 8(a - b)^{-1}(a - b - 2^{-n+1})^{-1}h^2 \sum_{G_n} u^2 \tag{125}$$

From the boundedness of $h^2 \sum_{G_n} u^2$ (Lemma 1) we now conclude that $\sum_n$ remains bounded as n increases.

LEMMA 3. Let G^* be any (open) square such that $\overline{G^*} \subset G$, and let $w \ (= w_n)$ denote any specified difference quotient of $u \ (= u_n)$. Then the sums $h^2 \sum_{G_n*} w^2$ remain bounded as n increases.

† The last equality follows from the fact that

$$4u(P) = u(P^N) + u(P^S) + u(P^E) + u(P^W)$$

Proof. If w is any one of the first-order difference quotients, u_x, $u_{\bar{x}}$, u_y, $u_{\bar{y}}$, the present lemma is contained in the preceding one. Suppose next that w is a second-order difference quotient, say $w = u_{x\bar{y}}$. According to the preceding lemma, the quantities $h^2 \sum\limits_{H_n} u_x{}^2$ are bounded, where H denotes a square such that $\overline{G^*} \subset H$, $\bar{H} \subset G$. Since u_x, like u, is harmonic, we may again apply the preceding lemma, with G replaced by H and u by u_x. We conclude that the sums $h^2 \sum\limits_{G_n{}^*} (u_{xx}{}^2 + u_{x\bar{x}}{}^2 + u_{xy}{}^2 + u_{x\bar{y}}{}^2)$, and a fortiori the sums $h^2 \sum\limits_{G_n{}^*} w^2$, remain bounded. Similarly, if w is a difference quotient of higher order, we introduce a sufficient number of nested squares, all containing G^* and contained in G, and repeat the above argument a sufficient number of times.

LEMMA 4. Let G^* denote any subset of G such that $\overline{G^*} \subset G$, and let $w\ (=w_n)$ denote either $u\ (=u_n)$ or any specified difference quotient of u. Then w is uniformly bounded and equicontinuous in G^*; i.e., there exists a quantity M independent of n such that the inequality $|w| < M$ holds at every point of G_n^*, and, for any $\epsilon > 0$, there exists δ independent of n such that $|w(P) - w(Q)| < \epsilon$ for any pair of points P, Q contained in G_n^* satisfying the inequality $\overline{PQ} < \delta$.

Proof. First we consider the particular case that G^* is an open square, of edge a, with vertices possessing dyadic rational coordinates. Let P

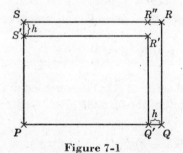

Figure 7-1

and Q denote two points possessing equal ordinates and belonging to G_n^* for all sufficiently large n. Referring to Fig. 7-1, we find by a simple computation, whose details are left as Exercise 31, that

$$h^2 \sum_{PQ'R'S'} w_{xy} = w(P) - w(Q) + w(R) - w(S) \tag{126}$$

and hence

$$|w(P) - w(Q)| \le h^2 \sum_{PQ'R'S'} |w_{xy}| + h \sum_{SR''} |w_x| \tag{127}$$

Summing (127) over all $a/h - 1$ possible choices of the line RS, we obtain

$$\left(\frac{a}{h} - 1\right)|w(P) - w(Q)| \leq \left(\frac{a}{h} - 1\right) h^2 \sum |w_{xy}| + h \sum |w_x| \quad (128)$$

where both sums are extended over the portion of G_n^* contained between the vertical lines through P and Q (including the former but not the latter). Applying the Schwarz inequality[1] to each sum, and noting that each sum is extended over $(a/h - 1) \cdot \overline{PQ}/h$ points, we obtain

$$|w(P) - w(Q)| \leq (h^2 \Sigma w_{xy}^2)^{1/2} a^{1/2} (\overline{PQ})^{1/2} + (a - h)^{-1/2}(\overline{PQ})^{1/2}(h^2 \Sigma w_x^2)^{1/2} \quad (129)$$

Extending both summations over all of G_n^* and taking account of Lemma 3, we conclude that

$$|w(P) - w(Q)| \leq C(\overline{PQ})^{1/2} \quad (130)$$

where C is independent of n. Thus the "horizontal equicontinuity" of the functions w_n is proved; "vertical equicontinuity" is established in the same manner, and it then follows immediately that (130) holds (with a new value of C) for arbitrary positions of P and Q in G_n^*. Thus equicontinuity is proved.

The boundedness of the functions w_n is now established by the following simple argument. Selecting P and Q, respectively, as points of G_n^* where w_n assumes its maximum and minimum values M_n and m_n, we find, by referring to (130), that the differences $M_n - m_n$ are bounded. If $\{M_{n_i}\}$ becomes arbitrarily large (positive or negative) for some subsequence $\{w_{n_i}\}$, the same must then be true of $\{m_{n_i}\}$ and, hence, of the sum $h^2 \sum_{G_{n_i}^*} w_{n_i}^2$, contrary to Lemma 3.

Having established the lemma in the particular case that G^* is a square, we dispose of the general case by covering $\overline{G^*}$ with a *finite* number of squares, each contained in G; the possibility of doing this is assured by the Heine-Borel theorem.

Now, by the Ascoli selection theorem, we can select a subsequence of the sequence $\{u_n\}$ which converges everywhere[2] in G, uniformly in every closed subset. From this subsequence we can extract a further subsequence on which the difference quotients $\{u_{n,x}\}$ also converge throughout G. Repeating this procedure a denumerable number of times and "diagonalizing" as in Sec. 1-1, we obtain a single subsequence $\{n_i\}$ such that the functions $\{u_{n_i}\}$ and each of their difference quotients converge throughout G, uniformly in every closed subset. (It will be shown later

---

[1] $(\Sigma|c_i|)^2 = (\Sigma|c_i| \cdot 1)^2 \leq (\Sigma c_i^2)(\Sigma 1^2) = $ (number of terms) $\cdot \Sigma c_i^2$.

[2] Each function u_n is defined only on a finite set of points, which varies with n, but it is evident that the definition of u_n may be extended to all of G without destroying the boundedness and equicontinuity.

that, in fact, the selection of subsequences is unnecessary.) We now confine attention to this subsequence.

LEMMA 5. The limit function U of the functions u_n is harmonic in G, and each sequence of difference quotients converges to the corresponding derivative of U, uniformly in every closed subset.

Proof. Let P and Q be two points of G having dyadic rational coordinates and lying on the same horizontal segment, as indicated in Fig. 7-2. Then, from the equality

$$u_n(Q) - u_n(P) = h \sum_{PQ'} u_{n,x} \tag{131}$$

and the uniform convergence of the sequences $\{u_n\}$ and $\{u_{n,x}\}$, we obtain, by passage to the limit,

$$U(Q) - U(P) = \int_P^Q U_x \, dx \tag{132}$$

where U_x denotes the limit of the sequence $\{u_{n,x}\}$. By continuity, we observe that the restriction that P and Q have dyadic rational coordinates

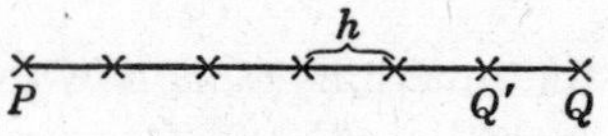

Figure 7-2

can be dropped, so that (132) holds for any horizontal line segment PQ lying in G. Fixing P and differentiating with respect to the abscissa of Q, we obtain, as was to be expected, that $\dfrac{\partial U}{\partial x}$ exists and equals U_x throughout G. Similarly, U possesses partial derivatives of all orders, which are the limits of the corresponding sequence of difference quotients. In particular, the quantity $\Delta U = \dfrac{\partial^2 U}{\partial x^2} + \dfrac{\partial^2 U}{\partial y^2}$ exists and is given by

$$\Delta U = \lim_{n \to \infty} (u_{n,x\bar{x}} + u_{n,y\bar{y}}) \tag{133}$$

Since $u_{n,x\bar{x}} + u_{n,y\bar{y}}$ vanishes at each point of B, it follows that ΔU must also vanish everywhere on B. Since ΔU is continuous in G and B is dense in G, it follows that U is harmonic.

Thus far, the only assumptions that have been made are that the domain G is bounded and that the boundary function f is continuous. In order to show that the function U obtained above does actually constitute, in some sense, a solution of the given Dirichlet problem, we must make some stronger assumptions. First, we require that the extension of the boundary function f to G which was employed in defining the sequence $\{u_n\}$ belongs to the space D. (Cf. Definition 5.) (According to a remark made in Sec. 5, this requirement does actually represent a restriction on f.)   Secondly, we assume that G is bounded by a single smooth convex curve Γ.   Actually, the latter assumption can be relaxed very considerably, and is introduced simply in order to avoid geometrical complications.

LEMMA 6.   The functions U and $U - f$ belong to the space D.

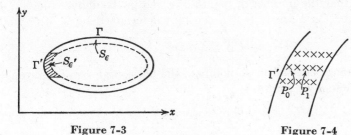

Figure 7-3   Figure 7-4

Proof.   Taking account of the definition of the functions u_n and the sets G_n, we obtain from Theorem 23 the inequality

$$h^2 \sum_{G_n} (u_{n,x}^2 + u_{n,y}^2) \le h^2 \sum_{G_n} (f_x^2 + f_y^2) \tag{134}$$

Letting n increase and taking account of Lemma 5, we obtain

$$\iint_{G'} \left[\left(\frac{\partial U}{\partial x} \right)^2 + \left(\frac{\partial U}{\partial y} \right)^2 \right] dx\, dy \le D(f) \tag{135}$$

where G' is any subdomain such that $\overline{G'} \subset G$.   Letting G' exhaust G, we conclude that $D(u)$ exists [and does not exceed $D(f)$].   Since D is a linear space, $U - f$ also belongs to D.

LEMMA 7.   Let S_ϵ denote the "ϵ boundary strip" of G; i.e., the subset of G consisting of all points whose distance from the boundary is less than ϵ.   Then, as ϵ approaches zero, the inequality

$$\iint_{S_\epsilon} (U - f)^2\, dx\, dy \le \gamma \epsilon^2 \tag{136}$$

holds for some constant γ.

Proof.   Let Γ' denote a portion of Γ such that the tangent at each point of Γ' makes an angle less than 30° with the vertical, and let horizontal lines be drawn through the end points of Γ', as indicated in Fig. 7-3. It will suffice to prove that (136) holds with S_ϵ replaced by S_ϵ', the portion of S_ϵ intercepted by the afore-mentioned lines, since, according to the restriction imposed on G, the strip S_ϵ can be expressed as the (overlapping) union of a *finite* number of portions to which the same argument can be applied.

Referring to Fig. 7-4, let P_0 be any boundary point of G_n, and let P_1 be any point of $G_n \cap S_\epsilon'$ lying on the same horizontal as P_0.   Letting

$v_n = u_n - f$, and noting that $v_n(P_0)$ vanishes, we obtain

$$v_n(P_1) = h\Sigma v_{n,x} \qquad (137)$$

where the summation is extended over the mesh points from P_0 to $P_1{}^W$, inclusive.   Applying the Schwarz inequality as in the proof of Lemma 4, and noting that the number of points involved in the above summation cannot exceed $C\epsilon/h$, where C is a constant determined by G, we obtain

$$v_n{}^2(P_1) \le C\epsilon h\Sigma v_{n,x}^2 \qquad (138)$$

Letting P_1 range over all mesh points of the horizontal segment through P_0 and summing, we obtain

$$\Sigma v_n{}^2 \le C^2\epsilon^2\Sigma v_{n,x}^2 \le C^2\epsilon^2\Sigma(v_{n,x}^2 + v_{n,y}^2) \qquad (139)$$

where all summations are extended over the entire segment.   We now sum (139) over all the horizontal "mesh segments" in S_ϵ' and obtain

$$h^2 \sum_{G_n \cap S_{\epsilon'}} v_n{}^2 \le h^2 C^2\epsilon^2 \sum_{G_n \cap S_{\epsilon'}} (v_{n,x}^2 + v_{n,y}^2) \qquad (140)$$

Now, from $v_{n,x}^2 = (u_{n,x} - f_x)^2 \le (u_{n,x} - f_x)^2 + (u_{n,x} + f_x)^2 = 2u_{n,x}^2 + 2f_x{}^2$ and the analogous inequality for $v_{n,y}^2$ we obtain

$$h^2 \sum_{G_n \cap S_{\epsilon'}} v_n{}^2 \le 2h^2 C^2\epsilon^2 \sum_{G_n \cap S_{\epsilon'}} (u_{n,x}^2 + u_{n,y}^2 + f_x{}^2 + f_y{}^2) \qquad (141)$$

Extending the summation on the right side of (141) to all of G_n, and then taking account of (134), we obtain

$$h^2 \sum_{G_n \cap S_{\epsilon'}} v_n{}^2 \le 4h^2 C^2\epsilon^2 \sum_{G_n} (f_x{}^2 + f_y{}^2) \qquad (142)$$

By uniform convergence,[1] we find that

$$\iint_{S_{\epsilon'}} (U - f)^2 \, dx \, dy \le 4C^2\epsilon^2 D(f) \qquad (143)$$

thus completing the proof.

LEMMA 8.   The inequality (136) may be replaced by the stronger assertion that

$$\lim_{\epsilon \to 0} \frac{1}{\epsilon^2} \iint_{S_\epsilon} (U - f)^2 \, dx \, dy = 0 \qquad (144)$$

Proof.   For ease in exposition, we first prove the lemma for the case that G is the unit disc, and then we indicate briefly how the argument

---

[1] Since uniform convergence is assured only on a compact subset of G, we employ the familiar device of replacing S_ϵ' on the left side of (142) by a slightly narrower strip S_ϵ'', letting $n \to \infty$, and then letting S_ϵ'' exhaust S_ϵ'.

extends to more general domains. Introducing polar coordinates, denoting $U - f$ as v, and writing $v(R,\theta) - v(r,\theta) = \int_r^R v_\rho \rho^{1/2} \cdot \rho^{-1/2} \, d\rho$, we obtain, by the Schwarz inequality,

$$[v(R,\theta) - v(r,\theta)]^2 \leq \left(\int_r^R v_\rho{}^2 \rho \, d\rho \right) \left(\int_r^R \frac{1}{\rho} \, d\rho \right)$$

$$\leq 2(1 - r) \int_r^1 v_\rho{}^2 \rho \, d\rho \qquad (\tfrac{1}{2} \leq r < R < 1) \quad (145)$$

Integrating with respect to θ, we obtain (letting $1 - r = \epsilon$)

$$\int_0^{2\pi} [v(R,\theta) - v(r,\theta)]^2 \, d\theta \leq 2\epsilon \int_0^{2\pi} \int_{1-\epsilon}^1 v_\rho{}^2 \rho \, d\rho \, d\theta \leq 2\epsilon D_\epsilon(v) \quad (146)$$

where $D_\epsilon(v)$ denotes the Dirichlet integral

$$\iint (v_x{}^2 + v_y{}^2) \, dx \, dy = \iint (v_\rho{}^2 + \rho^{-2} v_\theta{}^2) \rho \, d\rho \, d\theta$$

extended over the strip S_ϵ. Letting r (and hence R) approach unity, we conclude that $\lim_{r \to 1} \int_0^{2\pi} v^2(r,\theta) \, d\theta$ exists.[1] (Cf. Sec. 4-2.) If this limit, which we denote as C, were positive, we would have (for ρ sufficiently close to unity)

$$\int_0^{2\pi} v^2(\rho,\theta) \rho \, d\theta > \tfrac{1}{2} C \qquad (147)$$

and hence, by integration from $1 - \epsilon$ to 1,

$$\iint_{S_\epsilon} v^2 \, dx \, dy > \tfrac{1}{2} C \epsilon \qquad (148)$$

which contradicts (136). Therefore, C must vanish, and so, fixing r and letting R approach unity in (146), we obtain

$$\int_0^{2\pi} v^2(r,\theta) \, d\theta \leq 2(1 - r) D_{1-r}(v) \qquad (149)$$

Replacing r by ρ, multiplying by $\rho \, d\rho$, and integrating, we obtain

$$\iint_{S_\epsilon} v^2 \, dx \, dy \leq 2 \int_{1-\epsilon}^1 (1 - \rho) D_{1-\rho}(v) \rho \, d\rho \qquad (150)$$

Since the factors $1 - \rho$, $D_{1-\rho}(v)$, and ρ are dominated respectively by ϵ, $D_\epsilon(v)$, and unity, we obtain

$$\iint_{S_\epsilon} v^2 \, dx \, dy \leq 2 \int_{1-\epsilon}^1 \epsilon D_\epsilon(v) \, d\rho = 2\epsilon^2 D_\epsilon(v) \qquad (151)$$

which obviously implies (144). Thus, the proof is complete for the case that G is the unit disc.

---

[1] Of course, the fact that $v \in D$, which is assured by Lemma 6, is essential here.

For more general domains we replace the circles $r =$ constant and the radii $\theta =$ constant by the inner boundary curves of the strips S_ϵ and their orthogonal trajectories, respectively. There is then no difficulty in imitiating the argument presented above, provided that, as assumed earlier, the domain is smoothly bounded.

LEMMA 9.   The function $v\,(=\,U-f)$ belongs to $\mathring{D}$.   (Cf. Definition 6.)

Proof.   We present only the proof for the case that G is the unit disc, since, as in the preceding proof, the extension to more general domains is quite apparent.   Let the functions g_n and h_n be defined (for $n = 1, 2, 3, \ldots$) as follows:

$$g_n = \begin{cases} 1 & 0 \leq r < 1 - \dfrac{1}{n} \\[2mm] \cos^2 n\pi \left(1 - r - \dfrac{1}{n}\right) & 1 - \dfrac{1}{n} \leq r < 1 - \dfrac{1}{2n} \\[2mm] 0 & r \geq 1 - \dfrac{1}{2n} \end{cases} \qquad h_n = 1 - g_n$$

$$(152)$$

We shall show that v is the limit in both the H and D norms (cf. Definitions 4 and 5) of the sequence $\{vg_n\}$.   Since the functions of this sequence are readily seen to belong to $\mathring{D}$, the lemma will then be proved.   As for the H norm, we merely observe that

$$H(v - vg_n) = H(vh_n) \leq \iint_{S_{1/n}} v^2 \, dx \, dy \to 0 \tag{153}$$

As for the D norm, we observe that

$$D(v - vg_n) = D(vh_n) = \iint_{S_{1/n}} [(vh_n)_x{}^2 + (vh_n)_y{}^2] \, dx \, dy \tag{154}$$

Adding the quantity $(vh_{n,x} - v_x h_n)^2 + (vh_{n,y} - v_y h_n)^2$ to the integrand, we obtain

$$D(v - vg_n) \leq 2 \iint_{S_{1/n}} [v^2(h_{n,x}^2 + h_{n,y}^2) + h_n{}^2(v_x{}^2 + v_y{}^2)] \, dx \, dy \tag{155}$$

From the inequalities (cf. Exercise 32)

$$h_{n,x}^2 + h_{n,y}^2 \leq n^2\pi^2 \qquad h_n{}^2 \leq 1 \tag{156}$$

we then obtain

$$D(v - vg_n) \leq 2n^2\pi^2 \iint_{S_{1/n}} v^2 \, dx \, dy + 2D_{1/n}(v) \tag{157}$$

and, by Lemma 8, the right-hand side of (157) approaches zero as n increases.   The proof is thus complete.

We now observe that the argument employed in Theorem 21 applies to the problem under consideration here, the functions u and g appearing in Sec. 5 corresponding to the functions U and f, respectively. Thus, we conclude that the function U exhibits the desired behavior near the boundary of the domain. Furthermore, we can assert that the selection procedure described following Lemma 4 to secure a convergent subsequence of mesh functions was not actually necessary; for if the *entire* sequence $\{u_n\}$ did not converge everywhere, it would be possible (on account of the boundedness of the mesh functions at each fixed point of B) to choose two subsequences converging to different values at some point, and then, by applying the afore-mentioned selection procedure, we could obtain distinct functions, $U^{(1)}$ and $U^{(2)}$, both of which would provide solutions to the given Dirichlet problem, contrary to the fact that at most one solution can exist. Therefore, the method described here serves to furnish the solution as well as to prove its existence. Finally, we remark that the assumption that the Dirichlet integral $D(f)$ is finite (or even that f possesses any differentiability properties) can be eliminated by an approximation argument similar to that employed in Sec. 2.

We sum up the results of this section in the form of a theorem.

Theorem 24. Let a continuous function f be defined on the boundary of a bounded domain G, and let f be extended continuously to $\bar{G}$. If the boundary possesses suitable smoothness properties, then the sequence $\{u_n\}$ converges to a function U which provides the solution to the Dirichlet problem, and each sequence of difference quotients converges, uniformly in every compact subset of G, to the corresponding derivative of U.

EXERCISES

31. Prove (126).

32. Prove the first inequality of (156).

33. Work out in detail the approximation argument mentioned near the end of this section.

7. Conformal Mapping

Perhaps the most striking single application of the fact that every bounded plane simply connected domain is a Dirichlet domain is to be found in the following theorem, which is justly credited to Riemann, despite the fact that his proof was invalid.

Theorem 25. Riemann Mapping Theorem. Any two bounded[1] simply connected domains can be conformally mapped onto each other.

---

[1] By employing a suitable transformation, we can remove the requirement of boundedness; it is actually only necessary to exclude the trivial case that the domain consists of the entire plane.

Proof. It suffices to consider the case when one of the domains is the unit disc, for the general case then follows by mapping one of the given domains onto the disc and then mapping the disc onto the other given domain. Given the bounded simply connected domain G, we select any point P_0 of G and define on the boundary the continuous function $f(P) = \log \overline{P_0P}$. Since G is a Dirichlet domain, there exists a function u harmonic in G, continuous in $\bar{G}$, and coinciding on the boundary with f. It should be emphasized that $u(P)$ is *not* given by the function $\log \overline{P_0P}$ for $P \in G$, for this latter function is not harmonic at P_0. In fact, it is readily seen that the function $u(P) - \log \overline{P_0P}$ is the Green's function $G(P,P_0)$ of G with pole at P_0; cf. Sec. 6-7 and Exercise 6-14. Let v denote a harmonic conjugate of u—that is, a real function such that $u + iv$ is analytic throughout G. (Cf. Sec. 6-2.) The function v is determined to within an arbitrary additive constant, and its single-valuedness is assured by the monodromy theorem. Then the function $w(z) = u + iv - \log (z - z_0)$ is analytic in G except at z_0. (z_0 and z denote the complex numbers corresponding to P_0 and P, respectively.) The function

$$f(z) = e^{-w(z)} = (z - z_0)e^{-u-iv} \tag{158}$$

is therefore analytic (and single-valued) in G, has a simple zero at z_0, and does not vanish anywhere else. Since $|f(z)| = e^{-G(P,P_0)}$, it follows from Exercise 6-15 that $|f(z)| < 1$ throughout G, while from the behavior of $G(P,P_0)$ near the boundary it follows that $|f(z)|$ approaches unity there. Given any complex number C of modulus less than unity, we construct a smooth simple closed curve Γ which surrounds P_0 and is so close to the boundary of G that $|f(z)| > |C|$ everywhere on Γ.† By the principle of the argument, it follows that the equation $f(z) = C$ has the same number of solutions inside Γ as the equation $f(z) = 0$, namely, one. It follows that $f(z)$ maps G in a one-to-one manner onto the unit disc, and the proof is complete.

The freedom in choosing z_0 and the existence of an arbitrary additive constant in v show that an arbitrary point of G may be chosen to map into the origin, and that the argument of the derivative at the selected point may be prescribed. [Note that $\arg f'(z_0) = -v(P_0)$.] Going over to the general case, where neither domain is necessarily a disc, we conclude that any pair of points, together with a direction at each point, may be made to correspond. It is easily proved, finally, that these correspondences uniquely determine the mapping.

While the above proof is due to Riemann, his argument was defective, for, as explained at the beginning of Sec. 5, the solvability of the Dirichlet problem, which plays an essential role in the proof, had not been rigorously established at that time.

† We accept without proof the intuitively plausible fact that such a curve exists.

The above theorem makes no assertion concerning a correspondence of boundary points, and in general it is impossible to extend the mapping continuously from the domains to their boundaries. However, in the simplest case, which is highly important in practical applications, as well as in the further development of the theory of conformal mapping, the existence of a boundary correspondence is guaranteed by the following theorem, which we state without proof.

Theorem 26. Caratheodory's Extension of the Riemann Mapping Theorem. If two domains are each bounded by a simple closed curve, then any conformal mapping of the domains onto each other can be continuously extended so as to provide a one-to-one correspondence of the boundary points.

Suppose now that a domain G bounded by a simple closed curve is mapped conformally onto the unit disc. Then, in view of the above theorem and the fact that harmonicity is preserved under conformal mapping (cf. Sec. 6-2), it follows that every Dirichlet problem for G can be converted into a Dirichlet problem for the disc. By solving the latter problem with the aid of the Poisson formula (6-27) and then mapping from the disc back to G, we can obtain the solution to the original problem.

Turning now to doubly connected domains, we begin by showing that the most obvious generalization of Theorem 25 is not true; that is, two doubly connected domains cannot, in general, be mapped conformally onto each other. This will be proved by showing that two annuli cannot be mapped conformally onto each other unless they are similar. To show this, we suppose that the analytic function $w = f(z)$ provides a one-to-one mapping of the annulus $R_1 < |z| < R_2$ onto the annulus $R'_1 < |w| < R'_2$. Then the function $\log |f(z)|$ must be harmonic in the former annulus and must approach the constant values $\log R'_1$ and $\log R'_2$ as $|z|$ approaches R_1 and R_2, or R_2 and R_1, respectively. Thus we have a Dirichlet problem in which both the domain and the boundary values possess rotational symmetry, and so, by uniqueness, we know that

$$\log |f(z)| = a \log |z| + b \tag{159}$$

and hence
$$\left| \frac{f(z)}{z^a} \right| = \text{constant} \tag{160}$$

Since an analytic function of constant modulus must reduce to a constant, we conclude that

$$f(z) = \text{constant} \cdot z^a \tag{161}$$

Now, $f(z)$ can be single-valued only if a is an integer, and can furnish a one-to-one mapping only if this integer equals ± 1. Thus, the mapping must assume one of the forms

$$f(z) = \text{constant} \cdot z \qquad f(z) = \frac{\text{constant}}{z} \qquad (162)$$

from which we immediately conclude, in either case, that $R_1/R_2 = R_1'/R_2'$, as was to be proved.

The above example serves to suggest the correct generalization of Theorem 25, which we now state and prove.

Theorem 27. With each doubly connected domain G there is associated a unique number $\alpha > 1$ such that G can be mapped one to one and conformally onto an annulus whose outer and inner radii are in the ratio $\alpha:1$. Thus, two doubly connected domains can be mapped onto each other if and only if they have the same "modulus," $2\pi/\log \alpha$.

Proof. We may assume that G is bounded (cf. footnote on page 211) and that neither of the two boundary components reduces to a single point, for otherwise the present theorem follows trivially from Theorem 25.† By Theorem 21, there exists a harmonic function u approaching the values one and zero, respectively, at the outer and inner boundary components. Accepting for the present the fact—which will be proved later—that the harmonic conjugate v of u is multiple-valued, changing by a positive constant c each time the inner boundary component is surrounded in the positive sense, we see that the function

$$f(z) = \exp\left[\frac{2\pi}{c}(u + iv)\right] \qquad (163)$$

is analytic and single-valued in G. Furthermore, $1 < |f(z)| < \exp(2\pi/c)$, and $|f(z)|$ approaches the indicated bounds as z approaches the inner and outer boundary components, respectively. Since the argument of $f(z)$ ($= 2\pi v/c$) increases by 2π when the inner component is surrounded in the positive sense, it follows by applying the principle of the argument (in a manner entirely analogous to its use in the proof of Theorem 25) that $f(z)$ assumes each complex value C satisfying the inequalities $1 < |C| < \exp(2\pi/c)$ exactly once. Thus, the function $f(z)$ provides the desired mapping.

It remains, however, to justify the assertion made above concerning the multiple-valuedness of v. First consider the case that the outer boundary component consists of the unit circle. By using a linear transformation to map the unit disc onto a half plane and then applying the Schwarz reflection principle (cf. Sec. 6-8), we find that u is harmonic

---

† Add the isolated boundary point to G, thus producing a simply connected domain, which can be mapped onto the unit disc with the added point mapping into the origin; then G itself is mapped onto the unit disc minus its center, so that $\alpha = \infty$.

beyond the circle, so that the analytic function $u + iv$ may also be extended beyond the circle. On the circle, the outward normal derivative of u must be nonnegative, since, by the maximum principle, u is less than one everywhere in G. According to the Cauchy-Riemann equations, the tangential derivative of v in the positive sense must also be non-negative. If the latter derivative were zero everywhere on the circle, we would conclude (recalling that the tangential derivative of u is zero everywhere on the circle, since u is constant there) that the derivative

$$\frac{d}{dz}(u + iv)$$

also vanishes everywhere on the circle, and this would imply, by an elementary theorem concerning analytic functions, that $u + iv$ is constant, which is certainly false. Thus, v must actually increase when the outer boundary is traversed once in the positive sense. Since any other positive circuit about the inner boundary component must give the same change in v, the desired result is established in the case under consideration. Turning to the case of an arbitrary (bounded) doubly connected domain, it will suffice to map it onto a domain of the type that has just been considered. This may be accomplished if we can find a *simply connected* domain $\tilde{G}$ containing G whose boundary consists precisely of the outer boundary component of G, for we can then map $\tilde{G}$ onto the unit disc and confine attention to the part of the unit disc which is the image of G. Such a domain $\tilde{G}$ is easily found—one draws a simple closed curve lying in G which surrounds the inner boundary component, and then adds to G the interior of this curve.

Theorem 27 admits interesting generalizations to domains of higher connectivity. For example, if G is a triply connected domain (none of whose boundary components consists of a single point) we define, in analogy with the above procedure, harmonic functions u_1 and u_2 each assuming the value unity on one of the boundary components and zero on the other two components. Their respective harmonic conjugates v_1 and v_2 are determined, and then it is shown that real constants c_1 and c_2 can be chosen such that the function

$$f(z) = \exp\left[c_1(u_1 + iv_1) + c_2(u_2 + iv_2)\right] \tag{164}$$

is single-valued in G and furnishes a one-to-one mapping. From the constancy of u_1 and u_2 on each boundary component it follows that $|f(z)|$ approaches constant values on each component, and it is then shown, by applying the principle of the argument, that the image of G consists of an annulus from which a portion of a concentric circle has been removed. More generally, a domain of connectivity n (≥ 3) can be mapped conformally onto an annulus from which $n - 2$ concentric circular arcs have been removed. Again, the essential idea is to construct harmonic functions $u_1, u_2, \ldots, u_{n-1}$, each of which assumes the value unity on one of the boundary components and zero on the others. The respective conjugate functions $v_1, v_2, \ldots, v_{n-1}$ are constructed, and then suitable constants $c_1, c_2, \ldots, c_{n-1}$ are determined such that the function $\exp\left[\sum_{k=1}^{n-1} c_k(u_k + iv_k)\right]$ accomplishes the mapping described above.

By similar methods, it is possible to show that a given multiply connected domain can be mapped onto other types of domains ("canonical domains") possessing a simple geometrical structure. The formulation of a suitable Dirichlet problem may be used in each case as the starting point of the proof of the existence of the desired mapping. In particular, the functions $u_1, u_2, \ldots, u_{n-1}$, which are known as the "harmonic measures" of the boundary components on which they assume the value unity, play an essential role in determining the mappings onto many of the known classes of canonical domains.

EXERCISES

34. Let $u_1, u_2, \ldots, u_n$ denote the harmonic measures of the boundary components of a domain G (of connectivity n). Prove that $u_1 + u_2 + \cdots + u_n \equiv 1$, but that no nontrivial linear relationship can exist among any smaller number of these functions.

35. Let p_{ij} denote the change in v_i, the harmonic conjugate of u_i, resulting from a circuit around the jth boundary component. Prove that, with a suitable convention concerning the sense of the circuit, the equality $p_{ij} = D(u_i, u_j)$ holds, and hence that $p_{ij} = p_{ji}$.

36. Prove that $\displaystyle\sum_{i=1}^{n} p_{ij} = 0$, and that p_{ij} is negative except when $i = j$.

37. Prove that $\displaystyle\sum_{i,j=1}^{n-1} p_{ij}\lambda_i\lambda_j$ is positive for all (real) values of $\lambda_1, \lambda_2, \ldots, \lambda_{n-1}$ except $\lambda_1 = \lambda_2 = \cdots = \lambda_{n-1} = 0$. *Hint:* Consider $D\left(\displaystyle\sum_{i=1}^{n-1} \lambda_i u_i\right)$.

38. Prove that for each set of numbers $\gamma_1, \gamma_2, \ldots, \gamma_{n-1}$ there exists one and only one set of numbers $c_1, c_2, \ldots, c_{n-1}$ such that the function $\displaystyle\sum_{i=1}^{n-1} c_i v_i$ increases by γ_j when the jth boundary component is surrounded. *Hint:* Prove directly from Exercise 37 (or from the general theory of quadratic forms) that
$$\det(p_{ij}) \qquad (i, j = 1, 2, \ldots, n-1)$$
does not vanish.

39. Prove that the modulus of a doubly connected domain, as defined in the statement of Theorem 27, is equal to the Dirichlet integral of the harmonic measure of either boundary component.

8. THE HEAT EQUATION

In this chapter we shall consider a few problems dealing with the simplest of all parabolic equations, namely, the one-dimensional heat equation

$$u_{xx} = u_t \tag{1}$$

Under suitable physical assumptions and choice of units, this equation governs the distribution of temperature on a homogeneous thin rod occupying part or all of the x axis, the variable t denoting, of course, the time.

1. The Initial-value Problem for the Infinite Rod

Let a rod occupying the entire x axis be given, and suppose that at a given instant, which may be taken as $t = 0$, the temperature distribution is described by the function $f(x)$. On physical grounds, it appears plausible that there should exist one and only one solution of (1) satisfying the afore-mentioned initial condition, and this conjecture is supported by the heuristic reasoning employed in Sec. 3-2. We shall now see how the existence of such a solution can be rigorously established. The basic idea to be employed is that the Fourier transform (with respect to x) of the solution (assuming that the latter exists and does indeed possess a transform) can be determined quite explicitly, under reasonable assumptions on $f(x)$. By applying the inversion formula to the transform, we shall be led to an apparent solution, and finally it will be verified that the apparent solution is *actually* a solution of (1) satisfying the prescribed initial condition.

We temporarily assume, therefore, that a function $u(x,t)$ exists for all values of x and all positive values of t, satisfying equation (1) and the prescribed initial condition, the latter being understood in the following sense:

$$\lim_{t \to +0} u(x,t) = f(x) \tag{2}$$

For each positive value of t, we multiply $u(x,t)$ by $e^{-i\lambda x}$ and integrate with respect to x; applying (1-20) and taking account of (1), we readily find that the function $\mathfrak{F}(u)$, which is clearly a function of λ and t, satisfies the very simple differential equation

$$-\lambda^2 \mathfrak{F}(u) = \frac{\partial}{\partial t} \mathfrak{F}(u) \tag{3}$$

The symbol of partial differentiation is used in (3), since $\mathfrak{F}(u)$ depends on λ as well as t. This equation, which constitutes a notable simplification over (1), can be solved explicitly by separation of variables. We thus obtain

$$\mathfrak{F}(u) = ce^{-\lambda^2 t} \tag{4}$$

where c, the "constant" of integration, may depend on λ. To determine c, we make the plausible assumption that

$$\lim_{t \to +0} \mathfrak{F}(u) = \mathfrak{F}(\lim_{t \to +0} u) \tag{5}$$

Taking account of (2) and (4), we then obtain

$$c = \mathfrak{F}(f) \tag{6}$$

and hence

$$\mathfrak{F}(u) = \mathfrak{F}(f)e^{-\lambda^2 t} \tag{7}$$

or, taking account of Exercise 4,

$$\mathfrak{F}(u) = \mathfrak{F}(f)\mathfrak{F}((4\pi t)^{-\frac{1}{2}}e^{-x^2/4t}) \tag{8}$$

Finally, by invoking the "convolution theorem" (1-21), we obtain, as the presumed solution of our problem, the function

$$u(x,t) = \int_{-\infty}^{\infty} f(\xi)K(\xi,x,t)\,d\xi \tag{9}$$

where $K(\xi,x,t)$ is the "heat kernel" $(4\pi t)^{-\frac{1}{2}} \exp[-(\xi - x)^2/4t]$.

Now, disregarding the questionable manner in which (9) was derived, let us study this formula on its own merits. First we observe, by an explicit computation, that $f(\xi) \equiv 1$ yields the result $u(x,t) \equiv 1$ (cf. Exercise 1), which satisfies both the differential equation (1) and the initial condition (2). It is to be noted that even in this simple case the function $u(x,t)$ does not possess a Fourier transform, so that the procedure leading to (9) was entirely formal; nevertheless, this procedure has indeed led to a solution of (1) satisfying (2) as well. Turning now to more general choices of $f(\xi)$, it is readily seen that the integral (9) converges for all x and all $t > 0$ if $f(\xi)$ is continuous and bounded for all ξ, and under these assumptions it is readily shown, by justifying differentiation under the integral sign (cf. Exercise 2), that $u(x,t)$ satisfies (1). It remains to establish (2). For any x we subtract from (9) the equation

$$f(x) = \int_{-\infty}^{\infty} f(x) K(\xi,x,t) \, d\xi \tag{10}$$

[which follows from the foregoing remark concerning the case $f(\xi) \equiv 1$]. We obtain

$$u(x,t) - f(x) = \int_{-\infty}^{\infty} [f(\xi) - f(x)] K(\xi,x,t) \, d\xi \tag{11}$$

Given any $\epsilon > 0$, we can choose $\delta > 0$ so small that $|f(\xi) - f(x)| < \epsilon$ whenever $|\xi - x| \le \delta$, while for all other values of ξ we have the inequality $|f(\xi) - f(x)| \le 2M$, where M denotes an upper bound on $|f(\xi)|$, which we have assumed to exist. Taking account of the fact that $K(\xi,x,t)$ is always positive, we obtain from (11)

$$|u(x,t) - f(x)| < \epsilon \int_{x-\delta}^{x+\delta} K(\xi,x,t) \, d\xi + 2M \int_{|\xi-x|>\delta} K(\xi,x,t) \, d\xi$$
$$< \epsilon \int_{-\infty}^{\infty} K(\xi,x,t) \, d\xi + 2M \int_{|w|>\delta/2t^{1/2}} e^{-w^2} \, dw \tag{12}$$

[The substitution $w = (\xi - x)/2t^{1/2}$ is employed in obtaining the last integral.] The first quantity of the last pair is equal to ϵ, while the second quantity clearly approaches zero as $t \to +0$. Taking account of the arbitrariness of ϵ, we conclude that (2) is satisfied.

It should be noted that the explicit form of the kernel $K(\xi,x,t)$ was not actually exploited in establishing (2)—only the fact that this kernel constitutes, as $t \to +0$, a sequence of functions "concentrated" at the point x in the sense of Exercise 1-11.

A brief remark concerning uniqueness is in order; by linearity, the question of whether a solution of (1) satisfying (2) is unique is equivalent to the question of whether the only solution of (1) satisfying (2) in the case $f(x) \equiv 0$ is given by $u(x,t) \equiv 0$. The answer to this is negative, as is shown by the example

$$u(x,t) = K_x(0,x,t) = -x(16\pi t^3)^{-1/2} e^{-x^2/4t} \tag{13}$$

(Cf. Exercise 3.) For any fixed value of x other than zero, the quantity $t^{-3/2} \exp(-x^2/4t)$, and hence $u(x,t)$, approaches zero as t does, while $u(0,t) \equiv 0$. However, a more profound analysis[1] shows that the only *bounded* solution (in fact, the only solution having either an upper or a lower bound) of (1) with $f(x) \equiv 0$ is the trivial one $u \equiv 0$. [Note in (13)

---

[1] The interested reader may consult the paper by D. V. Widder in *Transactions of the American Mathematical Society*, vol. 55, 1944.

that u is unbounded as x and t approach zero under the constraint $x^2/4t = c$, c denoting any positive constant.]

EXERCISES

1. Prove that, when $f(\xi) \equiv 1$ in (9), $u(x,t) \equiv 1$.

2. Justify the assertion made in the text that the function $u(x,t)$ defined by (9) satisfies (1) under the assumption that $f(\xi)$ is continuous and bounded; more generally, assume, aside from continuity (or even measurability), only that $f(\xi)$ is "small" compared to $\exp \xi^2$ in the sense that there exist two constants A and B $(A > 0,$ $0 < B < 2)$ such that $|f(\xi)| < A \exp |\xi|^B$.

3. From the fact that $K(0,x,t)$ satisfies (1), show that the function (13) also satisfies the same equation.

4. Inserting the function $f(\xi) = \cos \xi$ into (9), show that $u(x,t)$ becomes equal to $g(t) \cos x$, where

$$g(t) = \int_{-\infty}^{\infty} K(\xi,0,t) \cos \xi \, d\xi$$

Then deduce from (1) and (2) that $g(t)$ must satisfy the differential equation $g'(t) + g = 0$ and the additional condition $g(0) = 1$. Thus, evaluate the integral

$$\int_{-\infty}^{\infty} \exp\left(\frac{-\xi^2}{4t}\right) \cos \xi \, d\xi$$

and, more generally, the integral

$$\mathcal{F}(\exp(-a\xi^2)) = \int_{-\infty}^{\infty} \exp(-a\xi^2) \cos \lambda\xi \, d\xi$$

a denoting a positive constant.

5. Prove that if $\int_{-\infty}^{\infty} |f(x)| \, dx$ exists, then $\int_{-\infty}^{\infty} u(x,t) \, dx$ also exists and is equal to $\int_{-\infty}^{\infty} f(x) \, dx$ for all t (>0). Give a physical interpretation of this result.

6. Prove that if $f(\xi)$ is bounded and has a mean value in the sense that

$$\lim_{y \to \infty} (2y)^{-1} \int_{-y}^{y} f(x) \, dx$$

exists, then $u(x,t)$ has, for each t (>0), the same mean value. Also prove that, for each x, $\lim_{t \to \infty} u(x,t)$ exists and is equal to the afore-mentioned mean value.

7. Let $f(x) \equiv -1$ for $x < 0$, $f(x) \equiv 1$ for $x > 0$. Prove that

$$u(x,t) = \left(\frac{2}{\pi}\right)^{1/2} \int_{0}^{x/2t^{1/2}} \cdot \exp\left(-\tfrac{1}{2}\xi^2\right) d\xi$$

and confirm the assertions of the preceding exercise.

8. Let $f(x)$ be any bounded continuous function defined for $x > 0$; extend the definition of $f(x)$ to all values of x by imposing the condition $f(-x) = -f(x)$ [so that, in particular, $f(0) = 0$]. Show that formula (9) yields a solution of (1) satisfying (2) even at $x = 0$, although $f(x)$ will, in general, be discontinuous there.

2. The Simplest Problem for the Semi-infinite Rod

In the remainder of this chapter we shall indicate, by presenting two examples in considerable detail, how the Laplace transform can be used to solve a wide variety of problems of heat conduction. In the present section we consider a very simple problem, which is well suited to serve as an introduction to the use of the Laplace transform.

Let us consider a semi-infinite rod whose end is maintained permanently at a fixed temperature, and suppose that at a given instant the temperature at all points of the rod (except the end) is uniform, but different from that at which the end is maintained. We seek to determine the distribution of temperature along the rod at any subsequent instant. By choosing the units of time, length, and temperature suitably, we may formulate the problem as follows:

$$u_{xx} = u_t \qquad (x > 0, t > 0) \tag{14}$$

$$\lim_{x \to +0} u(x,t) = 0 \quad (t > 0) \qquad \lim_{t \to +0} u(x,t) = 1 \quad (x > 0) \tag{15}$$

We note that this problem can be solved by the method of the preceding section, through the device explained in Exercise 8; in this way we are led to the problem of Exercise 7, and thus conclude that the problem under consideration here possesses the solution

$$u(x,t) = \left(\frac{2}{\pi}\right)^{\frac{1}{2}} \int_0^{x/2t^{\frac{1}{2}}} \exp\left(-\tfrac{1}{2}\xi^2\right) d\xi \tag{16}$$

However, we shall temporarily disregard this method of getting the solution, and instead investigate the problem directly by Laplace transform.

Assuming that the problem under consideration has a bounded solution $u(x,t)$ such that all the subsequent operations are permissible, we proceed as follows. For each fixed $x > 0$ we take the Laplace transform of both sides, obtaining[1]

$$\int_0^\infty e^{-st} u_{xx}\, dt = \int_0^\infty e^{-st} u_t\, dt \tag{17}$$

Moving the differential operator $\dfrac{\partial^2}{\partial x^2}$ before the integral sign on the left side of (17) and integrating by parts on the right, we obtain for the transform $U(x,s)$ of $u(x,t)$ the equation

$$U_{xx} = sU - 1 \tag{18}$$

---

[1] We might also attempt to transform (14) with respect to x instead of t; this would lead, in place of (18), to a differential equation of first rather than second order, but having the disadvantage of involving the unknown quantity $\lim\limits_{x \to 0} u_x$.

Although this is, like (14), a partial differential equation (since U, like u, is a function of two variables), it has the advantage of involving differentiation only with respect to one of the variables, so that it may be solved as an ordinary differential equation. For the general solution of (18) we obtain

$$U(x,s) = \frac{1}{s} + A \exp (s^{1/2}x) + B \exp (-s^{1/2}x) \tag{19}$$

where the "constants" A and B may depend on s.

Before proceeding further with the solution of the problem, we clarify the ambiguity in (19) arising from the presence of the radical $s^{1/2}$. Although $s^{1/2}$ is a two-valued function of the complex variable s, it may be rendered single-valued by restricting s to a simply connected domain not containing the origin and then assigning a definite value to the radical at any one point of the domain.[1] In particular, if s is confined to the half plane $\operatorname{Re} s > 0$, as is the case here,[2] we may, and shall, interpret $s^{1/2}$ unambiguously as that value which is positive, rather than negative, for positive values of s.

Returning to (19), we shall now determine A. Since $u(x,t)$ was assumed bounded, say $|u(x,t)| < M$, we have, for any positive value of s,

$$|U(x,s)| < M \int_0^\infty e^{-st} \, dt = \frac{M}{s} \tag{20}$$

If A were different from zero, $U(x,s)$ would be unbounded as a function of x. Thus $A = 0$. Next, performing in the equation

$$U(x,s) = \int_0^\infty u(x,t) \exp (-st) \, dt$$

the limiting operation under the integral sign, we obtain, taking account of the first condition of (15), the relation

$$0 = \frac{1}{s} + B \tag{21}$$

Thus we obtain the explicit formula

$$U(x,s) = \frac{1}{s} - \frac{1}{s} \exp (-s^{1/2}x) \tag{22}$$

---

[1] This is a simple illustration of the monodromy theorem of the theory of analytic functions.

[2] From (19) it follows that the integral $\int_0^\infty u(x,t) \exp (-st) \, dt$ cannot converge in a half plane containing the origin in its interior, for $U(x,s)$, *considered as a Laplace transform*, must be single-valued. Of course, this does not contradict the fact that $U(x,s)$, when analytically continued beyond the half plane of convergence, becomes a multiple-valued function.

[Strictly speaking, since $A = 0$ was established only for real values of s, (22) is also subject to this restriction. However, a similar argument could have been used for nonreal values of s, or, more directly, we may argue that since both sides of (22) are analytic in a half plane and identical on a curve in that region, they are identical throughout the half plane.]

In order to obtain $u(x,t)$, we apply the inversion formula, obtaining (for positive values of x and t)

$$u(x,t) = \lim_{\tau \to \infty} \frac{1}{2\pi i} \int_{\sigma-i\tau}^{\sigma+i\tau} \left[\frac{1}{s} - \frac{1}{s} \exp\left(-s^{\frac{1}{2}}x\right) \right] e^{st} \, ds \qquad (23)$$

σ being any positive constant, and the path of integration being the vertical line segment connecting $\sigma - i\tau$ to $\sigma + i\tau$. Since we know

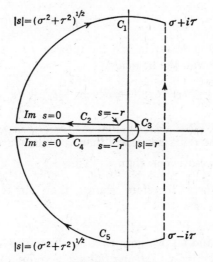

Figure 8-1

already (cf. Sec. 1-4) that the inverse transform of s^{-1} is given by the function $f(t) \equiv 1$ (for $t > 0$), it remains to invert $s^{-1} \exp\left(-s^{\frac{1}{2}}x\right)$. At this point, we exploit the fact that the latter function is analytic beyond the half plane of convergence. More precisely, it is analytic in the entire complex plane except the origin, so that if we cut the plane along the negative half of the real axis we obtain a function single-valued and analytic in the remainder of the plane. By the Cauchy integral theorem, the vertical path of integration indicated in (23) may be replaced by any other curve which connects the end points $\sigma \pm i\tau$ and avoids crossing the cut. In particular, it will be found convenient to employ the path of integration shown in Fig. 8-1.

First, we shall determine upper bounds on[1] $\left|\int_{C_1}\right|$ and $\left|\int_{C_5}\right|$ which suffice to show that $\int_{C_1}$ and $\int_{C_5}$ each approach zero as $\tau \to \infty$. Introducing polar coordinates on C_1, we obtain

$$\int_{C_1} = -i \int_{\alpha}^{\pi} \exp\left[-xR^{\frac{1}{2}} \exp\left(\frac{1}{2}i\theta\right) + tR \exp\left(i\theta\right)\right] d\theta \qquad (24)$$

and hence, by replacing the integrand by its absolute value,

$$\left|\int_{C_1}\right| \leq \int_{\alpha}^{\pi} \exp\left(-xR^{\frac{1}{2}} \cos \frac{1}{2}\theta + tR \cos \theta\right) d\theta \qquad (25)$$

Now, as θ varies between α and π, $tR \cos \theta \leq tR \cos \alpha = t\sigma$, and hence

$$\left|\int_{C_1}\right| \leq e^{t\sigma} \int_{\alpha}^{\pi} \exp\left(-xR^{\frac{1}{2}} \cos \frac{1}{2}\theta\right) d\theta$$

$$\leq e^{t\sigma} \int_{0}^{\pi} \exp\left(-xR^{\frac{1}{2}} \cos \frac{1}{2}\theta\right) d\theta \qquad (26)$$

Since $xR^{\frac{1}{2}} \cos \frac{1}{2}\theta$ is decreasing on the interval $0 \leq \theta \leq \pi$, we have, for any θ_0 in this interval, the inequality

$$\left|\int_{C_1}\right| \leq \theta_0 e^{t\sigma} \exp\left(-xR^{\frac{1}{2}} \cos \frac{1}{2}\theta_0\right) + (\pi - \theta_0) e^{t\sigma} \qquad (27)$$

Given $\epsilon > 0$, we choose θ_0 such that $\pi - \theta_0 < \epsilon \exp\left(-t\sigma\right)$, so that the second term on the right side is less than ϵ, while, for all sufficiently large R, the first term also is less than ϵ. Thus $\left|\int_{C_1}\right| < 2\epsilon$ for R sufficiently large, and we have therefore shown that $\lim_{\tau \to \infty} \int_{C_1} = 0$. Similarly, $\lim_{\tau \to \infty} \int_{C_5} = 0$.

Turning to C_2 and C_4, we note that on these lines $s^{\frac{1}{2}}$ is equal to $i|s|^{\frac{1}{2}}$ and $-i|s|^{\frac{1}{2}}$, respectively. Hence, taking account of the directions of integration, we get (by setting $\rho = -s = |s|$)

$$\int_{C_2} + \int_{C_4} = -2i \int_{r}^{R} \rho^{-1} \sin\left(x\rho^{\frac{1}{2}}\right) e^{-t\rho} d\rho$$

$$= -4i \int_{r^{\frac{1}{2}}}^{R^{\frac{1}{2}}} v^{-1} \sin xv \, e^{-tv^2} dv \qquad (28)$$

Thus, $\lim_{\substack{r \to 0 \\ R \to \infty}} \left(\int_{C_2} + \int_{C_4}\right)$ exists and equals $-4i \int_{0}^{\infty} v^{-1} \sin xv \exp\left(-tv^2\right) dv$.

(Note that the separate integrals $\int_{C_2}$ and $\int_{C_4}$ do *not* approach finite limits as $r \to 0$.)

[1] By $\int$ we mean $\int s^{-1} \exp\left(-s^{\frac{1}{2}}x + st\right) ds$ along the indicated curve; $\alpha = \arg\left(\sigma + i\tau\right) = \arctan \tau/\sigma$; $R = \left(\sigma^2 + \tau^2\right)^{\frac{1}{2}}$.

Finally, for C_3 we obtain

$$\int_{C_3} = \int_{C_3} s^{-1}\left[1 - \frac{xs^{\frac{1}{2}}}{1!} + \frac{(xs^{\frac{1}{2}})^2}{2!} - \cdots\right]\left(1 + \frac{st}{1!} + \frac{s^2t^2}{2!} + \cdots\right) ds$$

$$= \int_{C_3} [s^{-1} + s^{-\frac{1}{2}}\psi(s,x,t)]\, ds \quad (29)$$

where $\psi(s,x,t)$ is, for each fixed (positive) x and t, bounded as s approaches zero. Hence, the integral of the second term of the integrand approaches zero as r does, and we conclude that $\lim_{r\to 0}\int_{C_3} = 2\pi i$.

We have thus established the result

$$\lim_{r\to\infty} \frac{1}{2\pi i}\int_{\sigma-ir}^{\sigma+ir} s^{-1}\exp\left(-xs^{\frac{1}{2}} + st\right) ds = 1 - \frac{2}{\pi}\int_0^\infty v^{-1}\sin xv\, e^{-tv^2}\, dv \quad (30)$$

Taking account of (23), we obtain as the presumed solution of our problem

$$u(x,t) = \frac{2}{\pi}\int_0^\infty v^{-1}\sin xv\, e^{-tv^2}\, dv \quad (31)$$

We could now analyze (31) and show directly that it provides a solution to our problem. Instead of doing this, we shall reconcile (31) with (16); this will suffice to establish the desired result, for we know that (16) does, in fact, solve our problem. Let us differentiate both sides of (31) with respect to x; as in the preceding section, there is no difficulty in justifying differentiation under the integral sign, and we obtain

$$u_x = \frac{2}{\pi}\int_0^\infty e^{-tv^2}\cos xv\, dv = \frac{1}{\pi}\int_{-\infty}^\infty e^{-tv^2}\cos xv\, dv = (\pi t)^{-\frac{1}{2}}e^{-x^2/4t} \quad (32)$$

(The second integral is obtained by observing that the integrand is an even function of v, while the evaluation of the integral may be performed either by complex integration or by the method outlined in Exercise 4.) Observing that $u(0,t) = 0$, we obtain by integration

$$u(x,t) = (\pi t)^{-\frac{1}{2}}\int_0^x e^{-y^2/4t}\, dy \quad (33)$$

and the change of variable $\xi = y/(2t)^{\frac{1}{2}}$ now yields (16).

EXERCISE

9. Replace the second condition of (15) by the following: $\lim_{t\to +0} u(x,t) = f(x)$, $x > 0$. Generalize the method employed in the present section to this case, and then reconcile the result with (9). (Cf. Exercise 8.)

3. The Finite Rod

Let a rod of finite length be given, and suppose that at a given instant the temperature distribution is known. Also, suppose that thereafter the ends are maintained at prescribed temperatures, which may vary with time. On physical grounds it appears plausible that the subsequent temperature distribution along the rod is uniquely determined. We now proceed to investigate this question in some detail.

The problem under consideration may be formulated as follows:

$$u_{xx} = u_t \qquad (0 < x < 1, t > 0) \tag{34}$$

$$u(0,t) = f_1(t) \qquad u(1,t) = f_2(t) \qquad u(x,0) = g(x) \tag{35}$$

We shall assume that $f_1(t)$ and $f_2(t)$ are continuous for $t \geq 0$, that $g(x)$ is continuous on the closed interval $0 \leq x \leq 1$, and that $f_1(0) = g(0)$, $f_2(0) = g(1)$. Then it may be expected that a solution exists which is continuous in the closed region $t \geq 0$, $0 \leq x \leq 1$ of the x,t plane, and that this solution is uniquely determined. The uniqueness question is settled, as in the case of the Dirichlet problem for Laplace's equation (cf. Theorem 6-3), by establishing a suitable maximum principle for solutions of the heat equation (1).

Theorem 1. Maximum Principle. Let $u(x,t)$ be continuous in the closed rectangle $a \leq x \leq b$, $T_1 \leq t \leq T_2$ and satisfy (1) throughout the interior. Then the maximum of u is assumed at least at one point on the vertical sides ($x = a$, $x = b$) or on the bottom ($t = T_1$) of the rectangle.

Proof. If the assertion is false, the maximum is assumed at a point (x_0, t_0) such that $a < x_0 < b$, $T_1 < t_0 \leq T_2$. Let $v = u + \epsilon(x - x_0)^2$, where ϵ is chosen positive and so small that v, like u, does not assume its maximum on the vertical sides or the bottom. Then v must also assume its maximum elsewhere in the rectangle, say at (x_1, t_1). If (x_1, t_1) is an interior point of the rectangle, the equality $v_t = 0$ and the inequality $v_{xx} \leq 0$ would hold at this point. However, these conditions would be inconsistent with the identity

$$v_{xx} - v_t = u_{xx} - u_t + 2\epsilon = 2\epsilon \ (>0) \tag{36}$$

which holds by virtue of (1) and the definition of v. The remaining possibility, that (x_1, t_1) might be on the top side of the rectangle, cannot be ruled out by the same argument, for no differentiability assumptions are made on the boundary. However, this difficulty is circumvented by the device of replacing T_2 by a slightly smaller value, say T_3, where by "slightly smaller" is meant that u still does not assume its maximum (for the new rectangle) on the vertical sides or bottom. Then the maximum is surely attained on the top of the new rectangle (not in the interior, by the preceding argument). Since u and v are now differentiable along the top, we may still assert that $v_{xx} \le 0$ at the point where the maximum is attained, while the previous equality $v_t = 0$ is now replaced by $v_t \ge 0$. However, this pair of inequalities is again inconsistent with (36). Thus the proof is complete.

If we replace u by $-u$, we get a repetition of Theorem 1 with "maximum" replaced by "minimum." We now immediately obtain the following corollary, which plays exactly the same role here as does Theorem 6-3 in the theory of harmonic functions.

CoROLLARY. UNIQUENESS THEOREM. Given a rectangle and a continuous function defined on the vertical sides and bottom, there exists at most one function continuous in the (closed) rectangle, satisfying the equation (1) in the interior, and coinciding with the given function where the latter is defined.

Proof. If there were two solutions, their difference would vanish on the vertical sides and on the bottom; by the maximum and minimum principles, the difference would then vanish throughout the rectangle.

Having settled the uniqueness question, we turn to existence. Although the general set of boundary and initial conditions (35) can be handled together, it is more convenient to split the problem into several simpler ones, and then treat each one separately. Consider the heat equation in combination with each of the following sets of boundary and initial conditions, where $h(x)$ denotes the linear function coinciding with $g(x)$ at $x = 0$, $x = 1$.

$$u(0,t) = 0 \qquad u(1,t) = 0 \qquad u(x,0) = g(x) - h(x)$$
$$(37a)$$

$$u(0,t) = g(0) \qquad u(1,t) = g(1) \qquad u(x,0) = h(x) \qquad (37b)$$
$$u(0,t) = 0 \qquad u(1,t) = f_2(t) - g(1) \qquad u(x,0) = 0 \qquad (37c)$$
$$u(0,t) = f_1(t) - g(0) \qquad u(1,t) = 0 \qquad u(x,0) = 0 \qquad (37d)$$

If solutions can be found for each of these problems, their sum clearly constitutes a solution to the problem posed at the beginning of this section. The first of these four problems will be solved by reducing it to one of the type considered in Sec. 1, the solution to the second can be written down at sight, and solutions to the third and fourth will be obtained by the use of Laplace transform. We now proceed with the details.

Let $f(x)$ be defined for all real x as follows: $f(x) = g(x) - h(x)$, $0 \le x \le 1$, $f(-x) = -f(x)$, $f(x + 2) = f(x)$. Clearly $f(x)$ is bounded and continuous for all x, and so we may assert, as in Sec. 1, that the function

$$u_1(x,t) = \int_{-\infty}^{\infty} f(\xi)K(\xi,x,t)\,d\xi \tag{38}$$

is a bounded solution of (1) and that $\lim_{t \to 0} u_1(x,t) = f(x)$. From the symmetry properties of $f(\xi)$, it is evident that $u_1(0,t) = u_1(1,t) = 0$, so that $u_1(x,t)$, when restricted to the strip $0 \le x \le 1$, $t \ge 0$, solves the first of the four problems.

As for the second problem, it suffices to note that

$$u_2(x,t) = h(x) \tag{39}$$

constitutes a solution.

Turning to the third problem, and assuming that a solution $u_3(x,t)$ exists, that it admits for each value of x a Laplace transform $U(x,s)$ with respect to t, and that the formal rules of the transform calculus are applicable, we find that the transform satisfies the differential equation

$$U_{xx} = sU \tag{40}$$

with "boundary conditions"

$$U(0,s) = 0 \qquad U(1,s) = \check{\psi}(s) = \int_0^{\infty} e^{-st}\psi(t)\,dt \tag{41}$$

where, for convenience, we have set $f_2(t) - g(1) = \psi(t)$. We readily obtain

$$U(x,s) = \frac{\check{\psi}(s)\ \sinh\ xs^{\frac{1}{2}}}{\sinh\ s^{\frac{1}{2}}} \tag{42}$$

By invoking the Laplace inversion formula, we are led *formally* to the presumed solution

$$u_3(x,t) = \frac{1}{2\pi i} \int_{\sigma-i\infty}^{\sigma+i\infty} e^{st}U(x,s)\,ds \tag{43}$$

However, this formula may very well be meaningless, for $\check{\psi}(s)$ may fail to exist, as in the case $\psi(t) = \exp{(t^2)}$. This objection may be overcome by performing another formal operation. Assuming for the moment (we shall justify this assumption later) that there exists a function $w(x,t)$, defined for $0 \le x < 1$, $t \ge 0$, whose transform is equal to $\sinh xs^{\frac{1}{2}}/\sinh s^{\frac{1}{2}}$, then the convolution theorem (Exercise 1-27) suggests, because of the form of (42), that $u_3(x,t)$ is the convolution of $\psi(t)$ with $w(x,t)$. Thus, instead of (43), we are led to consider the formula

$$u_3(x,t) = \int_0^t \psi(\tau)w(x,\ t-\tau)\,d\tau \tag{44}$$

We note that (44), in addition to involving no assumption concerning the existence of the transform $\check{\psi}(s)$, is consistent with the uniqueness theorem established earlier in this section, for the value of u_3 at any instant $t = t_0$ (> 0) depends, according to (44), only on the definition of $\psi(t)$ in the interval $0 \leq t \leq t_0$, not on its behavior for $t > t_0$.

We now turn to the existence of the function $w(x,t)$ appearing in (44). For the present, the only properties of the function $\sinh xs^{\frac{1}{2}}/\sinh s^{\frac{1}{2}}$ that we shall use are that, for each fixed value of x, $0 \leq x < 1$, it is analytic in the half plane Re $s > 0$ and that, on any fixed vertical line Re $s = \sigma > 0$, its modulus is, for large (positive or negative) values of Im s ($= \tau$), given by

$$\left| \frac{\sinh xs^{\frac{1}{2}}}{\sinh s^{\frac{1}{2}}} \right| = e^{-(1-x)(|\tau|/2)^{\frac{1}{2}}}[1 + o(1)] \tag{45}$$

where $o(1)$ denotes a function of τ which approaches zero as τ becomes infinite. It follows that, for all values of x in the afore-mentioned range and for all values of t, it is meaningful to define a function $w(x,t)$ as follows:

$$w(x,t) = \frac{1}{2\pi i} \int_{\sigma-i\infty}^{\sigma+i\infty} e^{st} \frac{\sinh xs^{\frac{1}{2}}}{\sinh s^{\frac{1}{2}}} ds \tag{46}$$

Also, $w(x,t)$ is continuous in both variables, vanishes identically on the line $x = 0$ and on the segment $0 \leq x < 1$, $t = 0$. [The behavior of $w(x,t)$ as $x \to 1$ is not immediately apparent from (46), but will be determined later after an alternative definition of $w(x,t)$ is obtained.] Furthermore, any number of differentiations with respect to x and t under the integral sign in (46) are easily justified because of the rapid approach of the integrand to zero for large τ, as indicated by (45). Since a trivial computation shows that the integrand of (46) satisfies the heat equation, it then follows that the same is true of $w(x,t)$.

In order to investigate the behavior of $w(x,t)$ near the line $x = 1$, we find it helpful to evaluate the right side of (46) by the method of residues. We note that the integrand is single-valued, despite the appearance of the radical $s^{\frac{1}{2}}$. This follows from the fact that the Maclaurin expansion of the function $\sinh z$ contains only odd powers of z, so that both $\sinh xs^{\frac{1}{2}}$ and $\sinh s^{\frac{1}{2}}$ may be expressed as the product of $s^{\frac{1}{2}}$ and a series containing only integral powers of s. On the other hand, the presence of infinitely many poles necessitates a certain amount of caution. Noting that the singularities are all simple poles, situated at $s = -n^2\pi^2$ ($n = 1, 2, 3, \ldots$), we select the parabolas

$$\text{Re } s = -(n - \tfrac{1}{2})^2\pi^2 + \frac{(\text{Im } s)^2}{(2n - 1)^2\pi^2} \tag{47}$$

and define the closed contour C_n as the curve consisting of the portion of the vertical line employed in (46) lying inside the parabola (47), together with the portion of the parabola to the left of the line. (Cf. Fig. 8-2, where for the sake of clarity we have selected a particular value of n, namely, $n = 4$.) If we let I_n and V_n denote the vertical and parabolic portions of C_n, respectively, we obtain by the method of residues the following result:

$$\frac{1}{2\pi i} \int_{I_n + V_n} e^{st} \frac{\sinh xs^{1/2}}{\sinh s^{1/2}}\, ds = \sum_{k=1}^{n-1} 2\pi k(-1)^{k+1} \sin k\pi x\, e^{-k^2\pi^2 t} \qquad (48)$$

Letting $n \to \infty$, we readily find that $\int_{V_n} \to 0$. (Cf. Exercise 12.) Hence the left side of (48) approaches $w(x,t)$, and we obtain the series expansion

$$w(x,t) = \sum_{k=1}^{\infty} 2\pi k(-1)^{k+1} \sin k\pi x\, e^{-k^2\pi^2 t} \qquad (49)$$

For $t \geq t_0$, t_0 any positive constant, the series (49) is dominated by the convergent series of constants $\sum_{k=1}^{\infty} 2\pi k \exp(-k^2\pi^2 t_0)$, so that (49) converges, for all positive t and all x, to a continuous function. Since each term of (49) vanishes for $x = 1$, we conclude that $w(x,t)$, which was

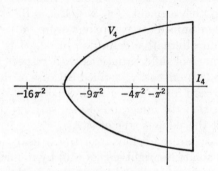

Figure 8-2

originally defined in the region $t > 0$, $0 \leq x < 1$, becomes continuous in the region $t > 0$, $0 \leq x \leq 1$ if defined to vanish on the line $t > 0$, $x = 1$.

Now, returning to (44), we immediately see that $u_3(x,t)$ is well defined, continuous, and a solution of the heat equation for $0 < x < 1$, $t > 0$, and continuous for $0 \leq x < 1$, $t \geq 0$, if defined to vanish for $x = 0$ and also for $t = 0$. It we formally set $x = 1$ in (44), we might conclude that u_3 vanishes, but this reasoning is invalid, on account of the highly singular behavior of $w(x,t)$ near the point $x = 1$, $t = 0$. In order to overcome this difficulty, we consider the particular case $\psi(t) = t^n$, n any positive integer. In this case $\psi(t)$ does indeed possess a transform, given by $\bar{\psi}(s) = n!s^{-n-1}$, and we can show without difficulty that (44) is entirely equivalent to (43).

Now, from (43) we can very easily deduce that $u_3(x,t)$ possesses the proper behavior near the line $x = 1$, namely, that $u_3(x,t)$ approaches $t_0{}^n$ as the point (x,t) approaches the point $(1,t_0)$, $t_0 \geq 0$. By superposition, we conclude that (44) furnishes the solution to the problem whenever $\psi(t)$ is a polynomial vanishing at $t = 0$. Now, for arbitrary continuous $\psi(t)$ vanishing at $t = 0$ and any $T > 0$ we can, by the Weierstrass approximation theorem, approximate $\psi(t)$ in the interval $0 \leq t \leq T$ by a sequence of polynomials $\{p_n(t)\}$, each of which also vanishes at $t = 0$. For each $p_n(t)$ we obtain by (44) a corresponding solution, $u_3{}^{(n)}(x,t)$. Hence, by construction, the functions $u_3{}^{(n)}(x,t)$ converge uniformly on the vertical sides and bottom of the rectangle $0 \leq x \leq 1$, $0 \leq t \leq T$, from which we conclude by the maximum principle that they converge uniformly in the interior as well. The limit function, which we term $u_3(x,t)$, clearly satisfies the prescribed conditions (37c). It remains only to show that $u_3(x,t)$ satisfies the heat equation in the interior. However, from the fact that $w(x,t)$ is continuous for $0 \leq x < 1$ and all values of t, it readily follows that the function $u_3(x,t)$ coincides with the function defined by (44), and there is no difficulty in justifying differentiation under the integral sign, from which it follows directly that $u_3(x,t)$ does satisfy (1).

Thus, a solution of (1) satisfying (37c) has been obtained. Replacing x by $1 - x$ and $f_2(t) - g(1)$ by $f_1(t) - g(0)$, we obtain a solution of (1) satisfying (37d). Thus, a complete solution has been obtained for the problem posed at the beginning of this section.

EXERCISES

10. Prove that the function $w(x,t)$ defined by (46) vanishes in the region $t < 0$, $0 \leq x < 1$.

11. Show that $w(x,t)$ may be interpreted as a limiting case of the temperature distribution resulting when there is applied to a rod initially at temperature zero a specified amount of heat at the right end over a very short time interval, after which this end is again maintained at zero temperature, while the other end is maintained permanently at zero temperature.

12. Prove the assertion made following (48) that $\int_{V_n} \to 0$.

13. Consider the "backward heat equation," $u_{xx} = -u_t$, with the conditions $u(0,t) = 0$, $u(x,0) = 0$, $u(1,t) = t$. Show that this problem leads to a function $U(x,s)$ which cannot be a Laplace transform. *Hint:* Observe the location of the singularities of $U(x,s)$.

9. GREEN'S FUNCTIONS AND SEPARATION OF VARIABLES

Among the most useful and frequently employed methods of solving certain types of problems in partial differential equations is the one known as "separation of variables." The reader is probably acquainted with the manner in which problems dealing with vibrating strings and circular membranes lead, by the use of the afore-mentioned method, to expansions in series of trigonometric functions and Bessel functions, respectively; similarly, the Dirichlet problem for a sphere leads to an expansion of the solution in a series of Legendre functions (spherical harmonics). We shall be only incidentally concerned with the particular functions arising in the study of such expansions. Rather, our primary interest will be in developing the theory underlying such expansions.

1. The Vibrating String

It is convenient to begin our considerations with a brief review (from a viewpoint different from that of Sec. 3-5) of the familiar problem of the vibrating string. Under suitable hypotheses and with suitable choice of physical units, the vibrations of a uniform string which is held fixed at both ends in the rest position are described by the one-dimensional wave equation

$$u_{xx} - u_{tt} = 0 \tag{1}$$

and the "boundary conditions"

$$u(0,t) \equiv u(1,t) \equiv 0 \tag{2}$$

At a given instant, which without loss of generality may be taken as $t = 0$, let the shape and velocity of the string be given by the functions $f(x)$ and $g(x)$, respectively.   Then the study of the subsequent motion of the string reduces to the determination of a function $u(x,t)$ defined and continuous for $0 \leq x \leq 1$, $t \geq 0$ and satisfying equations (1), (2), and the "initial conditions"

$$u(x,0) = f(x) \qquad u_t(x,0) = g(x) \tag{3}$$

Disregarding (3) temporarily, we seek a function which satisfies (1) and (2) and is expressible as the product of functions each depending on only one of the independent variables; i.e., we seek a solution of (1) having the form

$$u = X(x)T(t) \tag{4}$$

Substituting from (4) into (1), we obtain

$$TX''(x) = XT''(t) \tag{5}$$

which is equivalent to

$$\frac{X''}{X} = \frac{T''}{T} \tag{6}$$

Since the left and right sides of (6) are independent of t and x, respectively, they must both reduce to a constant, which we term $-\lambda$.   Thus, $X(x)$ and $T(t)$ satisfy the (ordinary) differential equations

$$X'' + \lambda X = 0 \tag{7a}$$
$$T'' + \lambda T = 0 \tag{7b}$$

In order that (2) be satisfied, it is necessary that

$$X(0) = X(1) = 0 \tag{8}$$

Trivially, the function $X \equiv 0$ satisfies (8) as well as the equation (7a), regardless of the value of the "separation constant" λ; however, a nontrivial function satisfying these two equations exists only if λ is one of the quantities $n^2\pi^2$, $n = 1, 2, 3, \ldots$, and for any one of these values of λ the corresponding solution must be of the form

$$X(x) = \text{constant} \cdot \sin n\pi x \tag{9}$$

The proof of these facts is left to the reader as Exercise 1.   The corresponding function $T(t)$ must then be of the form

$$T(t) = \text{constant} \cdot \cos n\pi t + \text{constant} \cdot \sin n\pi t \tag{10}$$

and (4) therefore assumes the form

$$u(x,t) = \sin n\pi x \,(\text{constant} \cdot \cos n\pi t + \text{constant} \cdot \sin n\pi t) \tag{11}$$

Since (1) and (2) are both linear and homogeneous, it follows that any finite combination of functions of form (11) will also satisfy the above pair of equations. It now appears plausible to attempt to form an *infinite* combination of all the particular solutions which have been obtained which will satisfy not only (1) and (2) but the initial conditions (3) as well. Temporarily disregarding questions relating to convergence, we assume a solution of the form

$$u = \sum_{n=1}^{\infty} \sin n\pi x (a_n \cos n\pi t + b_n \sin n\pi t) \tag{12}$$

and obtain from (3) the conditions

$$f(x) = \sum_{n=1}^{\infty} a_n \sin n\pi x \tag{13a}$$

and

$$g(x) = \sum_{n=1}^{\infty} n b_n \sin n\pi x \tag{13b}$$

We recognize that these last two equations represent "half-range" trigonometric expansions, and so we obtain for the coefficients the formulas

$$a_n = 2 \int_0^1 f(x) \sin n\pi x \, dx \qquad b_n = \frac{2}{n} \int_0^1 g(x) \sin n\pi x \, dx \tag{14}$$

Now, if the functions $f(x)$, $g(x)$ are such that the coefficients a_n, b_n approach zero sufficiently rapidly with increasing n, series (12) will indeed converge to a function continuous for $0 \leq x \leq 1$, $t \geq 0$ and will permit term-by-term differentiation, so that (1) and (2) will be satisfied. [For example, suppose that $f(x)$ and $g(x)$, when defined for all x as odd functions of period 2, possess continuous derivatives of fourth and third orders respectively. Then, by an elementary argument,[1] a_n and b_n cannot exceed Cn^{-4} for some sufficiently large constant C, and then there is no difficulty in justifying the term-by-term differentiations of (12) needed to show that (1) is satisfied.] Furthermore, from the general theory of trigonometric series, expansions (13a) and (13b) will actually converge to $f(x)$ and $g(x)$, respectively, so that the initial conditions (3) will be satisfied.

It should be stressed that the essential feature of the above method is that the family of functions $\{\sin n\pi x\}$ obtained by determining all non-trivial solutions of (7a) which satisfy (8) is "large" enough to permit any "reasonable" function to be expressed in the form of a linear combination of these particular functions. Now, while this fact is known, as indicated above, from the theory of trigonometric series, which may be developed

---

[1] Rewriting a_n in the form $\int_{-1}^{1} f(x) \sin n\pi x \, dx$ and integrating by parts four times, we obtain $a_n = (n\pi)^{-4} \int_{-1}^{1} f''''(x) \sin n\pi x \, dx$, and hence $|a_n| \leq Cn^{-4}$, where $C = 2\pi^{-4} \max |f''''(x)|$; similarly for b_n.

quite independently of equations (7a) and (8), the question arises whether the functions obtained by replacing this pair of equations by somewhat different equations would still possess the afore-mentioned property. As suggested by the introductory remarks to this chapter, a satisfactory theory concerning such systems of functions exists, and we shall begin by demonstrating in the next section, quite independently of the theory of trigonometric series, that expansions of the form (13a) for any "reasonable" function are indeed possible.

EXERCISES

1. Prove the assertions made in the sentence following (8).

2. Prove, without employing the explicit form of the general solution of (7a), that those values of λ (if any) for which (7a) admits a nontrivial solution satisfying (8) must be positive. *Hint:* Multiply both sides of (7a) by X, and then add $[X'(x)]^2$ to each side.

3. Obtain (12) from (1), (2), (3) by Laplace transform.

2. The Green's Function of the Operator $\dfrac{d^2}{dx^2}$

Let $f(x)$ be any function of class C^2 defined on the interval $0 \leq x \leq 1$ and vanishing at both end points. Then $f(x)$ is unambiguously determined by its second derivative, $f''(x)$; for if $g''(x) \equiv f''(x)$, it follows that $g(x) - f(x)$ must be a linear function, and hence, if f and g each vanish at both end points, their difference must vanish identically.

Now, we shall derive an integral formula expressing $f(x)$ in terms of its second derivative, which, for convenience, we temporarily denote by $-h(x)$. Integrating the equation

$$f''(x) = -h(x) \tag{15}$$

twice, we obtain

$$f(x) = -\int_0^x \int_0^t h(\xi)\, d\xi\, dt + c_1 x + c_2 \tag{16}$$

where c_1 and c_2 are arbitrary constants. The conditions $f(0) = f(1) = 0$ yield for these constants the following formulas:

$$c_1 = \int_0^1 \int_0^t h(\xi)\, d\xi\, dt \qquad c_2 = 0 \tag{17}$$

If we interchange the order of integration in (16) and (17), we obtain

$$f(x) = -\int_0^x \int_\xi^x h(\xi)\, dt\, d\xi + x\int_0^1 \int_\xi^1 h(\xi)\, dt\, d\xi \tag{18}$$

Upon performing the inner integration in each case, we obtain

$$f(x) = \int_0^x h(\xi)(\xi - x)\, d\xi + \int_0^1 h(\xi)(x - \xi x)\, d\xi \tag{19}$$

or, finally,

$$f(x) = \int_0^1 h(\xi)G(x,\xi)\,d\xi \tag{20}$$

where

$$G(x,\xi) = \begin{cases} (1-x)\xi & 0 \le \xi \le x \\ x(1-\xi) & x \le \xi \le 1 \end{cases} \tag{21}$$

Now, the function $G(x,\xi)$ was studied in Sec. 5-8 in order to illustrate the expansion theorem for a continuous symmetric kernel, and it was shown that its eigenfunctions are precisely the functions given by (9), so that any function expressible in form (20) [with some continuous $h(x)$] can also be expanded in a uniformly convergent series of the functions (9). As in Sec. 5-8, we stress here the method of proof, rather than the strength of the result, for it is well known that the requirement that $f(x)$ be of class C^2 is far more than is actually needed for this result to hold.

While the existence of a representation of form (20) of $f(x)$ is essential in the above considerations, the fact that $f(x)$ and $h(x)$ are related in the precise manner defined by (15) is of no importance. For example, let three continuous functions, $p(x)$, $q(x)$, $r(x)$, be defined on the interval $a \le x \le b$, and suppose that there exists a continuous symmetric function $K(x,\xi)$ such that every twice continuously differentiable function $f(x)$ vanishing at the ends of the interval can be expressed in the form

$$f(x) = \int_a^b h(\xi)K(x,\xi)\,d\xi \tag{22}$$

where

$$Lf = pf''(x) + qf'(x) + rf(x) = -h(x) \tag{23}$$

Then, exactly as in the above example [where $p \equiv 1$, $q \equiv r \equiv 0$, $a = 0$, $b = 1$, $K(x,\xi) = G(x,\xi)$], the function $f(x)$ will be expandible in a uniformly convergent series of eigenfunctions of $K(x,\xi)$. It will be shown in the following section that, under certain conditions on the functions p, q, and r, such a function $K(x,\xi)$ exists. This function will be termed the "Green's function" of the differential operator $p\dfrac{d^2}{dx^2} + q\dfrac{d}{dx} + r$; in particular, the function $G(x,\xi)$ considered above is the Green's function of the operator $\dfrac{d^2}{dx^2}$ (for the interval $0 \le x \le 1$).

EXERCISES

4. Prove that the function $G(x,\xi)$ may be characterized as follows:

(a) $G(0,\xi) \equiv G(1,\xi) \equiv 0$.

(b) $G(x,\xi)$ is continuous in x for each fixed ξ.

(c) $G_{xx}(x,\xi) = 0$, $x \ne \xi$.

(d) $\displaystyle\lim_{x \to \xi+0} G_x(x,\xi) - \lim_{x \to \xi-0} G_x(x,\xi) = -1$.

5. Prove that (20) is entirely equivalent to (15) *together with* the boundary conditions $f(0) = f(1) = 0$.

3. The Green's Function of a Second-order Differential Operator

In this section we shall consider the problem of representing a function $f(x)$ in the form (22), where $h(x)$ is related to $f(x)$ according to (23). As in Sec. 2, it is understood that $f(x)$ is of class C^2 in the closed interval $I: x_1 \leq x \leq x_2$, and vanishes at both ends.

We now impose on the functions $p(x)$, $q(x)$, $r(x)$ appearing in (23) the following restrictions:

1. $p(x)$, $q(x)$, and $r(x)$ are continuous in I,
2. $p > 0$ throughout I,
3. $p'(x) \equiv q(x)$, so that

$$pf''(x) + qf'(x) = [pf'(x)]' \tag{24}$$

4. The only solution of the equation

$$pu'' + qu' + ru = 0 \tag{25}$$

vanishing at both ends of I is the trivial one, $u \equiv 0$. (This condition is satisfied, for example, in the case considered in the preceding section.)

Now, let $f_1(x)$ denote the (unique) solution of (25) which vanishes at x_1 and whose first derivative there is equal to unity. Similarly, let $f_2(x)$ denote the solution satisfying the same conditions at x_2. Then the so-called "Wronskian determinant"

$$W(f_1, f_2) = \begin{vmatrix} f_1 & f_2 \\ f_1' & f_2' \end{vmatrix} = f_1 f_2' - f_2 f_1' \tag{26}$$

never vanishes in the interval.[1] Therefore, given any point ξ inside I, there exist uniquely determined quantities $c_1(\xi)$ and $c_2(\xi)$ such that

$$\begin{align} c_1 f_1(\xi) - c_2 f_2(\xi) &= 0 \\ c_1 f_1'(\xi) - c_2 f_2'(\xi) &= \frac{1}{p(\xi)} \end{align} \tag{27}$$

Indeed, denoting the constant (cf. footnote) value of $(pW)^{-1}$ by C, we obtain

$$c_1 = Cf_2(\xi) \qquad c_2 = Cf_1(\xi) \tag{28}$$

Now, consider the function $K(x, \xi)$ defined as follows:

$$K(x, \xi) = \begin{cases} Cf_2(\xi)f_1(x) & x_1 \leq x \leq \xi \\ Cf_1(\xi)f_2(x) & \xi \leq x \leq x_2 \end{cases} \tag{29}$$

---

[1] $(pW)' = f_1(pf_2')' - f_2(pf_1')' = f_1(-rf_2) - f_2(-rf_1) = 0$. Hence, $pW = $ constant, and therefore W vanishes everywhere or nowhere. To rule out the former possibility, we note that W reduces at x_1 to $-f_2(x_1)$, so that if W vanished there, $f_2(x)$ would be a nontrivial solution of (25) vanishing at both ends of I, contradicting condition 4.

$K(x,\xi)$ is evidently continuous in the closed square $x_1 \leq x$, $\xi \leq x_2$ and symmetric in its arguments; i.e.,

$$K(x,\xi) = K(\xi,x) \tag{30}$$

Let $u(x)$ and $v(x)$ be any two functions of class C^2 defined on an interval $x_3 \leq x \leq x_4$ contained in the afore-mentioned interval I. Then, denoting for brevity $(pu')' + ru$ by $L(u)$, we obtain

$$\int_{x_3}^{x_4} [uL(v) - vL(u)] \, dx = \int_{x_3}^{x_4} [u(pv')' - v(pu')'] \, dx$$
$$= \int_{x_3}^{x_4} (upv' - vpu')' \, dx = (upv' - vpu') \Big|_{x_3}^{x_4} \tag{31}$$

Now let ξ be any inner point of I and let us apply (31) twice, first to the interval $x_1 \leq x \leq \xi$ and then to the interval $\xi \leq x \leq x_2$, replacing $v(x)$ each time by $K(x,\xi)$. [Note that in each of these two intervals $K(x,\xi)$ is of class C^2, although in the entire interval I it is not even of class C^1.] We obtain, since $L(v) = 0$,

$$-\int_{x_1}^{\xi} L(u)K(x,\xi) \, dx = [upK_x(x,\xi) - K(x,\xi)pu'] \Big|_{x_1}^{\xi-0} \tag{32}$$

and $$-\int_{\xi}^{x_2} L(u)K(x,\xi) \, dx = [upK_x(x,\xi) - K(x,\xi)pu'] \Big|_{\xi+0}^{x_2} \tag{33}$$

Adding these two equations and observing, with the aid of a simple computation based on (29) and the definitions of C and of the functions $f_1(x)$ and $f_2(x)$, that

$$K_x(x,\xi) \Big]_{x=\xi+0} - K_x(x,\xi) \Big]_{x=\xi-0} = -\frac{1}{p(\xi)} \tag{34}$$

we obtain

$$-\int_{x_1}^{x_2} L(u)K(x,\xi) \, dx = -upK_x(x,\xi) \Big|_{x_1}^{x_2} + u(\xi) \tag{35}$$

We note that (35) furnishes the solution to the differential equation

$$L(f) = (pf')' + rf = -h(x) \tag{36}$$

with prescribed values of $f(x_1)$ and $f(x_2)$, namely,

$$f(\xi) = \int_{x_1}^{x_2} h(x)K(x,\xi) \, dx + fpK_x(x,\xi) \Big|_{x_1}^{x_2} \tag{37}$$

In particular, if $f(x)$ is required to vanish at both ends of I, (37) reduces to (22).

We conclude this section by giving a definition of the Green's function $K(x,\xi)$ of the operator L which, in contrast to (29), emphasizes essential properties of this function, rather than the specific manner in which it is constructed. $K(x,\xi)$ is now defined as follows: For each inner point ξ of I, it is a continuous function of x, vanishing at both end points, and satisfies, for all values of x except ξ, the equation $L(K) = 0$; $K_x(x,\xi)$ is discontinuous at ξ, the nature of the discontinuity being given by (34).

EXERCISES

6. If condition 3 immediately preceding (24) is omitted, (31) no longer holds. Show that, in its place, the following identity holds:

$$\int_{x_3}^{x_4} [uL(v) - vM(u)]\, dx = [quv + puv' - (pu)'v]\, \Big|_{x_3}^{x_4} \tag{38}$$

where M, the "adjoint" of the operator L, is given by

$$M(u) = pu'' + (2p' - q)u' + (p'' - q' + r)u \tag{39}$$

7. Prove that the adjoint of the operator M defined above is L.

8. Prove that M is uniquely determined by the requirement that, for arbitrary (sufficiently differentiable) functions u and v, the left side of (38) shall be equal to a "boundary term"—i.e., an expression involving only the values of u and v and their first derivatives at the ends of the interval.

9. Relate the present use of the term "adjoint" to its use in Chap. 5.

10. Let the operator L satisfy conditions 1 and 2 but not 3. Prove that there exists a positive function $\rho(x)$ defined on I such that the operator $\rho(x)L$ is self-adjoint.

11. Prove that, if $r(x) \le 0$ throughout I, then $K(x,\xi) > 0$ throughout the open square $x_1 < x$, $\xi < x_2$.

4. Eigenfunction Expansions

Let the differential operator L satisfy conditions 1 through 4 imposed in the preceding section, and let $\rho(x)$ be a given function continuous and positive in the interval I. Then the "boundary-value problem"

$$L(u) + \lambda\rho u = 0 \qquad u(x_1) = u(x_2) = 0 \tag{40}$$

admits, for each value of the parameter λ, the trivial solution $u \equiv 0$. However, as shown in Sec. 1 $\left(\text{where } L = \dfrac{d^2}{dx^2}, \rho \equiv 1, x_1 = 0, x_2 = 1\right)$, there may exist certain particular values of λ for which (40) admits a nontrivial solution. A comprehensive theory concerning the existence of these special values of λ (eigenvalues) and the properties of the corresponding nontrivial solutions of (40) (eigenfunctions) is most readily developed by converting (40) into an integral equation. According to the results of the preceding section, any solution of (40) satisfies the equation

$$u(\xi) = \lambda \int_{x_1}^{x_2} K(x,\xi)\rho(x)u(x)\, dx \tag{41}$$

[cf. (22)], where $K(x,\xi)$ is the Green's function of L (for the interval $x_1 \le x \le x_2$). Conversely, if u satisfies (41), it also satisfies (40) (i.e., the boundary conditions as well as the differential equation). If $\rho(x) \equiv 1$, we would conclude that the problem of determining eigenvalues and eigenfunctions for (40) is equivalent to that of determining the eigenvalues and eigenfunctions of $K(x,\xi)$. By a simple procedure, we shall show that even when $\rho(x)$ is not $\equiv 1$, it is possible to interpret the eigenvalues and eigenfunctions of (40) as being, aside from a simple factor, the eigenvalues and eigenfunctions of a symmetric kernel closely related to $K(x,\xi)$. Let

$$\rho^{\frac{1}{2}}u = \tilde{u} \qquad [\rho(x)\rho(\xi)]^{\frac{1}{2}}K(x,\xi) = \tilde{K}(x,\xi) \tag{42}$$

Then (41) assumes the form

$$\tilde{u}(\xi) = \lambda \int_{x_1}^{x_2} \tilde{K}(x,\xi)\tilde{u}(x)\,dx \tag{43}$$

Since $\tilde{K}(x,\xi)$ is clearly continuous and symmetric (and not $\equiv 0$), we know that it possesses a finite or denumerable set of eigenvalues λ_1, λ_2, λ_3, . . . and eigenfunctions $\tilde{u}_1$, $\tilde{u}_2$, $\tilde{u}_3$, ... Now, letting $f(x)$ denote any function of class C^2 vanishing at both ends of I, we conclude from (22) that

$$\rho^{\frac{1}{2}}(\xi)f(\xi) = \int_{x_1}^{x_2} L(f)\rho^{-\frac{1}{2}}(x)\tilde{K}(x,\xi)\,dx \tag{44}$$

By the expansion theorem of Chap. 5, $\rho^{\frac{1}{2}}(\xi)f(\xi)$ is expressible as a uniformly convergent series,

$$\rho^{\frac{1}{2}}(\xi)f(\xi) = \Sigma c_n\tilde{u}_n(\xi) \tag{45}$$

the coefficients c_n being given by the familiar formula

$$c_n = \int_{x_1}^{x_2} \rho^{\frac{1}{2}}(\xi)f(\xi)\tilde{u}_n(\xi)\,d\xi \tag{46}$$

[It is understood that the $\tilde{u}_n(\xi)$ are taken as an orthonormal set, $\int_{x_1}^{x_2} \tilde{u}_n(\xi)\tilde{u}_m(\xi)\,d\xi = \delta_{nm}$.] Letting $u_n(\xi) = [\rho(\xi)]^{-\frac{1}{2}}\tilde{u}_n(\xi)$, we obtain in place of (45) the entirely equivalent expansion

$$f(\xi) = \Sigma c_n u_n(\xi) \tag{47}$$

while (46) may be replaced by

$$c_n = \int_{x_1}^{x_2} \rho(\xi)f(\xi)u_n(\xi)\,d\xi \tag{48}$$

It may be noted that we have not indicated whether the summations in (45) and (47) are finite or infinite, but this matter is easily settled. If there were only a finite number of eigenvalues and eigenfunctions, the set of functions of class C^2 vanishing at both ends of I would constitute a finite-dimensional subspace of $L_2(x_1,x_2)$ (cf. Chap. 5). In fact, however, the afore-mentioned functions constitute a dense subset of $L_2(x_1,x_2)$, and so it follows that the functions $\tilde{u}_n(\xi)$ actually constitute an orthonormal basis of $L_2(x_1,x_2)$. (Cf. Exercises 13 and 14.)

We have, therefore, shown that the boundary-value problem (40) admits nontrivial solutions $\{u_n\}$ for a denumerable sequence $\{\lambda_n\}$ of values of the parameter λ. Any sufficiently smooth function vanishing at both ends of I admits the uniformly convergent expansion (47), the coefficients c_n being given by (48), provided that the functions $\bar{u}_n$ are normalized as indicated following (46).

Although the foregoing treatment employed the existence of a Green's function, which, in turn, was based on the assumption that (40) admits only the trivial solution for $\lambda = 0$ (i.e., that $\lambda = 0$ is not an eigenvalue), the final result involves neither the Green's function nor the eigenvalues. It therefore appears plausible that the proof might be modified so as to eliminate this assumption. This is indeed so, and is easily proved as follows: The eigenfunctions of (40) are not changed if L is replaced by $L + c\rho$, and the eigenvalues are merely reduced by c, where c is any constant. Let c be so chosen that $L + c\rho$ possesses a Green's function (cf. Exercise 18); then the above argument is applicable, and we obtain the fact that the functions $\{\bar{u}_n\}$ form an orthonormal basis in $L_2(x_1,x_2)$.

EXERCISES

12. Prove that the Green's function $K(x,\xi)$ cannot be degenerate.

13. Prove, as asserted following (48), that the set of functions of class C^2 vanishing at both ends of I constitutes a dense subset of $L_2(x_1,x_2)$. *Hint:* Approximate an arbitrary member of $L_2(x_1,x_2)$ in norm by a continuous function; then approximate the latter by a smooth function satisfying the end conditions.

14. Prove that the functions $u_n(\xi)$, like $\bar{u}_n(\xi)$, constitute a basis for $L_2(x_1,x_2)$. [Of course, the functions $u_n(\xi)$ will not, in general, constitute an *orthonormal* basis.]

15. Prove that, if $r \leq 0$, all the eigenvalues of (40) are positive.

16. Let λ and μ be any distinct numbers, $\lambda < \mu$, and let u and v be nontrivial solutions of the equations $Lu + \lambda\rho u = 0$, $Lv + \mu\rho v = 0$, respectively. Prove that between any two zeros of u there must exist at least one zero of v. Then use this result to prove that (40) can possess only a finite number of negative eigenvalues, so that the eigenvalues may be enumerated in ascending order, $\lambda_1 < \lambda_2 < \cdots$.

17. Prove that the eigenfunction u_1 corresponding to the lowest eigenvalue λ_1 (cf. preceding exercise) does not vanish in the interior of the interval I. *Hint:* First make the additional assumption $r(x) \leq 0$ and use Exercise 11; then show that this assumption may be dropped.

18. Prove, as asserted at the end of the present section, that the constant c can be so chosen that zero is not an eigenvalue of the operator $L + c\rho$. *Hint:* Exploit the fact that $L_2(x_1,x_2)$ is separable, while the real-number system is uncountable.

5. A Generalized Wave Equation

We shall now show how the method of separation of variables can be used to solve boundary-value problems involving equations more difficult than (1). Let $a(x)$, $b(x)$, $c(x)$ be defined and continuous on a bounded closed interval I: $x_1 \leq x \leq x_2$, and let α and β be two given constants.[1]   Let us consider the equation

$$au_{xx} + bu_x + cu = \alpha u_{tt} + \beta u_t \tag{49}$$

together with the "boundary conditions"

$$u(x_1,t) \equiv u(x_2,t) \equiv 0 \qquad (t \geq 0) \tag{50}$$

and "initial conditions"[2]

$$u(x,0) = f(x) \qquad u_t(x,0) = g(x) \tag{51}$$

Since (49) contains (1) as a particular case, we may describe the former as a "generalized wave equation."

Proceeding as in Sec. 1, we seek a function $X(x)T(t)$ satisfying (49) and (50), obtaining for $X(x)$ and $T(t)$ the ordinary differential equations:

$$aX'' + bX' + cX + \lambda X = 0 \tag{52}$$
$$\alpha T'' + \beta T' + \lambda T = 0 \tag{53}$$

where λ, as in (7), denotes a "separation constant."

Now, assuming that a is positive throughout I, we multiply (52) by a suitable positive function $\rho(x)$ such that the operator

$$\rho a \frac{d^2}{dx^2} + \rho b \frac{d}{dx} + \rho c$$

is self-adjoint (cf. Exercise 10).   Then, by the results of the preceding section, the eigenfunctions $\{\phi_n(x)\}$ of (52) form a basis in $L_2(x_1,x_2)$, so that the *formal* expansions

$$f(x) = \sum_{n=1}^{\infty} f_n\phi_n(x) \qquad g(x) = \sum_{n=1}^{\infty} g_n\phi_n(x) \tag{54}$$

will actually converge to $f(x)$ and $g(x)$ if, as we henceforth assume, the coefficients $\{f_n\}$, $\{g_n\}$ approach zero with sufficient rapidity as $n \to \infty$. Let the functions $T_n^{(1)}(t)$ and $T_n^{(2)}(t)$ be defined as the solutions of

$$\alpha T'' + \beta T' + \lambda_n T = 0 \tag{55}$$

satisfying the conditions

$$T_n^{(1)}(0) = 1 \qquad T_n^{(1)\prime}(0) = 0 \qquad T_n^{(2)}(0) = 0 \qquad T_n^{(2)\prime}(0) = 1 \tag{56}$$

respectively.   Then the series

$$u = \sum_{n=1}^{\infty} \phi_n(x)[f_n T_n^{(1)}(t) + g_n T_n^{(2)}(t)] \tag{57}$$

---

[1] At least one of these is assumed different from zero, for otherwise the variable t disappears from (49).   Actually, α and β could be allowed to depend on t without causing any essential change in the following considerations.

[2] If $\alpha = 0$, the second condition in (51) must be dropped.

certainly satisfy (50) and the first of the initial conditions (51). Now, in addition to the assumption previously made on the coefficients appearing in the expansions (54), we assume that they approach zero with sufficient rapidity to assure the convergence of (57) for all $t > 0$, and also to permit termwise differentiation of this series as many times as may be required. Then (57) will also satisfy (49) and the second initial condition of (51), and thus will provide a solution to the problem under consideration.

It may be noted that, though the functions $\{\phi_n(x)\}$ and the coefficients $\{f_n\}$ and $\{g_n\}$ are independent of α and β [depending only on the left side of (49) and on the data (51)], the convergence of the series (57) is influenced by α and β. Thus, when $x_1 = 0$, $x_2 = 1$, $a \equiv 1$, $b \equiv 0$, $c \equiv 0$, there is a marked contrast between the cases $\alpha = 1$, $\beta = 0$ and $\alpha = -1$, $\beta = 0$. In the former, we have the problem considered in Sec. 1, where we saw that very mild hypotheses on the functions $f(x)$ and $g(x)$ suffice to guarantee the existence of a solution. In the latter, on the other hand, we have the Laplace equation, and the functions $T_n^{(1)}(t)$ and $T_n^{(2)}(t)$ are given by $\cosh n\pi t$ and $(n\pi)^{-1} \sinh n\pi t$, respectively. Since these functions become large, either for fixed n and large t or vice versa, much more stringent conditions must be imposed to assure the convergence of the series (57), even for some small range of the variable t. This difference reflects the fact, pointed out in Chap. 3, that the Cauchy problem is meaningful for hyperbolic, but not for elliptic, equations.

EXERCISE

19. Consider the system of equations (49), (50), (51) with $a \equiv 1$, $b \equiv c \equiv 0$, $\alpha = 1$. Analyze the behavior of the functions $T_n^{(1)}(t)$ and $T_n^{(2)}(t)$ for positive and for negative values of the constant β. (A physical interpretation of β as a frictional coefficient is helpful.)

6. Extension of the Definition of Green's Functions

In this section we indicate briefly how the definition of the Green's function can be broadened, and the scope of the method of separation of variables correspondingly enlarged.

1. In the previous sections of this chapter the only boundary conditions that have been considered are the simplest possible ones, namely, that the functions under consideration vanish at both ends of the given interval. While the linearity of these conditions (i.e., the fact that $c_1 u_1 + c_2 u_2$ satisfy them whenever u_1 and u_2 do) is essential in the method of separation of variables, there is considerable flexibility in the nature of the conditions that can be handled. For example, suppose that in Sec. 1 the conditions (2) are changed to the following:

$$u(0,t) \equiv u_x(1,t) \equiv 0 \tag{2'}$$

The method of separation of variables leads in this case to the functions $\{\sin(n - \tfrac{1}{2})\pi x\}$, which are the eigenfunctions of the symmetric kernel

$$\tilde{G}(x,\xi) = \begin{cases} x & 0 \le x \le \xi \\ \xi & \xi \le x \le 1 \end{cases} \tag{58}$$

[or, more concisely, $\tilde{G}(x,\xi) = \min(x,\xi)$], just as in the previous case we were led to the functions $\{\sin n\pi x\}$, which are the eigenfunctions of the symmetric kernel $G(x,\xi)$ defined by (21). Furthermore, just as (20) furnishes the solution to (15) with conditions $f(0) = f(1) = 0$, the solution to (15) with conditions $f(0) = f'(1) = 0$ is given by

$$f(x) = \int_0^1 h(\xi)\tilde{G}(x,\xi)\,d\xi \tag{20'}$$

Also, $\tilde{G}(x,\xi)$ is uniquely determined by the conditions listed in Exercise 4 except for an obvious change in part (a). Therefore, the whole discussion of Sec. 1 carries over with straightforward modifications to the present problem.

We term $\tilde{G}(x,\xi)$, in analogy with $G(x,\xi)$, the Green's function of the operator $\dfrac{d^2}{dx^2}$ for the boundary conditions $f(0) = f'(1) = 0$. More generally, given a differential operator L, as defined by (23), and any pair of linear boundary conditions, we can associate with L, by an obvious modification of the procedure carried out in Sec. 3, a Green's function $K(x,\xi)$ for the given boundary conditions; as before, it is necessary and sufficient that the only solution of the equation $Lu = 0$ which satisfies the boundary conditions is the trivial one $u \equiv 0$. Actually, the assumption that L is self-adjoint is not essential insofar as the existence of $K(x,\xi)$ is concerned; however, as might be expected, the self-adjointness of L is essential for the symmetry of $K(x,\xi)$, without which a satisfactory theory of expansion in series of eigenfunctions cannot be obtained. One might expect that the self-adjointness of L is sufficient to assure the symmetry of $K(x,\xi)$, but this is not correct, as is indicated by the following example: Let $L = \dfrac{d^2}{dx^2}$, and let the boundary conditions be given by $f(0) + f'(1) = f(1) + 2f'(0) = 0$. Then an elementary computation furnishes for the Green's function the formula

$$K(x,\xi) = \begin{cases} 1 + \tfrac{1}{2}\xi - \tfrac{1}{2}\xi x & 0 \le x \le \xi \\ 1 + \tfrac{3}{2}\xi - \tfrac{1}{2}\xi x - x & \xi \le x \le 1 \end{cases} \tag{59}$$

and it is readily seen (by interchanging ξ and x in either of the above expressions and comparing the result with the other expression) that $K(x,\xi)$ is not symmetric.

2. We now indicate how Green's functions can be associated with differential operators of order higher than the second. We shall con-

sider only the simplest of all fourth-order differential operators, namely, $\frac{d^4}{dx^4}$, together with the linear boundary conditions

$$f(0) = f'(0) = f(1) = f'(1) = 0$$

Then, by an elementary computation entirely analogous to that performed in Sec. 2, we find that the (unique) solution of the equation

$$f''''(x) = h(x) \tag{60}$$

satisfying the afore-mentioned boundary conditions is given once again by (20), where $G(x,\xi)$ is now defined as follows:

$$G(x,\xi) = \begin{cases} \tfrac{1}{6}x^2(1 - \xi)^2(3\xi - x - 2\xi x) & 0 \le x \le \xi \\ \tfrac{1}{6}\xi^2(1 - x)^2(3x - \xi - 2\xi x) & \xi \le x \le 1 \end{cases} \tag{61}$$

The symmetry of $G(x,\xi)$ (which, of course, is termed the Green's function of the operator $\frac{d^4}{dx^4}$ for the given boundary conditions) assures us, exactly as is Sec. 2, that its eigenfunctions span $L_2(0,1)$. Thus, in complete analogy with the manner in which separation of variables was used successfully to solve the system consisting of equations (1), (2), and (3), we may now solve the system consisting of the following equations (which play the same role in the theory of vibrating beams as that played by the former system in the theory of vibrating strings):

$$u_{xxxx} + u_{tt} = 0 \tag{62}$$
$$u(0,t) \equiv u_x(0,t) \equiv u(1,t) \equiv u_x(1,t) \equiv 0 \tag{63}$$
$$u(x,0) = f(x) \qquad u_t(x,0) = g(x) \tag{64}$$

The eigenfunctions and eigenvalues of the Green's function are obtained in the present case as follows: We write out the general solution of the equation

$$X'''' - \lambda X = 0 \tag{65}$$

in the form[1]

$$X = a \sin \lambda^{1/4}x + b \cos \lambda^{1/4}x + c \sinh \lambda^{1/4}x + d \cosh \lambda^{1/4}x \tag{66}$$

When the boundary conditions are imposed on (66), we find, by an elementary computation, that the coefficients a, b, c, d must all vanish unless λ satisfies the transcendental equation

$$\cos \lambda^{1/4} \cosh \lambda^{1/4} = 1 \tag{67}$$

This equation is readily seen to have infinitely many real roots, for the left side vanishes when $\lambda = (2n - \tfrac{1}{2})^4\pi^4$ and exceeds unity when

---

[1] For $\lambda = 0$, formula (66) breaks down, but in this case the general solution of (65) is given by an arbitrary cubic polynomial, and it is easily shown that no such (non-trivial) function can satisfy the boundary conditions.

$\lambda = (2n\pi)^4$, $n = 1, 2, 3, \ldots$ [Of course, the existence of infinitely many eigenvalues is also assured by the fact that the eigenfunctions must span $L_2(0,1)$.]

3. Finally, we consider briefly the problem of associating a Green's function with a *partial* differential operator and of using it to justify the application of separation of variables to the solution of partial differential equations in three or more variables. Only a single example will be considered, but this should suffice to indicate how more difficult problems may be treated. It will be seen that, though the same basic ideas are employed as in the previous sections of the present chapter, a certain difficulty arises from the fact that the Green's function encountered here is unbounded.

We consider the two-dimensional wave equation

$$\Delta u = u_{xx} + u_{yy} = u_{tt} \tag{68}$$

together with the boundary condition

$$u \equiv 0 \qquad \text{on } x^2 + y^2 = 1 \qquad \text{for } t \geq 0 \tag{69}$$

and the initial conditions

$$u(x,y,0) = f(x,y) \qquad u_t(x,y,0) = g(x,y) \qquad (x^2 + y^2 < 1) \tag{70}$$

This system, an obvious analogue of the system (1), (2), (3), describes the vibration of an elastic circular membrane clamped in rest position along the circumference, the displacement and velocity of each point of the membrane being specified at $t = 0$. We seek nontrivial solutions of (68) which satisfy (69) and are of the form

$$u = v(x,y)T(t) \tag{71}$$

By substituting (71) into (68), we are immediately led to the pair of equations

$$\Delta v + \lambda v = 0 \tag{72a}$$
$$T'' + \lambda T = 0 \tag{72b}$$

Proceeding formally, we temporarily assume that (72a) admits, for a denumerable set $\{\lambda_n\}$ of values of λ, nontrivial solutions $\{v_n\}$ which vanish on the unit circle [so that (71) will satisfy (69)], and that these functions constitute a basis for the Hilbert space of functions quadratically integrable over the unit disc D. Then the functions $f(x,y)$ and $g(x,y)$ are expanded in series,

$$f(x,y) = \sum_{n=1}^{\infty} a_n v_n(x,y) \tag{73a}$$

$$g(x,y) = \sum_{n=1}^{\infty} b_n v_n(x,y) \tag{73b}$$

and the solution of the given problem is presumably represented by the series

$$u = \sum_{n=1}^{\infty} v_n(x,y)(a_n \cos \lambda_n^{1/2}t + b_n\lambda_n^{-1/2} \sin \lambda_n^{1/2}t) \tag{74}$$

It is now necessary to justify the statements made in the preceding paragraph. First, by comparing (72a) with (6-57) and taking account of the discussion at the end of Sec. 6-12, we conclude that any function which satisfies (72a) in D and vanishes on the circumference must satisfy the homogeneous integral equation

$$v(P) = \frac{\lambda}{2\pi} \iint_D v(Q)G(P,Q) \, d\xi \, d\eta \tag{75}$$

where $G(P,Q)$, the Green's function of D, is defined by (6-24). [The step from (72a) to (75) is justified by the Hölder continuity of v throughout D, which is certainly assured by the fact that v satisfies (72a).] Next we shall show, conversely, that any (quadratically integrable) solution of (75) satisfies (72a) and vanishes on the circumference of D. We shall use the following two facts, proofs of which are deferred to the following paragraph: (1) $\iint_D \iint_D G^2(P,Q) \, d\xi \, d\eta \, dx \, dy$ is finite, so that $G(P,Q)$ and each of its iterates may be considered as a completely continuous operator in $L_2(D)$ (cf. Sec. 5-6); (2) $G_{(2)}(P,Q)$, the first iterate of $G(P,Q)$, is continuous (jointly) in P and Q as these points vary throughout the closed disc $\bar{D}$. From (75) and the first of the above assertions, it follows that v also satisfies the integral equation

$$v(P) = \left(\frac{\lambda}{2\pi}\right)^2 \iint_D v(Q)G_{(2)}(P,Q) \, d\xi \, d\eta \tag{76}$$

It now follows from (76) and the second of the above assertions that v is continuous throughout $\bar{D}$; returning again to (75), we conclude, by the argument presented in Sec. 6-12, that v vanishes on the boundary of D and satisfies (72a) throughout D. In this manner we have established the complete equivalence of the integral equation (75) with the partial differential equation (72a), *together with* the condition that v shall vanish on the circumference. Now it was shown in Sec. 6-12 that every function sufficiently differentiable in D and vanishing on the circumference is representable in the form (6-97). Since such functions are dense in $L_2(D)$ (cf. Exercise 20), it follows from the theory of hermitian completely continuous operators presented in Sec. 5-4 that the eigenfunctions $\{v_n\}$ of the symmetric kernel $(2\pi)^{-1}G(P,Q)$ constitute an orthogonal basis of $L_2(D)$. Hence, if the sequences $\{a_n\}$, $\{b_n\}$ of coefficients appearing in (73a) and (73b) approach zero with sufficient rapidity, these series will actually converge throughout $\bar{D}$ to f and g, respectively. The coefficients may be obtained by the usual formulas, namely,[1]

$$a_n = (f,v_n) = \iint_D f(x,y)v_n(x,y)\,dx\,dy$$
$$b_n = (g,v_n) = \iint_D g(x,y)v_n(x,y)\,dx\,dy \tag{77}$$

Thus, as in the simpler problems considered earlier in this chapter, series (74) will furnish, under suitable restrictions on the functions $f(x,y)$ and $g(x,y)$, the solution to the given problem. It should be pointed out that the eigenvalues $\{\lambda_n\}$ are all positive, so that (74) actually involves trigonometric, rather than hyperbolic, functions of t. The proof runs as follows: If the nontrivial function v satisfies (72a) and vanishes everywhere on the circumference, then it must have a positive maximum or negative minimum (or both), and, in either case, the extremum would be assumed in D, not on the circumference. At a positive maximum we find from (72a) that Δv is opposite in sign to λ; since v_{xx} and v_{yy} must both be nonpositive at an interior maximum, it follows that λ is nonnegative. A similar argument applies for a negative minimum. The possibility that λ vanishes is ruled out by the maximum and minimum principles for harmonic functions. Thus, the only tenable conclusion is that λ must be positive.

It remains to prove the two statements that were temporarily assumed in the preceding paragraph. As for statement 1, we exploit the inequality

$$G^2(P,Q) \le 3(\log^2 \overline{OQ} + \log^2 \overline{PQ} + \log^2 \overline{PQ^*}) \tag{78}$$

which follows by applying the Schwarz inequality[2] to (6-24) (with $R = 1$). Introducing polar coordinates r, θ and ρ, ϕ for P and Q, respectively, we obtain

$$\iint_D \iint_D \log^2 \overline{OQ}\,d\xi\,d\eta\,dx\,dy = \int_0^{2\pi}\int_0^1\int_0^{2\pi}\int_0^1 (\log \rho)^2 \rho r\,d\rho\,d\phi\,dr\,d\theta \tag{79}$$

This integral obviously exists. To show that

$$\iint_D \left(\iint_D \log^2 \overline{PQ}\,d\xi\,d\eta\right) dx\,dy$$

exists, we replace the region D in the inner integral by the disc Δ_P of radius 2 with center at P. Since the latter disc contains D, we see that the integral under consideration is dominated by

$$\iint_D \left(\iint_{\Delta_P} \log^2 \overline{PQ}\,d\xi\,d\eta\right) dx\,dy$$

Since the inner integral exists and is independent of P, we conclude that

---

[1] The functions v_n are assumed normalized, so that $(v_n,v_m) = \delta_{nm}$.
[2] $(1 \cdot a + 1 \cdot b + 1 \cdot c)^2 \le (1^2 + 1^2 + 1^2)(a^2 + b^2 + c^2)$.

the dominated integral exists. Finally, we consider

$$\iint_D \left(\iint_D \log^2 \overline{PQ^*} \, dx \, dy \right) d\xi \, d\eta$$

In the inner integral we replace the region D by the circular sector S with vertex at Q^* which circumscribes D. (Cf. Fig. 9-1.) Introducing polar

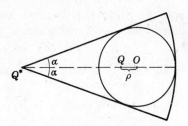

Figure 9-1

coordinates with origin at Q^*, we obtain

$$\iint_S \log^2 \overline{PQ^*} \, dx \, dy = \int_{-\alpha}^{\alpha} \int_0^{1+1/\rho} (\log r)^2 r \, dr \, d\theta$$
$$[\rho = \overline{OQ} = (\overline{OQ^*})^{-1}, \sin \alpha = \rho] \quad (80)$$

From the inequalities $\alpha \le \frac{1}{2}\pi \sin \alpha = \frac{1}{2}\pi\rho$ (cf. Exercise 22) and $1 + 1/\rho \le 2/\rho$ we find that the integral appearing in (80) is dominated by

$$\pi\rho \int_0^{2/\rho} (\log r)^2 r \, dr$$

which works out to

$$\pi\rho^{-1} \left(2 \log^2 \frac{2}{\rho} - 2 \log \frac{2}{\rho} + 1 \right)$$

The fourfold integral under consideration is therefore dominated by

$$\pi \int_0^{2\pi} \int_0^1 \left(2 \log^2 \frac{2}{\rho} - 2 \log \frac{2}{\rho} + 1 \right) d\rho \, d\phi$$

or

$$2\pi^2 \int_0^1 \left(2 \log^2 \frac{2}{\rho} - 2 \log \frac{2}{\rho} + 1 \right) d\rho$$

which is readily seen to be finite. Thus the proof of statement 1 is complete. Turning to statement 2, we write

$$G_{(2)}(P,Q) = \iint_D G(P,W)G(W,Q) \, ds \, dt \quad (81)$$

where s and t denote the rectangular coordinates of W. Letting P, Q, and W be represented, for convenience, by complex numbers $re^{i\theta}$, $\rho e^{i\phi}$,

$\tau e^{i\alpha}$, respectively, we rewrite (81) in the form

$$G_{(2)}(P,Q) = \int_0^{2\pi} \int_0^1 \log\left|\frac{1 - r\tau e^{i(\theta-\alpha)}}{re^{i\theta} - \tau e^{i\alpha}}\right| \cdot \log\left|\frac{1 - \rho\tau e^{i(\phi-\alpha)}}{\rho e^{i\phi} - \tau e^{i\alpha}}\right| \cdot \tau \, d\tau \, d\alpha \quad (82)$$

Without loss of generality (since P and Q play symmetric roles), we may assume that $r \le \rho$. We split the range of τ into three intervals: $0 \le \tau \le r$, $r \le \tau \le \rho$, $\rho \le \tau \le 1$. In the first interval, $G(P,W)$ and $G(W,Q)$ may be expanded into the following series:[1]

$$G(P,W) = -\log r + \sum_{n=1}^{\infty} n^{-1}\tau^n(r^{-n} - r^n) \cos n(\alpha - \theta) \quad (83a)$$

$$G(W,Q) = -\log \rho + \sum_{n=1}^{\infty} n^{-1}\tau^n(\rho^{-n} - \rho^n) \cos n(\alpha - \phi) \quad (83b)$$

(83b) is still valid in the second interval, but (83a) must be replaced by

$$G(P,W) = -\log \tau + \sum_{n=1}^{\infty} n^{-1}r^n(\tau^{-n} - \tau^n) \cos n(\alpha - \theta) \quad (83c)$$

In the third interval, (83c) is still valid, but now (83b) must be replaced by

$$G(W,Q) = -\log \tau + \sum_{n=1}^{\infty} n^{-1}\rho^n(\tau^{-n} - \tau^n) \cos n(\alpha - \phi) \quad (83d)$$

Splitting the disc D into three parts corresponding to the afore-mentioned intervals, and taking account of the orthogonality properties of the trigonometric functions appearing in the above expansions,[2] we find that (81) works out to the sum of the three following expressions:

$$\iint_{\tau \le r} = \pi\left[r^2 \log r \log \rho + \sum_{n=1}^{\infty} n^{-2}(2n + 2)^{-1}r^{2n+2}\right.$$
$$\left. \times (r^{-n} - r^n)(\rho^{-n} - \rho^n) \cos n(\theta - \phi)\right] \quad (84a)$$

---

[1] The following expansions are immediate consequences of the identity

$$\log|1 - z| = \operatorname{Re} \log(1 - z) = -\operatorname{Re}\left(z + \frac{z^2}{2} + \frac{z^3}{3} + \cdots\right)$$

valid for $|z| < 1$.

[2] $\int_0^{2\pi} \cos n(\alpha - \theta) \cos m(\alpha - \phi) \, d\alpha = \pi \cos n(\theta - \phi)$ if $n = m$, 0 if $n \ne m$, for positive integers n, m.

$$\iint_{r \leq \tau \leq \rho} = \pi \left\{ \left[\tau^2 \log \tau - \frac{\tau^2}{2} \right]_r^\rho + \sum_{n=1}^\infty n^{-2} \left[\frac{\tau^2}{2} - \frac{\tau^{n+2}}{n+2} \right]_r^\rho \right.$$

$$\left. \times r^n(\rho^{-n} - \rho^n) \cos n(\theta - \phi) \right\} \quad (84b)$$

$$\iint_{\rho \leq \tau \leq 1} = \pi \left[2 \int_\rho^1 \tau \log^2 \tau \, d\tau + \sum_{n=1}^\infty n^{-2} r^n \rho^n \cos n(\theta - \phi) \right.$$

$$\left. \times \int_\rho^1 (\tau^{-n} - \tau^n)^2 \tau \, d\tau \right] \quad (84c)$$

Thus, $G_{(2)}(P,Q)$ is expressed as the sum of three infinite series, each term of which depends continuously on the variables r, θ, ρ, ϕ. If each of these terms can be dominated by a constant in such a manner that all three of the dominating series converge, the continuity of $G_{(2)}(P,Q)$ will be established. In (84a), the term $r^2 \log r \log \rho$ is dominated by $(r \log r)^2$, since $r \leq \rho \leq 1$, and this dominating expression never exceeds e^{-2} in the interval $0 \leq r \leq 1$. Each term in the first summation is dominated by[1] $n^{-2}(2n + 2)^{-1} r^2 (1 - r^{2n})(1 - \rho^{2n})$, and a fortiori by the quantity n^{-2}. Since the series $\sum_{n=1}^\infty n^{-2}$ is convergent, the continuity of (84a) is established. Similar crude estimates suffice to establish the continuity of (84b) and (84c), and in this manner the continuity of $G_{(2)}(P,Q)$ in $\bar{D}$ is proved.

EXERCISES

20. Prove that the functions of class C^2 in the unit disc D, continuous in $\bar{D}$ and vanishing on the circumference, constitute a dense subset of $L_2(D)$.

21. The orthogonality of the functions v_n and v_m associated with distinct eigenvalues λ_n and λ_m follows from representation (75). An alternative approach would be to integrate the identity $v_n \Delta v_m - v_m \Delta v_n = (\lambda_n - \lambda_m) v_n v_m$ over D and apply Green's second identity and the fact that v_n and v_m vanish on the boundary. What is the difficulty associated with this approach?

22. Prove that the inequalities $2/\pi \leq \sin \alpha/\alpha \leq 1$ hold in the interval $0 \leq \alpha \leq \pi/2$.

23. Prove that $\iint_D \iint_D G^2(P,Q) \, dx \, dy \, d\xi \, d\eta$ is finite for any bounded domain D, where $G(P,Q)$ denotes the Green's function of this domain. *Hint:* Cover D by a disc Δ and compare the Green's functions of the two domains.

[1] Here we employ the inequalities $0 \leq r \leq \rho \leq 1$.

SOLUTIONS TO SELECTED EXERCISES

CHAPTER 1

1. The equality $\int_0^1 (\sin n\pi x - \sin m\pi x)^2 \, dx = 1$ holds for any two distinct positive integers n and m. Hence the inequality $|\sin n\pi x - \sin m\pi x| < 1$ cannot hold for all x, and so no uniformly convergent subsequence can exist. (A more delicate argument shows that not even a pointwise convergent sequence can exist.)

7. Let I denote the smallest closed interval, $a \leq x \leq b$, containing the compact set R. The set $I - R$, being open, consists of a finite or countable union of open intervals whose end points belong to R, so that $f(x)$ is defined at each of these end points. Extend $f(x)$ linearly into each of these open intervals. Then $f(x)$ is evidently defined and continuous in I. Finally, let $f(x) = f(a)$ for $x < a$, and $f(x) = f(b)$ for $x > b$. Then $f(x)$ has been extended continuously (and without altering the maximum and minimum) to the entire real axis, and a fortiori to every set containing R.

8. The Taylor series of $(1 - u)^{1/2}$ about $u = 0$ is of the form $1 - \sum_{n=1}^{\infty} c_n u^n$, where $c_n > 0$. If $\sum_{n=1}^{\infty} c_n$ diverged (or even converged to a value exceeding one), we could find an index N such that $\sum_{n=1}^{N} c_n > 1$, and then we could find (by continuity) a value of u below one such that $\sum_{n=1}^{N} c_n u^n > 1$. This would imply that $(1 - u)^{1/2}$ can assume negative values in the interval $|u| < 1$. [Of course, we are exploiting the fact that the Taylor series converges to $(1 - u)^{1/2}$ in this interval.] Thus, $\sum_{n=1}^{\infty} c_n$ converges, and so, given $\epsilon > 0$, we can find N such that $(1 - u)^{1/2}$ is approximated within ϵ by $1 - \sum_{n=1}^{N} c_n u^n$. Setting $u = 1 - x^2$, $|x| \leq 1$, we conclude that $|x|$ is approximated within ϵ by the polynomial $1 - \sum_{n=1}^{N} c_n (1 - x^2)^n$.

12. For fixed N, $\int_a^b \left(f - \sum_{n=0}^N c_n p_n \right)^2 dx$ is minimized by choosing $c_n = f_n$, and

the minimum value of the above integral is found to be $\int_a^b f^2 \, dx - \sum_{n=0}^N f_n^2$. (Cf.

Sec. 5-2.) Since the minimum is nonnegative, $\sum_{n=0}^N f_n^2 \le \int_a^b f^2 \, dx$; letting $N \to \infty$,

we conclude that $\sum_{n=0}^\infty f_n^2 \le \int_a^b f^2 \, dx$. Given $\epsilon > 0$, we can choose N and a poly-

nomial $p(x)$ of degree N such that $|f - p| < \epsilon$ throughout the interval; hence
$\int_a^b (f - p)^2 \, dx < \epsilon^2(b - a)$. Since the p_n are of degree n, p can be expressed as

$\sum_{n=0}^N c_n p_n$. Taking account of the minimizing property of the coefficients f_n explained

above, we conclude that $\int_a^b f^2 \, dx - \sum_{n=0}^N f_n^2 < \epsilon^2(b - a)$. Letting $N \to \infty$ and

taking account of the inequality obtained previously, we find that $0 \le \int_a^b f^2 \, dx -$

$\sum_{n=0}^\infty f_n^2 < \epsilon^2(b - a)$. Since ϵ is arbitrary, it follows that $\sum_{n=0}^\infty f_n^2 = \int_a^b f^2 \, dx$.

17. Let $f(x) = e^{-|x|} \operatorname{sgn} x$. It follows from Theorem 3 and Exercise 16 that the inversion formula (13) may be applied for all values of ξ. In particular, for $\xi = 0$, (13) assumes the form

$$\frac{1}{2\pi} \lim_{A \to \infty} \int_{-A}^A \frac{-2i\lambda}{1 + \lambda^2} \, d\lambda = 0$$

On the other hand, $\int_{-\infty}^\infty \frac{\lambda}{1 + \lambda^2} \, d\lambda$ does not exist, so that the use of a symmetric interval of integration is essential.

20. Assuming that $f(x)$ approaches zero at $\pm \infty$ and that integration by parts is justified,

$$\mathfrak{F}(f'(x)) = e^{-i\lambda x} f(x) \Big|_{-\infty}^\infty + i\lambda \int_{-\infty}^\infty f(x) e^{-i\lambda x} \, dx = i\lambda \mathfrak{F}(f)$$

21. Assuming that all integrals are meaningful and that interchange of order of integration is permissible,

$$\int_{-\infty}^\infty f * g e^{-i\lambda x} \, dx = \int_{-\infty}^\infty e^{-i\lambda x} \left[\int_{-\infty}^\infty f(\xi) g(x - \xi) \, d\xi \right] dx$$
$$= \int_{-\infty}^\infty f(\xi) e^{-i\lambda \xi} \left[\int_{-\infty}^\infty g(x - \xi) e^{-i\lambda(x-\xi)} \, dx \right] d\xi$$
$$= \int_{-\infty}^\infty f(\xi) e^{-i\lambda \xi} \left[\int_{-\infty}^\infty g(t) e^{-i\lambda t} \, dt \right] d\xi = \mathfrak{F}(f) \mathfrak{F}(g)$$

23. Since s may be replaced by $s - s_0$ in the following argument, we may confine attention to the case $s_0 = 0$. Employing the notation used in the proof of Theorem 4, we express, for σ ($= \operatorname{Re} s) > 0$, the difference $\bar{f}(s) - \bar{f}(0)$ in the form

$$s \int_0^\infty e^{-sx}[g(x) - \bar{f}(0)]\, dx$$

Note that $\bar{f}(0) = g(\infty)$ and that $|g(x)| < M$ for some M. Thus, for any $A > 0$,

$$|\bar{f}(s) - \bar{f}(0)| < |s| \int_0^A 2M\, dx + |s| \cdot \max_{x \geq A} |g(x) - \bar{f}(0)| \cdot \int_A^\infty e^{-\sigma x}\, dx < 2MA|s| +$$

$\left|\dfrac{s}{\sigma}\right| \cdot \max\limits_{x \geq A} |g(x) - \bar{f}(0)|$. Let s be confined to the wedge $|\arg s| \leq \theta < \frac{1}{2}\pi$, so that $|s/\sigma| \leq \sec \theta$. Given $\epsilon > 0$, choose A so large that $\max\limits_{x \geq A} |g(x) - \bar{f}(0)| < \frac{1}{2}\epsilon \cos \theta$, and then confine s to the intersection of the afore-mentioned wedge and the disc $|s| < \epsilon/4MA$. Then we obtain the inequality $|\bar{f}(s) - \bar{f}(0)| < \frac{1}{2}\epsilon + \frac{1}{2}\epsilon = \epsilon$.

28. Let L denote the period of $f(x)$. For σ ($= \operatorname{Re} s) > 0$, the integral (23) is

dominated by $\displaystyle\sum_{n=0}^\infty \int_{nL}^{(n+1)L} e^{-x\sigma} |f(x)|\, dx$, or $\displaystyle\int_0^L e^{-x\sigma} |f(x)|\, dx \cdot \sum_{n=0}^\infty e^{-nL\sigma}$. Since

the latter series converges, the integral (23) exists at least in the half plane $\sigma > 0$, while the example $f(x) \equiv 1$ shows that the abscissa of convergence may equal zero. For $\sigma > 0$, the convergence of the above series justifies rewriting $\bar{f}(s)$ in the form

$\displaystyle\int_0^L e^{-xs}f(x)\, dx \cdot \sum_{n=0}^\infty e^{-nLs}$, or $(1 - e^{-Ls})^{-1} \displaystyle\int_0^L e^{-xs}f(x)\, dx$. According to the proof

of Theorem 4, the latter integral defines an entire function of s, so that $\bar{f}(s)$ may be extended analytically to the entire plane except that simple poles will appear at the zeros of $(1 - e^{-Ls})$, namely, the points $2n\pi i/L$ ($n = 0, \pm 1, \pm 2, \ldots$), unless the value of the integral happens to be zero. [For example, if $f(x) = \sin x$, the transform $\bar{f}(s)$ is equal to $(1 + s^2)^{-1}$, so that only two of the zeros of $(1 - e^{-2\pi s})$ turn out to be poles of $\bar{f}(s)$.]

31. If the opposite inequality $g(x') > G(x')$ held at some other point of the interval, there would exist a third point x'' such that $g(x'') = G(x'')$. By the uniqueness portion of the theorem, g and G would coincide.

32. By the theorem of mean value, $|f(x,y_1) - f(x,y_2)| = |f_y(x,y_3)| \cdot |y_1 - y_2|$ for some value of y_3 between y_1 and y_2. Hence, $|f(x,y_1) - f(x,y_2)| \leq k|y_1 - y_2|$, if k is chosen as $\max\limits_R |f_y|$.

35. $|f(x,y) - f(\xi,\eta)| = |[f(x,y) - f(x,\eta)] + [f(x,\eta) - f(\xi,\eta)]| \leq |f(x,y) - f(x,\eta)| + |f(x,\eta) - f(\xi,\eta)| \leq k[|y - \eta| + |x - \xi|]$. This inequality obviously extends to a larger number of independent variables.

CHAPTER 2

7. From (5) we find that the relation $2z_x + 2z_y = 5$ must hold everywhere on the given line, and since $z_x z_y = 1$, it follows that $2z_x + 2/z_x = 5$, so that either $z_x \equiv 2$ or $z_x \equiv \frac{1}{2}$. If the former solution is chosen, the characteristic strip associated with

each point of the given line is found by solving the system $\dfrac{dx}{ds} = q,\ \dfrac{dy}{ds} = p,\ \dfrac{dz}{ds} = 2pq$,

$\dfrac{dp}{ds} = 0,\ \dfrac{dq}{ds} = 0$ with the conditions $x = 2t,\ y = 2t,\ z = 5t,\ p = 2,\ q = \frac{1}{2}$ for $s = 0$. We thus obtain $x = 2t + \frac{1}{2}s,\ y = 2t + 2s,\ z = 5t + 2s,\ p = 2,\ q = \frac{1}{2}$. Solving the first two equations of the latter system for s and t and substituting into the third, we obtain $z = 2x + \frac{1}{2}y$. Similarly, if we choose $z_x \equiv \frac{1}{2}$ along the given line, we obtain $z = \frac{1}{2}x + 2y$.

CHAPTER 3

7,8. The substitution $r = x + t$, $s = x - t$ converts (2a) into the form $u_{rs} = 0$, and so $u = a(r) + b(s) = a(x + t) + b(x - t)$. From this general solution, we obtain $u(x,0) = a(x) + b(x)$, $u_t(x,0) = a'(x) - b'(x)$. From the equations $a(x) + b(x) = f(x)$, $a'(x) - b'(x) = g(x)$, we obtain $a'(x) = \frac{1}{2}[f'(x) + g(x)]$ and hence $a(x) = \frac{1}{2}\left[f(x) + \int_0^x g(\xi)\, d\xi + c \right]$, where c denotes an arbitrary constant. From this we obtain $b(x) = \frac{1}{2}\left[f(x) - \int_0^x g(\xi)\, d\xi - c \right]$, and hence $a(x + t) + b(x - t) = \frac{1}{2}\left[f(x + t) + f(x - t) + \int_{x-t}^{x+t} g(\xi)\, d\xi \right]$. When $g(\xi) \equiv 0$, this simplifies to (45).

12. Applying integration by parts twice to the second derivatives and once to the first derivatives, we find that $\iint vLu\, dx\, dy = \iint u[(av)_{xx} + 2(bv)_{xy} + (cv)_{yy} - (dv)_x - (ev)_y + f]\, dx\, dy +$ boundary integral.

14. Let Q and Q' be fixed points, and let $u(P) = R_{L^*}(P;Q)$, $v(P) = R_L(P;Q')$. Apply (51) to the rectangle formed by the horizontal and vertical lines through Q and Q'. Since $Lu = L^*v = 0$, the right side of (51) must vanish. When account is taken of equations (53a) to (53d) and the corresponding conditions for u, the vanishing of the right side of (51) is found to be equivalent to $u(Q') = v(Q)$, or $R_{L^*}(Q';Q) = R_L(Q;Q')$.

15. On the vertical characteristic the assigned values of u determine q, and so (22) may be considered as an ordinary differential equation $\dfrac{dp}{dy} = g(y,p)$. Since the value of p at the intersection of the characteristics is determined by the assigned values of u on the horizontal characteristic, a unique solution $p(y)$ is determined. Similarly, q can be found at each point of the horizontal characteristic.

16. Let the data be of the form $u(x,y_0) = \phi(x)$, $u(x_0,y) = \psi(y)$ [so that $\phi(x_0) = \psi(y_0)$]. Let $v = u - \phi(x) - \psi(y) + \phi(x_0)$. Then v must vanish on the two given characteristics and must satisfy an equation similar to (22). The iterative procedure employed in the proof of Theorem 1 applies without any change to the present problem. As the point (x,y) (cf. Fig. 3-1) approaches the horizontal characteristic, the segment PM does not shrink to a point, and so, in contrast to the noncharacteristic problem, q need not vanish on the horizontal characteristic; similarly, p need not vanish on the vertical characteristic.

20. Referring to Fig. 3-3, we note that $\xi^2 = t^2 + r^2 - 2rt \cos \theta$, and so $\xi\, d\xi = rt \sin \theta\, d\theta$. The area dS of the zone determined by colatitudes θ and $\theta + d\theta$ is given by $2\pi t^2 \sin \theta\, d\theta$, and so $dS = 2\pi tr^{-1}\xi\, d\xi$, or $\frac{1}{2}r^{-1}\xi\, d\xi = (4\pi t)^{-1}\, dS$. Multiplying by ϕ and integrating, we conclude that the right sides of (74) and (76) are equal.

22. $u = x^2t + 2xyt + 3z^3t + 3zt^3 + t^3/3$.

24. The Legendre transformation reduces the given equation to the form $\omega = \xi^2 + \eta^2$. Thus $x = \omega_\xi = 2\xi$, $y = \omega_\eta = 2\eta$, $u = x\xi + y\eta - \omega = x \cdot \frac{1}{2}x + y \cdot \frac{1}{2}y - (\frac{1}{4}x^2 + \frac{1}{4}y^2) = \frac{1}{4}x^2 + \frac{1}{4}y^2$.

25. The Legendre transformation, if applied formally, would yield $\omega = 0$, $x = \omega_\xi = 0$, $y = \omega_\eta = 0$, so that x and y could not vary. Thus, all solutions of the given equation must be developable. This may be seen without considering the Legendre transformation, for the equation $u_{xx}u_{yy} - u_{xy}^2 = 0$ is an easy consequence of the given equation.

CHAPTER 4

2. *Existence of basis:* Let f_1, f_2, . . . , f_n be a set such that representation (1) is possible. If $f_1 = o$, discard it, then eliminate the first element (if any) which is expressible as a linear combination of the preceding elements. Repeat this procedure as long as possible. The elements that are not eliminated constitute a basis.

Uniqueness of number of elements in basis: Let the dimension n be momentarily redefined as the *smallest* possible number of elements in a basis. For $n = 1$ uniqueness is trivial; apply induction on n.

Next, suppose n given independent elements f_1, f_2, . . . , f_n fail to form a basis. Choose a basis ϕ_1, ϕ_2, . . . , ϕ_n, and then apply the elimination procedure described above to the (ordered) set f_1, f_2, . . . , f_n, ϕ_1, ϕ_2, . . . , ϕ_n. None of the f's are eliminated by this procedure; hence at least one ϕ must be retained. This would furnish a basis consisting of at least $n + 1$ elements, contradicting uniqueness. The "only if" assertion follows from uniqueness.

If f_1, f_2, . . . , f_n form a basis, the equality $\Sigma\alpha_i f_i = \Sigma\beta_i f_i$, or $\Sigma(\alpha_i - \beta_i)f_i = o$, implies, by independence, $\alpha_i - \beta_i = 0$. Thus the coefficients are uniquely determined.

8. If $\|f - f_n\| \to 0$, $\|g - f_n\| \to 0$, then $\|f - g\| = \|(f - f_n) + (f_n - g)\| \leq \|f - f_n\| + \|g - f_n\| \to 0$. Hence $\|f - g\| = 0$, $f = g$.

10. $\|f\| = \|(f - g) + g\| \leq \|f - g\| + \|g\|$; hence $\|f\| - \|g\| \leq \|f - g\|$. Interchanging f and g, we get $\|f\| - \|g\| \geq - \|f - g\|$. Thus, $\big|\|f\| - \|g\|\big| \leq \|f - g\|$.

12. M is obviously a manifold. Consider the sequence $\{f_n\}$, where $f_n = \{1, \frac{1}{2}, . . . , 1/n, 0, 0, 0, . . .\}$. Then $\{f_n\}$ converges to the element $\{1, \frac{1}{2}, \frac{1}{3}, . . .\}$, which does not belong to M.

13. (a) For convenience let $a = 0$, $b = \pi$. Let $f_n(x) = n^{1/2} \sin nx$ for $0 \leq x \leq \pi/n$, $f_n(x) = 0$ for $\pi/n < x \leq \pi$. Then $f_n(x) \to 0$ pointwise, but, since $\|f_n\|^2 = \frac{1}{2}\pi$, the sequence does not converge in norm to the zero element ($f \equiv 0$).

(b) Define suitable functions $f_n(x)$ vanishing outside a subinterval and having a triangular graph of fixed height within the subinterval. Let the subintervals repeatedly move across the interval as their lengths approach zero.

18. By homogeneity of the norm, it suffices to prove that $\|\Sigma\lambda_k f_k\|$ has a positive lower bound under the restriction $\Sigma|\lambda_k| = 1$. Suppose the contrary; then there exist n subsequences $\{\lambda_k^{(i)}\}$ ($k = 1, 2, . . . , n$) such that $\Sigma\lambda_k^{(i)}f_k \to o$, $\Sigma|\lambda_k^{(i)}| = 1$. Since each sequence $\{\lambda_k^{(i)}\}$ is bounded in absolute value, there exists a subsequence of $\{i\}$ on which each $\{\lambda_k^{(i)}\}$ converges, say to λ_k. Then $\Sigma|\lambda_k| = 1$, while (by continuity of the norm) $\|\Sigma\lambda_k f_k\| = 0$, or $\Sigma\lambda_k f_k = o$. The last equality contradicts the independence of the f's.

31. If $\{f_n\}$ is bounded, then a suitable subsequence of $\{Tf_n\}$ is convergent; applying S to each term of the subsequence, we conclude that $\{STf_n\}$ is also convergent. Therefore, ST is completely continuous. The sequence $\{Sf_n\}$ is bounded; hence a suitable subsequence of $\{TSf_n\}$ is convergent. Therefore, TS is completely continuous.

32. First, consider $n = 2$. Then the product may be written as $I - K$, where $K = K_1 + K_2 - K_1K_2$ is completely continuous by the two preceding exercises. For $n > 2$, a trivial induction may be used.

34. $S_1 = S_1 I = S_1(TS_2) = (S_1 T)S_2 = IS_2 = S_2$.

38. Let $\|K\| = \alpha$ and $\|K^m\| = \beta < 1$. Then series (27) is dominated by the convergent series $1 + \alpha + \alpha^2 + \cdots + \alpha^{m-1} + \beta + \beta\alpha + \cdots + \beta\alpha^{m-1} + \beta^2 + \beta^2\alpha + \cdots + \beta^2\alpha^{m-1} + \cdots$. The proof presented in the text that $I - K$ is invertible if $\|K\| < 1$ carries over without change, since only the existence of a convergent series which dominates series (27) is needed.

41. Let $a = 0$, $b = \pi$, $K(x,y) = \frac{2}{\pi} \sum_{n=1}^{\infty} n^{-2} \sin nx \sin ny$. By uniform convergence,

$$\int_0^\pi K(x,y)f(y) \, dy = \sum_{n=1}^{\infty} n^{-2}f_n \sin nx, \quad \text{where} \quad \pi f_n = 2\int_0^\pi f(y) \sin ny \, dy. \text{ By ele-}$$

mentary theory of Fourier series, $\int_0^\pi K(x,y)f(y) \, dy \equiv 0$ if and only if $f \equiv 0$. On the other hand, $(40')$ is not solvable if $g(0) \neq 0$, since each term in the above series vanishes at $x = 0$.

42. Similar to convergence proof employed in Theorem 1-6.

CHAPTER 5

1. If $(f,g) = 0$, the Schwarz inequality is certainly satisfied. If (f,g) is real and positive, we obtain $\|f - \lambda g\|^2 = \|f\|^2 - 2\lambda(f,g) + \lambda^2\|g\|^2 \geq 0$ for all real values of λ, and so $|(f,g)|^2 \leq \|f\|^2\|g\|^2$. If (f,g) is not zero and not positive, we replace f in the above argument by cf, where $|c| = 1$ and $(cf,g) > 0$.

4. Let $\phi_1, \phi_2, \ldots, \phi_n$ be any orthonormal basis in the given space and let each element $\sum_{i=1}^{n} \alpha_i \phi_i$ be mapped into the element $\{\alpha_1, \alpha_2, \ldots, \alpha_n\}$ of U_n.

5. The "hermitian form" $\sum_{i,j=1}^{n} (f_i,f_j)\lambda_i\bar{\lambda}_j$ can be written as $\left\| \sum_{i=1}^{n} \lambda_i f_i \right\|^2$, and so is always real-valued and nonnegative. If the elements are independent, this form assumes positive values except when $\lambda_1 = \lambda_2 = \cdots = \lambda_n = 0$, and if the elements are not independent, the form assumes the value zero for nontrivial choices of the λ's as well. According to the theory of hermitian forms, the determinant of the coefficients must be positive and zero in these two cases, respectively.

7. The set of all elements with only a finite number of nonzero components, each of which has rational real and imaginary parts, is easily seen to be dense in l_2 and denumerable. The second part is obvious.

12. If $\{f_n\}$ is bounded, say $\|f_n\| < M$, then $\{T^*f_n\}$ is also bounded, and so $\{TT^*f_n\}$ contains a convergent subsequence. Confining attention to the corresponding subsequence of $\{f_n\}$, we obtain $\|T^*(f_n - f_m)\|^2 = (TT^*(f_n - f_m), f_n - f_m) \leq \|f_n - f_m\| \cdot \|TT^*(f_n - f_m)\| \leq 2M \cdot \|TT^*f_n - TT^*f_m\| \to 0$. Thus $\{T^*f_n\}$ is a Cauchy sequence; since H is complete, $\{T^*f_n\}$ is convergent, and so T^* is completely continuous.

13. Let $\{\phi_n\}$ be an orthonormal basis in a separable Hilbert space, and let $T_nf = \sum_{k=1}^{n} (f,\phi_k)\phi_k$. For each f, $T_nf \to f$, so that $\{T_n\}$ converges strongly to the identity operator I; each T_n is completely continuous, but I is not. (Note that $\|I - T_n\| = 1$, so that the sequence $\{T_n\}$ does not converge in norm to I.)

14. From the hint, we easily conclude that $(Tf,g) = 0$ for every pair of elements f,g. Choose f and then let $g = Tf$. We obtain $(Tf,Tf) = 0$, and so $Tf = o$ and $T = 0$.

17. The necessity is proved by (30) and the remarks preceding (30). To prove sufficiency, let T be idempotent and hermitian, and let M be the subspace spanned by the range of T—i.e., the smallest subspace containing all elements of the form Tf. If $f \in M^\perp$, then $(Tf,Tf) = (T^2f,f) = (Tf,f) = 0$ (since $f \in M^\perp$ and $Tf \in M$), and so $Tf = o$. If f is in the range of T, then $f = Tg$ for some element g, and so $f = T^2g =$

$T(Tg) = Tf$. By continuity, $f = Tf$ holds for all elements in M. (That is, the range of T is precisely M, not a proper subset.) Taking account of the projection theorem and of the linearity of T, we conclude that $T = P_M$.

19. If $Tf = \mu f, f \neq o$, we obtain $\mu = (Tf,f)/(f,f)$. Since (Tf,f) is real, by Theorem 10, μ is real. If $Tf = \mu f$, $Tg = \lambda g$, then $(\mu - \lambda)(f,g) = (\mu f,g) - (f,\lambda g) = (Tf,g) - (f,Tg) = (Tf,g) - (Tf,g) = 0$. Thus, $\mu \neq \lambda$ implies that $(f,g) = 0$.

20. (a) and (b) are obvious. (c) For any element f, $(T^{-1}f,f) = (T^{-1}f,TT^{-1}f) = (TT^{-1}f,T^{-1}f)$, since T is hermitian. Thus $(T^{-1}f,f) = (f,T^{-1}f)$, and so $(T^{-1}f,f)$ is real for each f. By Theorem 10, T^{-1} is hermitian. (d) $(ST)^* = T^*S^* = TS$. Thus $(ST)^* = ST$ if and only if $ST = TS$.

21. (a) If $PQ = QP$, then PQ is hermitian by the preceding exercise, and also $(PQ)^2 = PQPQ = PPQQ = PQ$, so that PQ is idempotent. By Exercise 17, PQ is a projection. Conversely, if PQ is a projection it must be hermitian, and so $PQ = QP$ by the preceding exercise. (b) If $PQ = O$, then QP also equals O, by part (a) (since O is a projection). Then $(P + Q)^2 = P^2 + Q^2 + PQ + QP = P + Q + O + O = P + Q$. Thus $P + Q$ is idempotent, and it is hermitian by preceding exercise. Thus $P + Q$ is a projection by Exercise 17. Conversely, if $P + Q$ is a projection, $P + Q = (P + Q)^2 = P^2 + Q^2 + PQ + QP = P + Q + PQ + QP$, and so $PQ + QP = O$. Multiplying on the right by P, we get $PQP + QPP = PQP + QP = O$. Multiplying on the left by P we get $PPQ + PQP = PQ + PQP = O$. Subtracting, we get $PQ = QP$; but since $PQ + QP = O$, PQ and QP each equal O. Conversely, $PQ = O$ implies that $QP = O$ by part (a). Thus $(P + Q)^2 = P^2 + Q^2 + PQ + QP = P + Q$, and so $P + Q$ is idempotent. Since it is hermitian by the preceding exercise, it is a projection by Exercise 17. (c) $P - Q$ is a projection if and only if $I - (P - Q)$, or $(I - P) + Q$, is a projection. By part (b), $(I - P)Q = O$, or $PQ = Q$, is necessary and sufficient.

24. (a) In l_2 let $K\{\alpha_1, \alpha_2, \alpha_3, \ldots\} = \{\alpha_2, 0, 0, \ldots\}$. (b) If K is hermitian and $Kf \neq o$, then $0 < (Kf,Kf) = (K^2f,f)$, and so $K^2f \neq o$. By repeating this argument, we see that K^4f, K^8f, $K^{16}f$, ... must all be distinct from o, and this could not happen if K were nilpotent.

25. By Theorem 11, $\|K^2\| = \sup_{\|f\|=1} |(K^2f,f)| = \sup_{\|f\|=1} (Kf,Kf) = \sup_{\|f\|=1} \|Kf\|^2 = (\sup_{\|f\|=1} \|Kf\|)^2 = \|K\|^2$. Similarly, $\|K^4\| = \|K^2\|^2 = \|K\|^4$, and, in general, $\|K^{2^n}\| = \|K\|^{2^n}$. If $\|K^k\| < \|K\|^k$, let m be any power of 2 exceeding k. Then $\|K^m\| = \|K^kK^{m-k}\| < \|K\|^k \cdot \|K\|^{m-k} = \|K\|^m$, which contradicts the previous result. [Note that this includes part (b) of the preceding exercise.]

26. If K is normal, $\|Kf\|^2 = (Kf,Kf) = (K^*Kf,f) = (KK^*f,f) = (K^*f,K^*f) = \|K^*f\|^2$. Conversely, if $\|Kf\| \equiv \|K^*f\|$, then $(Kf,Kf) \equiv (K^*f,K^*f)$, or $(K^*Kf,f) \equiv (KK^*f,f)$. Thus, $(Tf,f) \equiv 0$, where $T = KK^* - K^*K$. By Theorem 11, $\|T\| = 0$, and so $KK^* = K^*K$.

31. Let $f(x) \in L_2$ possess the Fourier expansion $f(x) \sim \sum f_n e^{2\pi i n x}$; then $\|f\|^2 = \sum |f_n|^2$ and $Kf = f_0 + \frac{1}{2}f_1 e^{2\pi i x}$, so that $\|Kf\|^2 = |f_0|^2 + \frac{1}{4}|f_1|^2 = \|f\|^2 - \frac{3}{4}|f_1|^2 - \sum_{n \neq 0,1} |f_n|^2$. Thus, for fixed $\|f\|$, $\|Kf\|$ is maximized by choosing each f_n except f_0 equal to zero, and in this case $\|Kf\| = \|f\|$. Thus $\|K\| = 1$. The equality $\||K\|| = (\frac{5}{4})^{\frac{1}{2}}$ is given by an elementary computation.

38. Set $x = y = \frac{1}{2}$ in (154).

39. For $0 \le x \le y$ we may write $K_{(2)}(x,y) = \int_0^x t^2(1 - x)(1 - y)\, dt + \int_x^y xt(1 - t)(1 - y)\, dt + \int_y^1 xy(1 - t)^2\, dt$. This works out to $\frac{1}{6}x(1 - y)(2y - x^2 - y^2)$.

For $x > y$ we merely interchange x and y in this expression. By elementary computation we obtain $\iint K_{(2)}^2(x,y)\, dx\, dy = 1/9{,}450$. Hence $\sum_{n=1}^{\infty} n^{-8} = \dfrac{\pi^8}{9{,}450}$.

CHAPTER 6

3. $u_{x_k x_k} = \sum_{i,j=1}^{n} u_{x_i' x_j'} \dfrac{\partial x_i'}{\partial x_k} \dfrac{\partial x_j'}{\partial x_k} = \sum_{i,j=1}^{n} a_{ik} a_{jk} u_{x_i' x_j'}$. Summing over k, we obtain

$\sum_{k=1}^{n} u_{x_k x_k} = \sum_{i,j=1}^{n} \left(\sum_{k=1}^{n} a_{ik} a_{jk} \right) u_{x_i' x_j'}$. An orthogonal transformation is characterized

by the equalities $\sum_{k=1}^{n} a_{ik} a_{jk} = \delta_{ij}$. Therefore, the triple summation reduces to

$\sum_{i,j=1}^{n} \delta_{ij} u_{x_i' x_j'}$, or $\sum_{i=1}^{n} u_{x_i' x_i'}$.

6. When $f(r)$, $r = \left(\sum_{i=1}^{n} x_i^2 \right)^{1/2}$, is substituted for u, (1) assumes the form $f'\,\Delta r +$

$f'' \sum_{i=1}^{n} r_{x_i}^2 = 0$, or $f''/f' + (n-1)/r = 0$. Integrating once we obtain $f' = \text{con-}$

stant $\cdot r^{1-n}$. Integrating again, we obtain (7a) for $n = 2$, (7b) for $n > 2$.

7. Taking account of (11b) and of the constancy of v on Γ, we find that the right side, and hence the left side, of (12) vanishes. Since the integrand of the left side is continuous and nonnegative, it must vanish throughout G. Therefore, $v_x \equiv v_y \equiv 0$, and hence v is constant throughout $\bar{G}$.

9. $\pi R^2 u_x(Q) = \iint_G u_x\, dx\, dy = \int_\Gamma u\, dy$ [by (8)]. Hence, $\pi R^2 |u_x(Q)| \le \max |u| \cdot$
$\int_\Gamma |dy| = M \cdot 4R$, or $|u_x(Q)| \le 4M/\pi R$. Establishing a new rectangular coordinate system (x', y') by rotating the axes so that the x' axis has the direction of the gradient of u, we obtain $[(u_x^2(Q) + u_y^2(Q)]^{1/2} = |u_{x'}(Q)| \le 4M/\pi R$. (The fact that the Laplace equation is invariant under rotation plays an essential role.)

10. (20a) follows directly by observing that the integral given in the hint is nonnegative and works out (using the second corollary) to $K - \pi R^2 u^2(Q)$. It can also be proved by applying the Schwarz inequality to $\pi R^2 u(Q) = \iint_G u\, dx\, dy$ to obtain

$$\pi^2 R^4 u^2(Q) \le \left(\iint_G 1^2\, dx\, dy \right) \cdot \left(\iint_G u^2\, dx\, dy \right) = \pi R^2 K$$

(20b) follows by applying this result to the disc G' with center at P internally tangent to G and observing that $\iint_{G'} u^2\, dx\, dy \le \iint_G u^2\, dx\, dy = K$.

12. The circular symmetry shows that u must have the form (7a), and the given data yield for a and b the values

$$a = \frac{c_2 - c_1}{\log R_2 - \log R_1} \qquad b = c_1 - \frac{c_2 - c_1}{\log R_2 - \log R_1}$$

Letting $R_1 \to 0$, we obtain $a \to 0$, $b \to c_1$, $u \to c_1$. Thus, as the inner circumference shrinks to a point its influence disappears.

13. Let P be any point of the punctured disc, let Γ' be any circle concentric with Γ and not containing or enclosing P, and let v be defined in the annulus as the harmonic function with values M on Γ' and m on Γ. By the maximum and minimum principles, $v \pm u \geq 0$, and hence $|u(P)| \leq v(P)$. Letting Γ' shrink and taking account of Exercise 12, we conclude that $|u(P)| \leq m$. Hence $M \leq m$; on the other hand, $M \geq m$, by the definition of these quantities.

15. Apply the maximum principle to the region obtained by deleting from the domain a small disc containing Q.

16. Apply (10) to the region obtained by deleting small discs with centers at Q_1 and Q_2, respectively, taking u and v as the two Green's functions. The desired result is then obtained by shrinking the discs.

24. Subtract the two expansions and employ the fact that the sum of a convergent power series vanishes identically only if all coefficients vanish.

29. For simplicity let Γ be the unit circle. (Otherwise minor modifications are needed.) Then, for any z (except 0) inside Γ,

$$f(z) = \frac{1}{2\pi i} \int_\Gamma \frac{f(\zeta)\, d\zeta}{\zeta - z} \qquad 0 = \frac{1}{2\pi i} \int_\Gamma \frac{f(\zeta)\, d\zeta}{\zeta - 1/\bar{z}}$$

Subtracting, setting $\zeta = e^{i\theta}$, $z = \rho e^{i\phi}$, and performing some elementary simplifications, we obtain

$$f(\rho e^{i\phi}) = \int_0^{2\pi} f(e^{i\theta}) P(\rho, \phi, 1, \theta)\, d\theta$$

Taking the real part of both sides and noting that P is real, we obtain (26b).

30. The necessary condition is simply a repetition of (11b). Assume that u solves the given Neumann problem. Then the harmonic conjugate v satisfies (by Cauchy-Riemann equations) $\dfrac{\partial v}{\partial s} = g(s)$, which determines v on Γ to within an arbitrary additive constant. [Note that the condition $\int_\Gamma g(s)\, ds = 0$ assures the single-valuedness of v in the case that Γ is a single curve.] Thus we obtain a Dirichlet problem for v. If v is found, *its* harmonic conjugate, with sign changed, solves the Neumann problem.

32. Let u be the given function, and U the harmonic function coinciding on the boundary with u. Then $u - U$ vanishes on the boundary and also possesses the "one-circle mean-value property." This suffices to permit the conclusion that $u - U$ vanishes throughout D.

35. If $u \geq 0$, let $R \to \infty$ in (48). If $u \geq c$ (or $u \leq c$), apply the same argument to $u - c$ (or $c - u$).

41. For $0 \leq r < h$, the left side of (96) can be rewritten as

$$\frac{\pi}{h^2 - r^2} \int_0^{2\pi} [r(1 + \cos 4\theta) - h(\cos \theta + \cos 3\theta)] P(r, 0, h, \theta)\, d\theta$$

Now, the Dirichlet problems with boundary values $(1 + \cos 4\theta)$ and $(\cos \theta + \cos 3\theta)$ on the circle of radius h are solved by $1 + (r/h)^4 \cos 4\phi$ and $(r/h) \cos \phi + (r/h)^3 \cos 3\phi$, respectively, which reduce for $\phi = 0$ to $[1 + (r/h)^4]$ and $[r/h + (r/h)^3]$. [Cf. (36).] Thus, the left side of (96) equals $\pi(h^2 - r^2)^{-1}\{r[1 + (r/h)^4] - h[r/h + (r/h)^3]\}$, or $-\pi r^3 h^{-4}$

Similar arguments apply when $0 \leq r < -h$ and when $r > |h|$.

42. Choosing unit density in the disc $r < R$ and taking account of rotational symmetry, we find that the potential $u(r)$ satisfies the equation $(ru')' + 2\pi r = 0$ in the disc. Integrating twice, we obtain $u = -\frac{1}{2}\pi r^2 + c_1 \log r + c_2$, but c_1 must vanish since u remains finite at the origin. The potential at the origin is readily computed, and this gives the value of c_2. Thus (89) is obtained. For the exterior of the disc we express u in the form (7a), and evaluate a and b by matching the "interior" and "exterior" formulas for u and u' on the circumference. This leads to the result previously obtained for u outside the disc, namely, $u = -\pi R^2 \log r$.

CHAPTER 7

3,4,5. The inequality (1) may be squared if $u(P) \geq 0$, and the equality (6-18) may be squared in any case, to yield, with the aid of the Schwarz inequality,

$$u^2(P) \leq \frac{1}{2\pi} \int_0^{2\pi} u^2(P + re^{i\theta}) \, d\theta$$

If $u(P) < 0$ and equality does not hold in (1), the squaring of both sides of (1) and the subsequent derivation of the inequality

$$u^2(P) < \frac{1}{2\pi} \int_0^{2\pi} u^2(P + re^{i\theta}) \, d\theta$$

may lead to an incorrect result. *Example:* The function $x^2 + y^2 - 1$ is subharmonic everywhere, but its square is *super*harmonic in the disc $x^2 + y^2 < \frac{1}{2}$; this may be seen either by direct computation or by applying the result of Exercise 11.

6. Consider the function max $(1, x^2 + y^2)$.

17. By rotational symmetry, $u = a \log r + b$, but a must vanish since u is bounded. Hence u reduces to a constant. According to Theorem 5, u demonstrates "proper" behavior near each *regular* boundary point, and so $u \equiv 1$.

19. Choose any exterior point O. If there is a unique boundary point Q closest to O, the function (21) is a barrier, and so Q is regular. If there are two or more such boundary points, choose any one of them, say Q, and then select a point O' between O and Q. Then (21), with O replaced by O', furnishes a barrier, and so Q is regular in this case also.

21. By Exercise 11, each of the given functions is subharmonic. Since $(x^2 + y^2)^{1/n} \to 1$ for each point (x,y) of the punctured disc, $u \geq 1$. However, $u \leq 1$ by the maximum principle. Hence, $u \equiv 1$.

29. Let v^+ be the solution of the Dirichlet problem with boundary values max $(0,v)$, and let v^- be the solution with boundary values max $(0,-v)$. Then $u = v^+ + v^-$, $v = v^+ - v^-$. (These equalities are correct on the boundary, and then, by Theorem 6-3, they remain correct in the domain.) Then $D(v) - D(u) = D(v^+ - v^-) - D(v^+ + v^-) = -4D(v^+,v^-)$, which by (6-9) is equal to

$$-4 \int v^- \frac{\partial v^+}{\partial n} \, ds$$

At boundary points where $v \geq 0$, v^- vanishes, and hence so does $v^- \dfrac{\partial v^+}{\partial n}$. At points where $v < 0$, $v^+ = 0$; since $v^+ \geq 0$ everywhere inside (by minimum principle),

$\dfrac{\partial v^+}{\partial n} \leq 0$ at such points. Hence $v^- \dfrac{\partial v^+}{\partial n} \leq 0$ everywhere on the boundary, and therefore $D(v) - D(u) \geq 0$. [A slight extension of the argument shows that the strict inequality $D(v) - D(u) > 0$ holds whenever v assumes both positive and negative values on the boundary.]

31. Since $h^2 w_{xy}(P) = w(P^{NE}) + w(P) - w(P^N) - w(P^E)$, the indicated summation covers each inner point of the rectangle $PQRS$ four times, twice with each sign, while each boundary point which is not a vertex is counted twice, once with each sign. Each vertex is counted only once, and examination of signs yields (126).

35. By (6-9) we get $D(u_i, u_j) = \displaystyle\int u_j \dfrac{\partial u_i}{\partial n} \, ds$. Since u_j vanishes on all boundary components except Γ_j and equals one on that component, $D(u_i, u_j) = \displaystyle\int_{\Gamma_j} \dfrac{\partial u_i}{\partial n} \, ds$. By the Cauchy-Riemann equations, this equals the increase in v_i resulting from a circuit of Γ_j (or any curve in G deformable into Γ_j).

37. By Exercise 35, the indicated Dirichlet integral is equal to the given quadratic form, while by Exercise 34 the integral is positive unless all λ's vanish.

39. Since by Exercise 30 the Dirichlet integral is conformally invariant, it suffices to prove the statement for an annulus of radii 1 and α. The harmonic measure of the outer circumference is given by $u = \log r / \log \alpha$, so that $u_x^2 + u_y^2 = (r \log \alpha)^{-2}$,

$$D(u) = \int_0^{2\pi} \int_1^\alpha (r \log \alpha)^{-2} r \, dr \, d\theta = 2\pi / \log \alpha.$$

CHAPTER 8

1. The substitution $\eta = (\xi - x)/2t^{1/2}$ reduces (9) in this case to the form $u(x,t) = \pi^{-1/2} \displaystyle\int_{-\infty}^\infty e^{-\eta^2} \, d\eta = 1$.

4. The substitution $\eta = \xi - x$ reduces (9) in this case to the form $u(x,t) = \cos x \displaystyle\int_{-\infty}^\infty (\cos \eta) K(\eta, 0, t) \, d\eta = g(t) \cos x$. Substituting the latter expression into (1) yields $g'(t) + g(t) = 0$, while the initial condition yields $g(0) = 1$. Thus $u(x,t) = e^{-t} \cos x$, or $\displaystyle\int_{-\infty}^\infty \exp\left(\dfrac{-\eta^2}{4t}\right) \cos \eta \, d\eta = 2e^{-t}(\pi t)^{1/2}$. Setting $\eta = |\lambda| x$ and $t = \lambda^2/4a$, we obtain $\displaystyle\int_{-\infty}^\infty \exp(-ax^2) \cos \lambda x \, dx = \mathfrak{F}(\exp(-ax^2)) = (\pi/a)^{1/2} \exp(-\lambda^2/4a)$.

5. $\displaystyle\int_{-\infty}^\infty u(x,t) \, dx = \int_{-\infty}^\infty \left[\int_{-\infty}^\infty f(\xi) K(\xi, x, t) \, d\xi\right] dx = \int_{-\infty}^\infty f(\xi) \left[\int_{-\infty}^\infty K(\xi, x, t) \, dx\right]$ $d\xi = \displaystyle\int_{-\infty}^\infty f(\xi) \, d\xi$. [By Exercise 1 and the symmetry of the kernel, $\displaystyle\int_{-\infty}^\infty K(\xi, x, t) \, dx \equiv 1$.] Physically, this result asserts that the total amount of heat is conserved.

6. It may be assumed that the mean value of $f(\xi)$ is zero, for otherwise we may simply subtract off the mean value. Letting $\displaystyle\int_0^\xi f(\eta) \, d\eta = F(\xi)$ and employing integration by parts, we rewrite (9) in the form $u(x,t) = -\displaystyle\int_{-\infty}^\infty F(\xi) K_\xi(\xi, x, t) \, d\xi = \displaystyle\int_{-\infty}^\infty F(\xi) K_x(\xi, x, t) \, d\xi$. Then $\displaystyle\int_{-y}^y u(x,t) \, dx = \int_{-\infty}^\infty F(\xi)[K(\xi, y, t) - K(\xi, -y, t)] \, d\xi$ $= \displaystyle\int_{-\infty}^\infty F(\xi)[K(\xi, y, t) - K(-\xi, y, t)] \, d\xi = \int_{-\infty}^\infty [F(\xi) - F(-\xi)] K(\xi, y, t) \, d\xi$. Given $\epsilon > 0$, we can find positive numbers A, M such that $|F(\xi) - F(-\xi)| < \epsilon|\xi|$ for $|\xi| > A$ and $|F(\xi) - F(-\xi)| < M$ for $|\xi| \leq A$. The last integral is then dominated by $M \int_{-A}^{A} K(\xi,y,t) \, d\xi + \epsilon \int_{|\xi| \geq A} |\xi| K(\xi,y,t) \, d\xi$, and hence by

$$\int_{-\infty}^{\infty} (M + \epsilon|\xi|) K(\xi,y,t) \, d\xi$$

This integral is reduced by the substitution $\eta = \xi - y$ to the form

$$\int_{-\infty}^{\infty} (M + \epsilon|y + \eta|) K(\eta,0,t) \, d\eta$$

which is dominated by $\int_{-\infty}^{\infty} [M + \epsilon(|y| + |\eta|)] K(\eta,0,t) \, d\xi$, and this integral works out

to $M + \epsilon[|y| + (t/\pi)^{1/2}]$. Thus, $\limsup\limits_{y \to \infty} \left| (2y)^{-1} \int_{-y}^{y} u(x,t) \, dx \right| \leq \frac{1}{2}\epsilon$ and hence

$\lim\limits_{y \to \infty} (2y)^{-1} \int_{-y}^{y} u(x,t) \, dx$ exists and equals zero.

From the formula $u(0,t) = \int_{-\infty}^{\infty} F(\xi) K_x(\xi,0,t) \, d\xi = (16\pi t^3)^{-1/2} \int_{0}^{\infty} \xi[F(\xi) - F(-\xi)]$

$\exp(-\xi^2/4t) \, d\xi$, we obtain $|u(0,t)| < (16\pi t^3)^{-1/2} \left[\int_{0}^{A} M\xi \exp\left(\frac{-\xi^2}{4t}\right) d\xi + \int_{A}^{\infty} \epsilon \xi^2 \right.$

$\left. \exp\left(\frac{-\xi^2}{4t}\right) d\xi \right] < (16\pi t^3)^{-1/2} \int_{0}^{\infty} (M\xi + \epsilon\xi^2) \exp\left(\frac{-\xi^2}{4t}\right) d\xi = (16\pi t^3)^{-1/2}[2Mt +$

$(4\pi t^3)^{1/2}\epsilon]$. Letting $t \to \infty$ and taking account of the arbitrariness of ϵ, we conclude that $u(0,t)$ approaches zero. Similarly, for any other fixed value of x, say x_0, $\lim\limits_{t \to \infty} u(x_0,t) = 0$. It is merely necessary to replace x in the above argument by

$x - x_0$ and to observe that $\lim\limits_{y \to \infty} (2y)^{-1} \int_{x_0-y}^{x_0+y} f(\xi) \, d\xi$ exists and equals zero, as in

the case $x_0 = 0$; this is an immediate consequence of the boundedness of $f(\xi)$, for

$(2y)^{-1} \left[\int_{x_0-y}^{x_0+y} f(\xi) \, d\xi - \int_{-y}^{y} f(\xi) \, d\xi \right] = (2y)^{-1} \left[\int_{y}^{x_0+y} f(\xi) \, d\xi - \int_{-y}^{x_0-y} f(\xi) \, d\xi \right]$,

and the latter expression is dominated by $|x_0/y| \cdot \max |f(\xi)|$. [The function $u(x,t) \equiv x$, corresponding to $f(\xi) \equiv \xi$, shows that the boundedness condition on $f(\xi)$ cannot be dropped.]

12. By adding $i\tau$ ($\tau = \operatorname{Im} s$) to both sides of (47), we find that the relationship $s = [(2n - 1)^{-1}\pi^{-1}\tau + i(n - \frac{1}{2})\pi]^2$ is satisfied on V_n, so that $s^{1/2} = \pm [(2n - 1)^{-1}\pi^{-1}\tau + i(n - \frac{1}{2})\pi]$. From this we immediately obtain

$$|\sinh s^{1/2}| = \sinh [(2n - 1)^{-1}\pi^{-1}|\tau|] > \frac{1}{2} \exp [(2n - 1)^{-1}\pi^{-1}|\tau|]$$

and $|\sinh xs^{1/2}| < \exp [x(2n - 1)^{-1}\pi^{-1}|\tau|]$, so that the integrand is dominated by $f_n(\tau) = 2 \exp [-(n - \frac{1}{2})^2\pi^2 t + (2n - 1)^{-2}\pi^{-2}\tau^2 t - (1 - x)(2n - 1)^{-1}\pi^{-1}|\tau|]$. Since $ds = [i + 2\tau(2n - 1)^{-2}\pi^{-2}] \, d\tau$ and $|\tau|$ is bounded by $c_n = (2n - 1)\pi[\sigma + (n - \frac{1}{2})^2\pi^2]^{1/2}$ as s traverses V_n, the inequality $|ds| < C|d\tau|$ holds, where C is independent of n. Thus, it will suffice to prove that $\int_{0}^{c_n} f_n(\tau) \, d\tau$ approaches zero with increasing n. The substitution $r = (2n - 1)^{-1}\pi^{-1}\tau$ reduces the integral to the more convenient form

$$\frac{2\pi \int_{0}^{[\sigma + (n - 1/2)^2\pi^2]^{1/2}} \exp [tr^2 - (1 - x)r] \, dr}{(2n - 1)^{-1} \exp [\pi^2(n - \frac{1}{2})^2 t]}$$

The numerator and denominator each become infinite with n; treating n as a continuous variable and employing L'Hôpital's rule, we find that the fraction approaches zero with increasing n.

13. Proceeding formally, we obtain $U(x,s) = \sin xs^{1/2}/s^2 \sin s^{1/2}$. This function has poles at the points $n^2\pi^2$ $(n = 1, 2, \ldots)$, and so cannot be a Laplace transform.

Taking account of the reversal of sign, we may assert that it is, in general, not possible to find a temperature distribution $u(x,t)$ defined for all *negative* values of t and satisfying (1) such that $u(x,0)$ coincides with a specified function $f(x)$. This result is the mathematical counterpart of the physical phenomenon of irreversibility.

CHAPTER 9

2. Multiplying both sides of (7a) by X and adding $(X')^2$, we obtain $2(XX')' + \lambda X^2 = (X')^2$. Integrating over the interval and taking account of (8), we obtain $\lambda \int_0^1 X^2 \, dx = \int_0^1 (X')^2 \, dx$. Since both integrals must be positive, λ is also positive.

9. If the inner product (f,g) of two functions is defined as $\int_{x_1}^{x_2} f(x)g(x) \, dx$, then (38) assumes the form $(u,Lv) = (Mu,v)$ for (sufficiently differentiable) functions u,v vanishing at both ends.

10. We seek a positive solution of the equation $(\rho p)' = \rho q$; such a function is given by $p^{-1}(x) \exp \int_{x_1}^x p^{-1}(\xi)q(\xi) \, d\xi$.

11. Recalling the definition of $f_1(x)$ given in the text, we first prove that $f_1'(x)$ never vanishes in the interval I. Assuming the contrary, let x_3 be the smallest value of x at which $f_1'(x)$ vanishes. Then, in the subinterval $x_1 < x < x_3$, f_1' is positive, and so f_1 increases monotonely from zero; since $(pf_1')' = -rf_1$, and the latter quantity is nonnegative in the subinterval, the quantity pf_1' cannot decrease from its value at x_1, namely, $p(x_1)$. This is inconsistent with the vanishing of $f_1'(x_3)$. It follows that f_1 increases monotonely throughout I, and so, in particular, $f_1(x_2) > 0$. This proves that the Green's function $K(x,\xi)$ exists, and also that it either increases or decreases monotonely in the interval $x_1 \leq x \leq \xi$; by the same type of argument, $K(x,\xi)$ shows the opposite monotone behavior on the interval $\xi \leq x \leq x_2$, and so $K(x,\xi)$ is either entirely positive or entirely negative throughout the interior of I and attains an extremum at ξ. The possibility of a minimum at ξ is ruled out by (34), and we conclude that $K(x,\xi)$ is positive for $x_1 < x < x_2$.

15. Multiplying (40) by u and then adding pu'^2 to both sides, we obtain $(puu')' + (r + \lambda\rho)u^2 = pu'^2$. Integrating over I and taking account of the vanishing of u at the end points, we obtain $\lambda \int \rho u^2 \, dx = \int pu'^2 \, dx + \int (-r)u^2 \, dx$. For a nontrivial function u the first two integrals must be positive and the third nonnegative (since $r \leq 0$); hence λ must be positive. (This argument is a simple extension of that employed in Exercise 2.)

16. Let x_3 and x_4 denote a pair of successive zeros of u. Since $uL(v) - vL(u) = (\lambda - \mu)\rho uv$, we obtain, from (31) and the vanishing of u at the end points, the equality

$$(\mu - \lambda) \int_{x_3}^{x_4} \rho uv \, dx = vpu' \Big|_{x_3}^{x_4}$$

To show that v must vanish at least once in the open interval $x_3 < x < x_4$, it evidently suffices to assume that u is positive there, and then rule out the possibility that v is also positive throughout this interval. Under the afore-mentioned assumption, $u'(x_3) > 0$, $u'(x_4) < 0$. If v were positive in the interval, vpu' would have to be nonnegative and nonpositive, respectively, at x_3 and x_4, and so each side of the above equation would be nonpositive. Recalling that $\mu > \lambda$, we conclude that ρuv, and hence v itself, cannot be positive throughout the interval.

Now, if there were infinitely many negative eigenvalues, it would immediately follow from the above result that each eigenfunction must possess infinitely many zeros. By a simple argument based on Rolle's theorem, there would exist for each eigenfunction a point in I at which the function and its first derivative both vanish, and this would imply the identical vanishing of the function.

17. Assuming momentarily that $r(x) \leq 0$, we know from Exercise 11 that $K(x,\xi)$ exists and is positive in the open square $x_1 < x$, $\xi < x_2$, and the same is true for $\tilde{K}(x,\xi)$. From Exercise 15 we know that λ_1, the lowest eigenvalue, is positive, and hence lowest in absolute value, and so, according to the theory developed in Sec. 5-4, the corresponding eigenfunction u_1 (normalized so that $\int_{x_1}^{x_2} \rho u_1{}^2 dx = 1$) is characterized by the fact that, among all functions $f(x)$ so normalized, it maximizes $|\iint \tilde{K}(x,\xi)f(x)f(\xi)\,dx\,d\xi|$. If u_1 assumed both positive and negative values, the inequality $|\iint \tilde{K}(x,\xi)f(x)f(\xi)\,dx\,d\xi| > |\iint \tilde{K}(x,\xi)u_1(x)u_1(\xi)\,dx\,d\xi|$ would hold for $f(x) = |u_1(x)|$. Thus we may assume that $u_1(x) \geq 0$ throughout I. If u_1 vanished at any interior point of I, u_1' would also have to vanish there, and this would imply the identical vanishing of u_1. Hence, u_1 vanishes only at the ends of I.

The assumption $r(x) \leq 0$ may be eliminated by the device of rewriting (40) in the form $(pu')' + (r - c\rho)u + (\lambda + c)\rho u = 0$, c being a positive constant sufficiently large that $r - c\rho \leq 0$.

19. For convenience let $x_1 = 0$, $x_2 = 1$ (as in Sec. 1). Then $\lambda_n = n^2\pi^2$, and (55) assumes the form $T'' + \beta T' + n^2\pi^2 T = 0$. The roots of the associated quadratic are $\frac{1}{2}[-\beta \pm (\beta^2 - 4n^2\pi^2)^{1/2}]$. If $\beta^2 < 4n^2\pi^2$, the roots are complex conjugates with real part $-\frac{1}{2}\beta$. Thus, any solution is expressible as the product of a sinusoidal function and the exponential function $\exp(-\frac{1}{2}\beta t)$. For $\beta > 0$ this factor provides a damping effect on the sinusoidal function, while for $\beta < 0$ it provides an exponentially *increasing* factor. If $\beta^2 \geq 4n^2\pi^2$, the foregoing discussion must be modified, but this inequality can hold for only a finite number of values of n.

20. Similar to Exercise 13.

21. The differentiability properties of the eigenfunctions on the boundary would have to be investigated in order to justify the use of the Green's identity.

23. The cancellation of logarithmic singularities at Q and the fact that $\tilde{G}(P,Q)$, the Green's function of Δ, is positive on the boundary of D enable us to assert (by the minimum principle) that $\tilde{G}(P,Q) > G(P,Q) > 0$ for any pair of points P, Q in D. Thus $\iint_D \iint_D G^2$ is dominated by $\iint_\Delta \iint_\Delta \tilde{G}^2$, and the latter is known to be finite.

SUGGESTIONS FOR FURTHER STUDY

1. B. B. Baker and E. T. Copson, "The Mathematical Theory of Huyghens' Principle," Oxford University Press, New York, 1939.
2. S. Banach, "Théorie des opérations linéaires," Chelsea Publishing Company, New York, 1955. (Reprint.)
3. H. Bateman, "Partial Differential Equations," Dover Publications, New York, 1944. (Reprint.)
4. S. Bergman, "The Kernel Function and Conformal Mapping," American Mathematical Society, New York, 1950.
5. S. Bergman and M. Schiffer, "Kernel Functions and Elliptic Differential Equations in Mathematical Physics," Academic Press, Inc., New York, 1953.
6. C. Caratheodory, "Variationsrechnung und partielle Differentialgleichungen erster Ordnung," Teubner Verlagsgesellschaft, mbH, Stuttgart, 1935.
7. H. S. Carslaw, "Introduction to the Mathematical Theory of the Conduction of Heat in Solids," Dover Publications, New York, 1945. (Reprint.)
8. H. S. Carslaw and J. C. Jaeger, "Operational Methods in Applied Mathematics," Oxford University Press, New York, 1941.
9. R. V. Churchill, "Fourier Series and Boundary Value Problems," McGraw-Hill Book Company, Inc., New York, 1941.
10. E. A. Coddington and N. Levinson, "Theory of Ordinary Differential Equations," McGraw-Hill Book Company, Inc., New York, 1955.
11. R. Courant and D. Hilbert, "Methoden der mathematischen Physik," vols. I and II, Springer-Verlag, Berlin, 1931 and 1937. (A slightly revised version of vol. I in English has been published by Interscience Publishers, Inc., New York.)
12. G. Doetsch, "Theorie und Anwendungen der Laplace-transformation," Dover Publications, New York, 1943. (Reprint.)
13. G. F. D. Duff, "Partial Differential Equations," University of Toronto Press, Toronto, 1956.
14. N. Dunford and J. Schwartz, "Linear Operators," Interscience Publishers, Inc., New York, 1958.
15. B. Friedman, "Principles and Techniques of Applied Mathematics," John Wiley & Sons, Inc., New York, 1956.
16. J. Hadamard, "Lectures in Cauchy's Problem in Linear Partial Differential Equations," Dover Publications, New York, 1952. (Reprint.)

17. E. L. Ince, "Ordinary Differential Equations," Dover Publications, New York, 1944. (Reprint.)

18. O. D. Kellogg, "Foundations of Potential Theory," Dover Publications, New York, 1953. (Reprint.)

19. W. V. Lovitt, "Linear Integral Equations," Dover Publications, New York, 1950. (Reprint.)

20. C. Miranda, "Equazioni alle Derivate parziale di Tipo ellitico," Springer-Verlag, Berlin, 1955.

21. Z. Nehari, "Conformal Mapping," McGraw-Hill Book Company, Inc., New York, 1952.

22. J. v. Neumann, "Mathematische Grundlagen der Quantenmechanik," Dover Publications, New York, 1943. (Reprint.)

23. I. G. Petrovsky, "Lectures on Partial Differential Equations," Interscience Publishers, Inc., New York, 1954.

24. I. G. Petrovsky, "Lectures on the Theory of Integral Equations," Graylock Press, Rochester, N.Y., 1957.

25. F. Riesz and B. Sz.-Nagy, "Functional Analysis," Frederick Ungar Publishing Co., New York, 1955.

26. I. N. Sneddon, "Elements of Partial Differential Equations," McGraw-Hill Book Company, Inc., New York, 1957.

27. A. J. W. Sommerfeld, "Partial Differential Equations in Physics," Academic Press, Inc., New York, 1949.

28. M. H. Stone, "Linear Transformations in Hilbert Space," American Mathematical Society, New York, 1932.

29. J. Tamarkin and W. Feller, "Partial Differential Equations," Brown University, Providence, R.I., 1941. (Mimeographed.)

30. E. C. Titchmarsh, "The Theory of Functions," Oxford University Press, New York, 1939.

31. E. C. Titchmarsh, "Introduction to the Theory of Fourier Integrals," Oxford University Press, New York, 1937.

32. E. C. Titchmarsh, "Eigenfunction Expansions Associated with Second-order Differential Equations," Oxford University Press, New York, 1946.

General references: [3], [11], [13], [23], [26], [27], [29]
References for particular topics:
   Chapter 1: Sec. 3: [9], [11, vol. I, chap. 2], [31]
             Sec. 4: [8], [12]
             Sec. 5: [10], [17]
             Sec. 6: [25], [30]
   Chapter 2: [6], [11, vol. II, chap. 2], [26, chap. 2]
   Chapter 3: [1] (for Sec. 8), [11, vol. II, chaps. 3, 5, 6], [16]
   Chapters 4 and 5: [2], [11, vol. I, chap. 3], [14], [19], [22], [24], [25], [28]
   Chapters 6 and 7: [4] (for Sec. 7-7), [5], [11, vol. II, chaps. 4, 7], [18], [20], [21], [26, chap. 4]
   Chapter 8: [5], [7], [8], [9], [12], [26, chap. 6]
   Chapter 9: [9], [11, vol. I, chaps. 5, 6], [15], [32]

INDEX

269